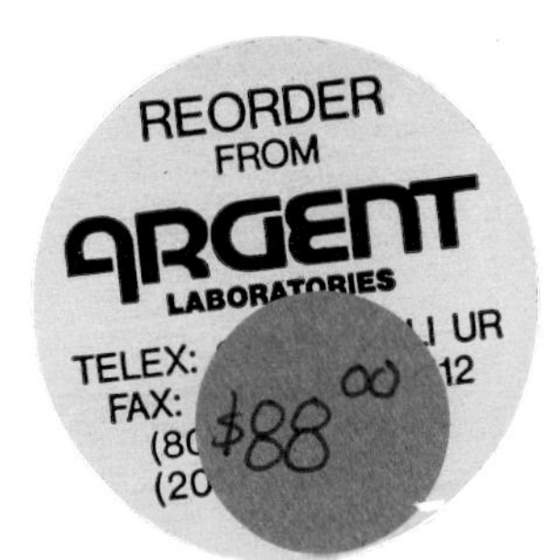
REORDER
FROM
ARGENT
LABORATORIES
TELEX:
FAX:
$88^{00}

D1604817

Sea Bass

CHAPMAN & HALL FISH AND FISHERIES SERIES

Amongst the fishes, a remarkably wide range of fascinating biological adaptations to diverse habitats has evolved. Moreover, fisheries are of considerable importance in providing human food and economic benefits. Rational exploitation and management of our global stocks of fishes must rely upon a detailed and precise insight of the interaction of fish biology with human activities.

The *Chapman & Hall Fish and Fisheries Series* aims to present authoritative and timely reviews which focus on important and specific aspects of the biology, ecology, taxonomy, physiology, behaviour, management and conservation of fish and fisheries. Each volume will cover a wide but unified field with themes in both pure and applied fish biology. Although volumes will outline and put in perspective current research frontiers, the intention is to provide a synthesis accessible and useful to both experts and non-specialists alike. Consequently, most volumes will be of interest to a broad spectrum of research workers in biology, zoology, ecology and physiology, with an additional aim of the books encompassing themes accessible to non-specialist readers, ranging from undergraduates and postgraduates to those with an interest in industrial and commercial aspects of fish and fisheries.

Applied topics will embrace synopses of fishery issues which will appeal to a wide audience of fishery scientists, aquaculturists, economists, geographers and managers in the fishing industry. The series will also contain practical guides to fishery and analysis methods and global reviews of particular types of fisheries.

Books already published and forthcoming are listed below. The Publisher and Series Editor would be glad to discuss ideas for new volumes in the series. . .

Available titles

1. **Ecology of Teleost Fishes**
   Robert J. Wootton
2. **Cichlid Fishes**
   Behaviour, ecology and evolution
   Edited by Miles A. Keenlyside
3. **Cyprinid Fishes**
   Systematics, biology and exploitation
   Edited by Ian J. Winfield and Joseph S. Nelson
4. **Early Life History of Fish**
   An energetics approach
   Ewa Kamler

5. **Fisheries Acoustics**
   David N. MacLennan and E. John Simmonds
6. **Fish Chemoreception**
   Edited by Toshiaki J. Hara
7. **Behaviour of Teleost Fishes**
   Second edition
   Edited by Tony J. Pitcher
8. **Genetics and Fish Breeding**
   Colin R. Purdom
9. **Fish Ecophysiology**
   J. Cliff Rankin and Frank B. Jensen
10. **Fish Swimming**
    John J. Videler
11. **On the Dynamics of Exploited Fish Populations**
    Raymond J.H. Beverton and Sidney J. Holt
    (Facsimile reprint)
12. **Sea Bass**
    Graham D. Pickett and Michael G. Pawson
13. **Fish Bioenergetics**
    Malcolm Jobling
14. **On the Sex of Fish and Gender of Scientists**
    Daniel Pauly
15. **Hake**
    Fisheries, products and markets
    Edited by Jürgen Alheit and Tony J. Pitcher

Forthcoming titles

Fisheries Ecology
Second edition
Edited by T.J. Pitcher and P. Hart

Impact of Species Change in the African Lakes
Edited by T.J. Pitcher

Environmental Biology of Fishes
M. Jobling

Electric Fishes
P. Moller

Adult sea bass (*Dicentrarchus labrax*).

Sea Bass

Biology, exploitation and conservation

Graham D. Pickett and Michael G. Pawson

MAFF Fisheries Laboratory
Lowestoft, UK

CHAPMAN & HALL

London · Glasgow · Weinheim · New York · Tokyo · Melbourne · Madras

Published by Chapman & Hall, 2–6 Boundary Row, London SE1 8HN

Chapman & Hall, 2–6 Boundary Row, London SE1 8HN, UK

Blackie Academic & Professional, Wester Cleddens Road, Bishopbriggs, Glasgow G64 2NZ, UK

Chapman & Hall GmbH, Pappelallee 3, 69469 Weinheim, Germany

Chapman & Hall Inc., One Penn Plaza, 41st Floor, New York NY10119, USA

Chapman & Hall Japan, Thomson Publishing Japan, Hirakawacho Nemoto Building, 6F, 1-7-11 Hirakawa-cho, Chiyoda-ku, Tokyo 102, Japan

Chapman & Hall Australia, Thomas Nelson Australia, 102 Dodds Street, South Melbourne, Victoria 3205, Australia

Chapman & Hall India, R. Seshadri, 32 Second Main Road, CIT East, Madras 600 035, India

First edition 1994

Typeset in 10/12 Photina by Acorn Bookwork, Salisbury, Wilts

Printed in Great Britain by St. Edmundsbury Press, Bury St. Edmunds, Suffolk

ISBN 0 412 40090 1

A catalogue record for this book is available from the British Library

Library of Congress Catalog Card Number: 93-74442

Printed on permanent acid-free text paper, manufactured in accordance with the proposed ANSI/NISO Z 39.48-199X and ANSI Z 39.48-1984

To Donovan and Betty Kelley
for their friendship, generosity, enthusiasm and dedication
to conserving the sea bass

Contents

Series foreword

Among the fishes, a remarkably wide range of biological adaptations to diverse habitats has evolved. As well as living in the conventional habitats of lakes, ponds, rivers, rock pools and the open sea, fish have solved the problems of life in deserts, in the deep sea, in the cold Antarctic, and in warm waters of high alkalinity or of low oxygen. Along with these adaptations we find the most impressive specializations of morphology, physiology and behaviour. For example, we can marvel at the high speed swimming of the marlins, sailfish and warm-blooded tunas, air-breathing in catfish and lungfish, parental care in the mouth-brooding cichlids and viviparity in many sharks and toothcarps.

Moreover, fish are of considerable importance to the survival of the human species in the form of nutritious, delicious and diverse food. Rational exploitation and management of our global stocks of fishes must rely upon a detailed and precise insight of their biology.

The *Chapman & Hall Fish and Fisheries Series* aims to present timely volumes reviewing important aspects of fish biology. Most volumes will be of interest to research workers in biology, zoology, ecology and physiology but an additional aim is for the books to be accessible to a wide spectrum of non-specialist readers ranging from undergraduates and postgraduates to those with an interest in industrial and commercial aspects of fish and fisheries.

Sea Bass comprises the 12th volume in the *Chapman & Hall Fish and Fisheries Series*. The authors are both experts on this species from the UK Lowestoft Fisheries Laboraotry and have recently helped steer a Sea Bass management plan through the councils of the European Community. The book contains a thorough and broad coverage of the essential features of European Sea Bass life history, growth, migrations, spawning and fisheries.

Although Sea Bass fisheries are of modest size in relation to large-scale commercial fisheries for cods, herrings and mackerel, this work focuses our attention on the value, impact and sustainability of such small-scale inshore fisheries. Small-scale fishing activities can form a significant part of the local economy and culture, and economic demand for luxury table fish by restaurant and leisure trade in the developed world can be a salient factor in the survival of local stocks.

Although the book is primarily concerned with the European Sea Bass, interesting comparisons with the North American striped bass are included. Moreover, principles developed for assessing the economic and cultural value of sport fisheries for bass in relation to the protection and conservation of vulnerable inshore stocks can be applied to a wide range of fish species of this type world-wide: the parallel that springs to mind most readily concerns endangered stocks of salmonids in the Pacific Northwest of North America.

Pickett and Pawson's *Sea Bass* illustrates how management and conservation of small-scale inshore fisheries depends critically upon insight of the processes driving the biology, ecology and exploitation regime. Consequently, in addition to providing a helpful source of reference for European Sea Bass, the synthesis provided by this book should help to establish the ground rules for such valuable work elsewhere.

Professor Tony J. Pitcher
Editor, *Chapman & Hall Fish and Fisheries Series*
Director, Fisheries Centre, University of British Columbia, Vancouver, Canada

Preface

Marine fisheries science in north-western Europe has been dominated over the last 20 years by the need to assess and manage the exploitation of internationally important commercial fish stocks. Fisheries perceived to have a more parochial interest have often been neglected, but the decline in distant-water and middle-distance fishing fleets, and intensification of effort on resources nearer at hand, has belatedly raised the profile of inshore fisheries. The provision of appropriate management advice is, however, complicated by their mixed and seasonal nature and because commercial catching interest can conflict with those of recreational anglers and, sometimes, aquaculture and stock enhancement initiatives. It is likely, moreover, that human activities and insults to the environment will have their greatest impact on fisheries that take place close to the shore.

With the above perspective, the Ministry of Agriculture, Fisheries and Food's Directorate of Fisheries Research set up a coastal fisheries group at the Lowestoft Laboratory in 1981. This group, along with shellfish and salmonid scientists, developed a research programme to investigate the small-boat fisheries of England and Wales. It soon became apparent that the sea bass, *Dicentrarchus labrax* (Linnaeus), was an important constituent of catches along many parts of the coast, and that there was a vigorous campaign (by anglers) to give the species greater protection, in the face of what some perceived to be a massive stock decline. Unfortunately – though not for us, the authors – the scientific knowledge of the biology of sea bass, and of the options and implications of appropriate management for its fishery, was somewhat sparse. As a consequence, a project was devised which aimed to elucidate the life history of the sea bass, to assess its stock structure and dynamics, and to describe its fishery and the interrelations both between catching methods and with other resources around the UK. This work has now attained its primary objective; a precautionary management package for the bass fishery was implemented in 1990.

We have written this book to draw together current knowledge about the European sea bass and its fishery, and to provide an insight to the approaches taken to achieve this. Some well-known aspects of fisheries science are presented in a review style focused on sea bass, but those parts of our investigations which are specific to the sea bass and to inshore fisheries have been dealt with more comprehensively. Our intention is to inform and instruct and, we hope, to provoke thought and speculation upon which further research can be based.

G.D. Pickett
M.G. Pawson

Acknowledgements

The authors wish to thank the following people who contributed to the work described in this book:

for the use of unpublished data: Mike Dunn, Simon Potten, John Lancaster and, particularly, David Garrod, Director of Fisheries Research, MAFF, Lowestoft;

for technical help in field and laboratory investigations; Dave Anderson, Kevin Benham, Paul Bonner, John Bridger, Neville Burt, Simon Dowson, Derek Eaton, Teresa Eaton, Keith Fell, Ed Gillespie, Garry Howlett, Jim Knights, Dave Lewis, Ian Mayer, Steve Mustow, John Rawle, Bill Riley, Brian Robinson, Peter Robinson, Pat Scholes, Miles Thomas, Bob Turner, Jon Ware, Stephen Warnes;

for allowing us to use their boats and share their tea (and egg sandwiches); Alec and Dave Baldacchino, Bill Brown, Ted Chappel, David Chapple, Brian Cooper, Richard Ede, Dick Langley, Ian MacKenzie, Mick Selling, David Smale, Nick Smith, Colin Thomas, Dickie Thomas, the late Mike Trott, John Watkins, Baron Woodward and many others;

for information and advice over the years; Dick Drennan, Barry Edwards, Pete Edwards, Colin George, Len Hawke and other members of MAFF's Sea Fisheries Inspectorate, Rex Aldous, Simon Bossey, Jim Howell, Steve Ozanne, John Rhydderch, Joss Wiggins, Harry Worden and other officers of Sea Fisheries Committees;

for all those who have collected bass scales, filled in logbooks, returned tags or allowed us to measure their fish, especially Malcolm Brindle, Alan Burchett, Kevin Mankelow and David Warnes;

for continuing to support and encourage research on bass; Andrew Grimm, Paul Llewellyn, Peter Reay, John Rylands and Sue Shackley;

for all the above, and for their particular interest in the bass; Bob Cox and Donovan Kelley;

and, finally, for their painstaking and constructive criticism of earlier manuscripts of this book; Simon Jennings and Steve Lockwood.

General introduction

The European sea bass (*Dicentrarchus labrax*, Linnaeus, 1758) is a fish that never ceases to excite the imagination. At the time of writing, it is very much in vogue in north-west Europe, featuring regularly in angling articles, cookery books and television programmes, and on gourmet menus. During the 1980s the sea bass has been the subject of increasing commercial interest and consequently of scientific research, both as a new species for cultivation and as an exploited resource in need of conservation and management of its fishery. In the United Kingdom, the sea bass has emerged from the obscurity of a minor commercial species in the 1970s to being the prized target of a large part of the coastal fishery. Its value has increased with rising demand, and it usually retails at a price above that of wild Atlantic salmon (*Salmo salar* L.). The sea bass also shares with the salmon the distinction of being the object of both fervent sporting interest and gastronomic esteem.

The sea bass is often spoken of passionately by those who hunt for it, sell it and eat it. Even the most dispassionate scientist cannot help but admire this fish, especially when confronted with living specimens at sea or in the laboratory. It is a beautiful, bristling predator that has a distinctive behaviour and biology and, on this merit alone, deserves the research attention it is currently receiving.

The sea bass is well known throughout much of the Mediterranean and coastal Europe and has long been valued as a food fish in France, Italy and Spain. It is hard to understand why it took so long to be appreciated in the United Kingdom. Yet it is in the UK, and in Ireland, that the sea bass has been esteemed as a sporting quarry since the early 19th century, by anglers who regard it as a gamefish of the sea.

It would be too simple to explain that the sea bass is regarded as being special because it is good to eat, with a high demand and a high price, and that its exploitation provides employment and generates wealth. It is true that fish become emotive subjects when livelihoods or, indeed, recreational amenities are at stake, but there is far more to the sea bass story than money. In order to appreciate fully the relative importance of the sea bass to humans we must examine the reasons why it is valued. Many socio-economic factors are involved and some of these have recently been studied. Moreover, it is not just the fish, but the whole ethos of time, place and catching method, that has attracted many, especially anglers, to fish for sea bass.

In the UK, anglers once had almost exclusive use of the sea bass resource, and in the period 1945–70, bass sport fishing was popularized by angling writers of the day. In many cases they wrote about fishing from the shore, from the lonely surf beaches of Inch Strand (Ireland) and St Brides Bay and Rhossili Bay (South Wales), which were known to many who had never even visited them. All are beautiful places, wild and exciting when wind and tide create a strong, running sea. Catching a large, silvery bass from the surf is probably one of angling's ultimate experiences. For many years this idyll typified bass fishing for those who could only read about this branch of sea angling. In reality, there was also plenty of boat fishing for bass being practised by those with access to vessels in the right places, although the standard of sport, if not the size of the catches, was thought by some to be of an inferior variety.

After the Second World War, following seven years of negligible fishing activity and with a greatly reduced fishing fleet, beach anglers had plenty of space and fish. By the 1960s, more British anglers owned cars and could afford holidays away from home, and the well-known but relatively unused beaches, where bass fed and could be caught close inshore, came under increased fishing pressure. Those anglers who wanted easier fishing went afloat and much publicity was given to places where good catches were taken. Ireland began to attract anglers intent on taking bass-fishing holidays to places such as Dingle, Rosscarbery, Youghal and Splaugh Rock. Boat fishing marks in English waters also became famous: the Eddystone and Manacles reefs, Hartland Point and Anchor Stone are a few that are still fondly remembered. In Brittany, the Pointe de Raz and Etel probably strike a similar chord with French anglers. All these places conjure up a vision of a swirling sea with sea birds diving to catch small fish that have been driven to the surface by marauding shoals of feeding bass.

The sea bass itself, whether caught from shore or boat, is an excellent sporting fish, hard fighting and attractive, defiant and aggressive to the last spiny gasp. It has been regarded as the prime sea sport fish in the UK since the dawn of serious sea angling in the 1820s. Holbrow wrote in 1909 that 'the sport afforded by bass is little inferior to salmon fishing', an opinion which remains shared by many to the present day. In the sea, the bass stands out from the usual run of fish that sea anglers catch around England and Wales. Perhaps only the grey mullets (*Crenimugil labrosus* and *Liza* spp.) offer such sport, weight for weight, once hooked. Bass are much easier to catch on hooks, however, and they are regarded by many as much better than mullet to eat.

To many commercial fishermen in France and Britain, the bass now often represents the difference between making a reasonable living and going out of business. Owing to overfishing, stocks of many of the main commercial species have declined dramatically, and, with increasingly restrictive catch-quota controls under the European Community's (EC's) Common Fisheries Policy, fishermen have turned to other accessible, non-quota fish, such as the sea bass. Some had no experience of bass fishing, and they had to adapt their methods, a few with outstanding success; others came to commercial bass fishing having started as anglers and then realizing the financial potential of their catch. There are many ways of catching sea bass and some techniques have developed almost into an art form. Consequently, there is a large element of job satisfaction in bass fishing. But these fish are all to easily converted from 'bars of silver' to gold, and competition for the resource is now great. Even some of the offshore fishing vessels now direct their efforts at bass for part of the season, and then there are protests from the less mobile inshore fishing communities.

Bass landings now help to support a growing number of fish merchants and specialist middlemen. At one time, most bass that were landed passed through the large auctions at major ports. Today, a host of small firms has appeared in the UK, concentrating on high-value species such as lobsters (*Homarus gammarus*), scallops (*Pecten maximus*), salmon and, of course, bass. A lively export trade exists between EC countries, in which the UK is a net exporter of sea bass, but they are imported to the UK from France in the winter when home supplies are low. The UK and France export bass to Italy and Spain and, because exploitation of wild stocks of bass has been unable to satisfy the demand for the fish around the Mediterranean, it is now being cultivated (along with sea breams, Sparidae) in several countries. Fish farms have appeared, in various forms, from Biscay to Israel, and production of sea bass is likely to continue to rise.

Sea bass often feature on menus in restaurants in Spain, Italy, Greece and France, particularly in the Midi and, since the early 1980s, in restaurants in major cities and coastal towns in southern England. Top chefs extol its virtues; for Floyd (1985) 'bass is king' and Anton Mossiman (former head chef at the Dorchester Hotel, London) regards it as the choice fish for summer barbecues. But this is not new: Smitt (1892) reports it being sought in Parisian markets for its firm, white flesh; the ancient Greeks and Romans valued it highly, and the latter even cultivated it.

It is obvious from the foregoing that many people have an interest in the European sea bass, and its continued abundance is a matter of concern to them, whether they depend on it for income or pleasure. There are many parallels with fisheries for related bass species in the United States, particularly those for the striped bass (*Morone saxatilis*, Walbaum), which has been studied for longer and in much greater depth than the European species. It may not be easy to identify the particular issues which led to fisheries scientists intensifying their efforts to understand the life history of the striped bass, with the aim of better management of its fishery, but with the European sea bass, one event stands out.

On 10 April 1981, Statutory Instrument No. 535 (Great Britain–Parliament, 1981) was signed by the Minister of Agriculture, Fisheries and Food and the Secretaries of State respectively concerned with the sea-fishing industry in Wales and Scotland. This announced that, from 1 May 1981, the minimum size at which the European sea bass could be legally landed in the UK waters by British fishermen would be raised from 26 to 32 cm total length. This minimum landing size (MLS) was to be increased further, from 32 to 38 cm, in May 1983. This legislative action was the Government's response to calls from many sea anglers, and some commercial fishermen, to conserve bass stocks, which were perceived to be suffering from increased exploitation, in particular, the development and expansion of inshore gill net fisheries around the coasts of England and Wales.

The choice of an increased MLS as the regulatory device arose from the results of research by Holden and Williams (1974), scientists working for the Ministry of Agriculture, Fisheries and Food (MAFF), at the Fisheries Laboratory, Lowestoft. Their study, undertaken in response to pressure from anglers who were concerned about apparent decreases in catch rates of large bass in some localities in the 1960s, indicated that bass mortality levels overall were actually quite low. The authors suggested that the rather local distribution and limited movements of bass, as indicated by tagging studies in England and in Ireland (Kennedy and Fitzmaurice, 1972), probably exaggerated the apparent impact of increasing fishing pressure on some parts of the bass populations. They concluded that the stock as a whole could be regarded as underexploited and that the most likely cause of decline in the bass population was its failure to breed successfully in many years and to produce sufficient juveniles to sustain catch rates. The most plausible biological explanation was that the bass around the United Kingdom are at the northern limit of the species' geographic range and, as a consequence, environmental and climatic conditions might be expected, on average, to be less than propitious for the survival of eggs, larvae and juvenile bass.

It seemed, therefore, that short of a favourable climatic change, the most sensible course of action would be to increase protection for those juveniles that were produced, and to ensure as far as possible that they had a good chance of surviving to enter the spawning stock. Published data on bass growth and maturity (Kennedy and Fitzmaurice, 1972) indicated an average size at first spawning of around 38 cm total length (TL) for females. Holden and Williams' analysis also suggested that, if bass could reach a size of 38 cm before first entering the fishery, this would improve and stabilize yields. When this course of action was first proposed, in 1974, the commercial fishery for bass in England and Wales (bass catches were negligible around Scotland) was considered to be insignificant compared with that of its recreational rod-and-line fishery. Similarly, the UK market for sea bass was extremely limited, with little or no infrastructure. Consequently, an MLS of 26 cm, which corresponds to a 2–3-year-old fish (the average age in MLS regulations for many commercially important North European species), was introduced in 1976. The Government's later decision, to implement a 38 cm MLS for bass, followed discussions with 'interested parties' in 1980. It is instructive to examine more closely the viewpoints of those various parties.

By 1980, a lucrative market for sea bass had been developed in mainland Europe, and the scale and variety of bass-catching operations were increasing all around the southern half of England and Wales. Representatives of the commercial fishery protested that a one-step increase from 26 to 38 cm in the minimum permissible takable size for bass, would almost completely annihilate those fisheries which had come to rely increasingly upon fish of between 30 and 40 cm. Such a large increase would, in any case, have caused considerable short-term losses of catch and earnings in some parts of the fishery. Some argued that there was a lack of evidence for heavy exploitation, and that additional protection (beyond the 26 cm MLS) was unwarranted. However, participants in the bass fishery in the Thames Estuary and around Cornwall, for example, were becoming aware that pressure on bass stocks was increasing. Because there was money to be made by catching and selling bass, fishing effort had been displaced from inshore fisheries for other species from which fishermen could no longer sustain their livelihoods. These people, along with anglers in general, lobbied for an MLS of 38 or 40 cm, or even higher. The outcome of this debate was the announcement of a two-stage increase already mentioned.

There is, however, another party which is interested in the outcome of fisheries-management deliberations: the fisheries scientists. Although the sea bass had been a minor species in the hierarchy of European fisheries issues, the fishery in which it is taken around Britain has assumed much more prominence since 1980 than hitherto. In England and Wales particularly, the majority of people involved in commercial fishing use day-boats within territorial waters (the 12 mile zone). Accordingly, the problems of assessing and managing small-boat fisheries, in which several resource species are caught in a variety of gears – often being used consecutively by the same boat at the appropriate seasons – is therefore worthy of attention. In view of the growing importance of being able to provide good scientific advice for the management of the bass fishery, which at that time appeared to be wholly an inshore issue of strictly national interest, a further research programme was implemented to provide the following information:

1. A description of the fishery in which bass are taken, detailing the distribution and characteristics of the fleet and catching methods, and an evaluation of effort and catches, both commercial and recreational;

2. The biological knowledge required to give management advice; population distribution, movements, stock identity, growth, recruitment, reproductive biology, etc.;
3. An assessment of the impact of exploitation on the bass stock(s), abundance trends, mortality rates, yield per recruit and general population dynamics;
4. An evaluation of the options and constraints for management of the fishery, identification of conservation objectives and development of a strategy to achieve them.

It would be useful at this stage to say something about the fisheries biologist's approach to the provision of management advice for the bass fishery, or any other. In the first place, it is recognized that, without a fishable resource, there can be no exploitation, no yield and no livelihood for fishermen (or fisheries scientists!) or sport for anglers. The immediate priority, therefore, was to give advice on the status of the bass population and to define the 'safe biological limits' within which management should aim to operate. It is important that sufficient appropriate data be collected systematically, interpreted objectively and presented in a way that can be readily understood by administrators, Ministers and exploiters alike. If this can be achieved, it is more likely that any forthcoming management measures will be consistent with the scientific data, practicable in implementation and, above all, effective when introduced.

The remaining problem, then, is to identify a management objective upon which general agreement can be reached, and which satisfies, as far as possible, the needs of those who have a specific interest in the sea bass fishery; essentially the commercial fishermen and bass anglers. It is important, therefore, to identify these groups and to try to evaluate their respective attitudes to the various potential management regimes and their likely effects on the fishery. As part of this procedure, in 1986, MAFF commissioned an economic evaluation of both the recreational and commercial sectors of the bass fishery with the Centre for Marine Resource Economics (CEMARE) at Portsmouth Polytechnic (now the University of Portsmouth). A more direct approach, however, is to contact and work with the fishery. In this way, first-hand information is forthcoming and opinions are freely given, against which judgements can be made at a more formal level on options and tactics for management. If this contact can be achieved at an early stage of the investigation, it is more likely that the views expressed will be forthright and without prejudice, compared with those given once positions have hardened in response to imminent decisions on regulation of the fishery.

Along similar lines, we would argue that fisheries-dependent data (catch returns and tagged fish reporting, for example) are less likely to be biased when the fishery's participants understand and agree with the use to which the information will be put. As a consequence, a data-collection strategy should be explicit and seek to impose as little as possible on the conduct of a fishery. The approach, therefore, has been to work and consult with those involved in the bass fishery, with an expressed aim of assessing the status of the exploited stock and, if necessary, recommending measures to sustain and possibly improve its potential for commercial and recreational utilization around the UK. The major part of the field work of MAFF's Directorate of Fisheries Research (DFR) has been carried out using chartered commercial fishing vessels, with some involvement of research vessels and scientists from other research institutes in those areas, such as reproductive biology, which are of fundamental scientific interest but were of less relevance to the immediate management issues.

Throughout the 1970s and 1980s, there has been considerable progress in the development of extensive and intensive farming of bass in south-western Europe and along the Mediterranean coasts, which has not been without effect on the supply-and-demand aspects of the market. Reports of this work have dominated the scientific literature where the sea bass is concerned. Although some fundamental biological knowledge has undoubtedly been gained, it is presented in publications in the context of applied aquaculture. It is possible that work directed to this end might have actually suppressed research on the reproductive biology and early life history of wild bass populations. This is in marked contrast to the picture for Atlantic salmon, where the impetus to enhance and improve survival of wild stocks has focused attention for a century or more on the species' natural spawning behaviour and juvenile lifestyle. Admittedly, this is much easier to observe and investigate in the salmon than in almost any marine fish. It is only relatively recently that farming of marketable salmon – as opposed to obtaining fertilized eggs from wild spawners and incubating and rearing-on before returning the progeny to the wild – has developed, with its imperative of better stock control, genetic manipulation and economically efficient husbandry.

In contrast to species like the Atlantic salmon and the striped bass, which are also pursued for fun and for food (and profit), investigations on the European bass have a very short history, in the fisheries context. Luckily, it has not yet become necessary to divert research effort to examine the effects of degradation of the bass's genetic identity or of its environment, and there has been relatively little debate on the appropriate allocation of the yield to the bass fishery. These issues can all too easily cast an obscuring shadow across one's perception of what knowledge is required to enable mankind to continue to enjoy the use of a renewable resource.

It is becoming more widely accepted in Europe that fisheries – that is, the resource and its exploiters – and not fish stocks *per se*, are the basic management unit. The collection of scientific data and their use in formulating management advice must therefore follow a policy which implicitly recognizes the importance of biological and fishery interactions. During the course of this work we have had an ideal opportunity to investigate the biology of the sea bass, to study its fishery around the UK and to meet and work with many of those who fish for it. The behaviour of the catcher is in many ways as interesting as that of the catch itself, and in that sense this book is about people and the animal they exploit. We hope that the way in which the material is presented will nourish the idea that both biological knowledge and an understanding of the ways in which fisheries behave are required in any successful application of science to fisheries management.

Although this book's three parts may be read independently, they are strongly interrelated and we refer across chapters where aspects are covered in more detail elsewhere. This book is not intended to be a complete review of all the scientific literature available on *D. labrax*, but we hope that in this description of the biology, exploitation and conservation of the bass in England and Wales, the broad spectrum of current knowledge is revealed.

Part One

Biology and Ecology

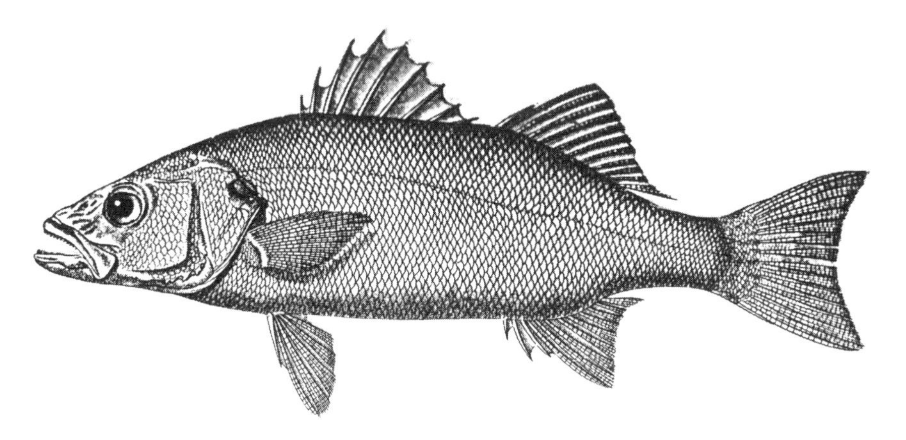

Bar Commun. From Cuvier and Valenciennes, 1828.

Chapter one

Nomenclature, taxonomy and morphology

1.1 NOMENCLATURE

The earliest records of the European sea bass date back to the ancient Greeks, when Aristotle named it 'sea wolf' or *Labrax*, the specific name which is still used today. Pliny (according to Cuvier and Valenciennes, 1828) gave it the Roman name *Lupus*, 'wolf', used both on account of the sea bass's voracity and its habit of hunting in groups. The word bass is derived from the Anglo-Saxon word boers, and appears in several forms, along with derivatives of wolf, in the following list of national names for this fish.

French bar, loup (wolf)
German seebarsch (sea perch)
Dutch zeebaars (sea perch)
Portuguese robalo (bass)
Spanish lubina (wolf)
Catalonian llobarro, llop (wolf)
Italian spigola (gleaner) branzino (bass)
Greek/Turkish lavrek (big-mouth)
Moroccan albar (bass)
Danish bars (bass)
Swedish havsaborre (sea perch)
Norwegian hav-abar (sea perch)
Yugoslavian lubin (wolf)
Russian lavraki (big-mouth)

Bace and basse are old English derivations of boers, which were still used in some areas of England earlier this century, as were the following local names for bass: sea perch, salmon bass, sea dace, white salmon and white mullett, with gapemouth, school bass, schoolie and checker being reserved for juvenile fish. Draenog and gannog are old Welsh names for bass. For the general text of this book, bass will be used as the common name of the European sea bass.

1.2 TAXONOMY

The present accepted scientific name for the bass in Europe is *Dicentrarchus labrax* (Linnaeus). Recent publications in the United States of America refer to it as *Morone labrax* (e.g. Waldman, 1986), the name also used in Europe for a hundred years until the late 1960s. The taxonomic affinities for bass, which are generally accepted in Europe, are as follows.

Order Perciformes
Suborder Percoidei
*Family Serranidae
Genus *Dicentrarchus*
Species *labrax*

The Order Perciformes (perch-like fishes) is the largest known order of fishes, numbering around 7500 identified species. The suborder Percoidei (Fig. 1.1) includes the families Serranidae (sea basses, sea perches and groupers), Percidae (perches, walleyes and darters), Sciaenidae (drums), Centropomidae (snooks), Carangidae (jacks, scads and pompanos), Pomatomidae (bluefish) and Centrarchidae (sandfishes, including large-mouth and smallmouth bass).

The Serranidae are widely distributed throughout temperate and tropical seas, although only three member species, other than *D. labrax*, have been recorded around the United Kingdom, and then rarely. These are the stone basse or wreckfish (*Polyprion americanus*), the dusky perch (*Epinephelus gigas*) and the comber (*Serranus cabrilla*). Most of the North American species which are close relatives of *D. labrax* live in fresh or brackish water.

The taxonomic status of the bass and related species in Europe and the USA has been the subject of much dispute and confusion over the last 200 years. There have been various interpretations of the systematics of the Perciformes since the time of Linnaeus in the mid 18th century, and some 19th century authors made fundamental mistakes concerning diagnostic features and in interpreting other authors' work. We hope that this is clarified in the following brief review of the systematics of the bass and closely related species.

Labrax was originally proposed as a generic name for bass by Klein in 1749, but Linnaeus later relegated this to a specific name by using *Perca* as the generic name. This was taken from the family Percidae, to which he assigned the European bass, which thus became *Perca labrax* (Linn. 1766). Following Linnaeus, various names were proposed and used until Cuvier and Valenciennes (1828), who reorganized ichthyological systematics, renamed the genus to which the bass belongs as *Labrax*, and gave it the specific name

*May be found under Moronidae in some listings.

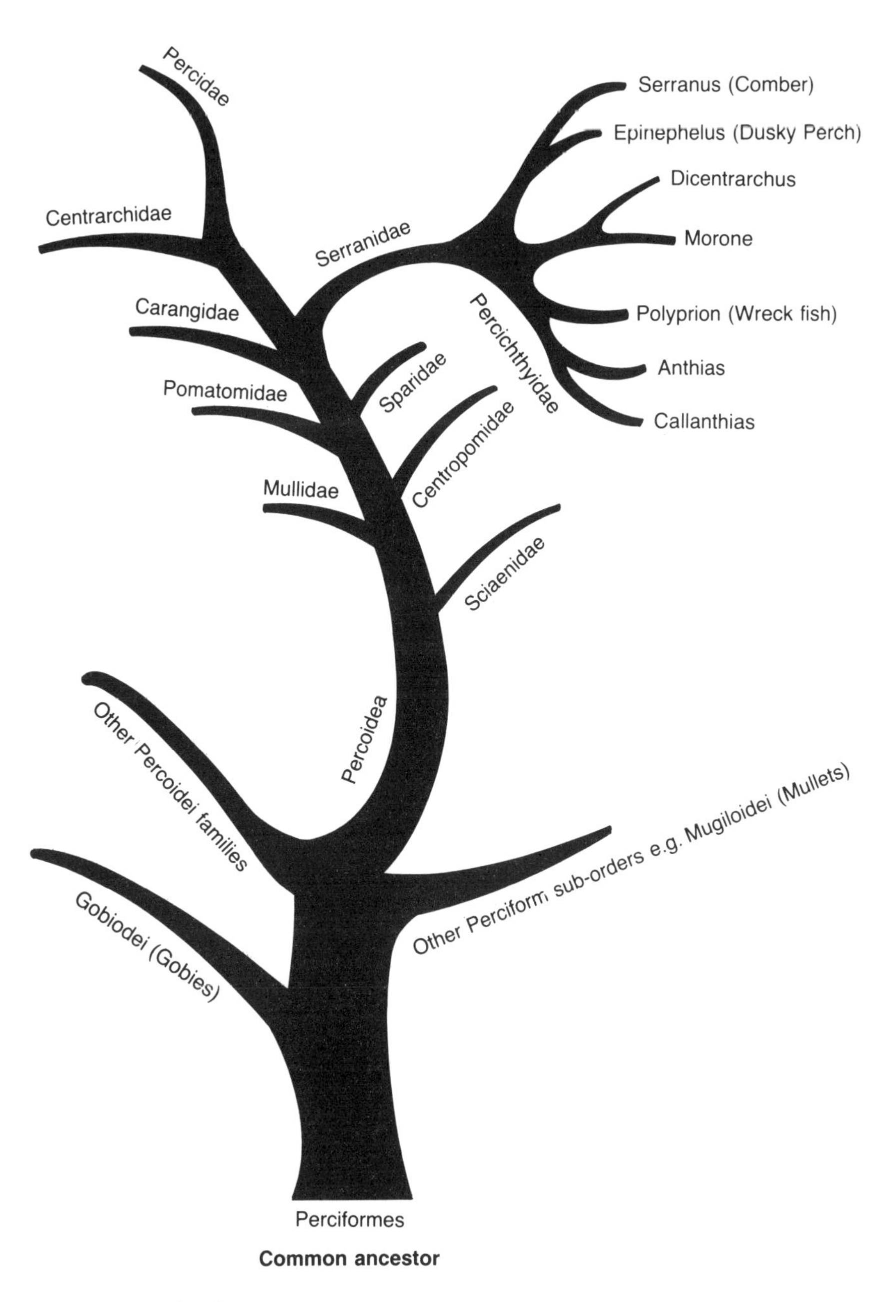

Fig. 1.1 Bass family tree.

Labrax lupus. They listed seven species in this genus, one of which was *Labrax elongata* (Cuvier), which was said to be found off northern Egypt and was first described by Geoffroy St Hilaire in 1809. In a major review of the Percoidei, however, Boulenger (1890) regarded *L. elongata* as synonymous with *Morone labrax*, because of errors which he detected in the works of Gill (1861). Gill had given credibility to *L. elongata* without even seeing a specimen, and had called it *Dicentrarchus elongata* on the basis that Geoffroy St Hilaire's illustration showed two anal-fin spines. According to Whitehead and Wheeler (1966), however, Geoffroy St Hilaire's original description and drawing of *D. elongata* did not contain sufficient detail of diagnostic features (Fig. 1.2) to enable it to be distinguished from *labrax* and, because the latter has three anal-fin spines, even the name *Dicentrarchus* has a doubtful origin.

Boulenger (1890) gave the generic name *Morone* priority for six species, of which four were American and two were Afro–European: *Morone labrax*, the bass, and a close relative, the Afro–European spotted bass, *Morone punctatus*.

The American members of the genus were the anadromous striped bass (*M. saxatilis*, Walbaum, 1792) and the freshwater white bass (*M. chrysops*, Rafinesque, 1820), white perch (*M. americana*, Gmelin, 1788) and yellow perch (*M. mississippiensis*, Jordan and Eigenmann, 1887). There was some dispute over the naming of each species, however, because the generic name *Roccus* was used by some taxonomists. This was originally proposed in 1815 by an American, Mitchill, who considered *Morone* and *Roccus* to be two separate genera, based on the (wrong) assumption that each had a different position of the pelvic fins on the body.

Thus, over the years, the two European sea basses have attracted all three generic names (*Morone*, *Dicentrarchus* and *Roccus*), and sufficient confusion has been caused that some authors (e.g. Audousset, 1978) still recognize the existence of a third European species, *D. elongata* (Pickett, 1989).

In an attempt to clarify the situation, Whitehead and Wheeler (1966) reviewed the generic names of the six substantiated bass species, and provided a key to them based on observed differences in the dentition of the basihyal (tooth plates on the floor of the mouth). They proposed that the six species fell into three groups each containing two species, corresponding to the genera *Morone* (white perch and yellow perch), *Roccus* (striped bass and white bass) and *Dicentrarchus* (sea bass and spotted bass). Subsequently, European authors seem to have accepted this suggestion, whereas Americans have generally stuck to *Morone* for all six species. Waldman (1986) based this latter argument on the characteristic possession, in all six species, of lateral tooth plates at the base of the tongue. These features had not been described in detail when Whitehead and Wheeler produced their review, and Waldman suggests that allocating

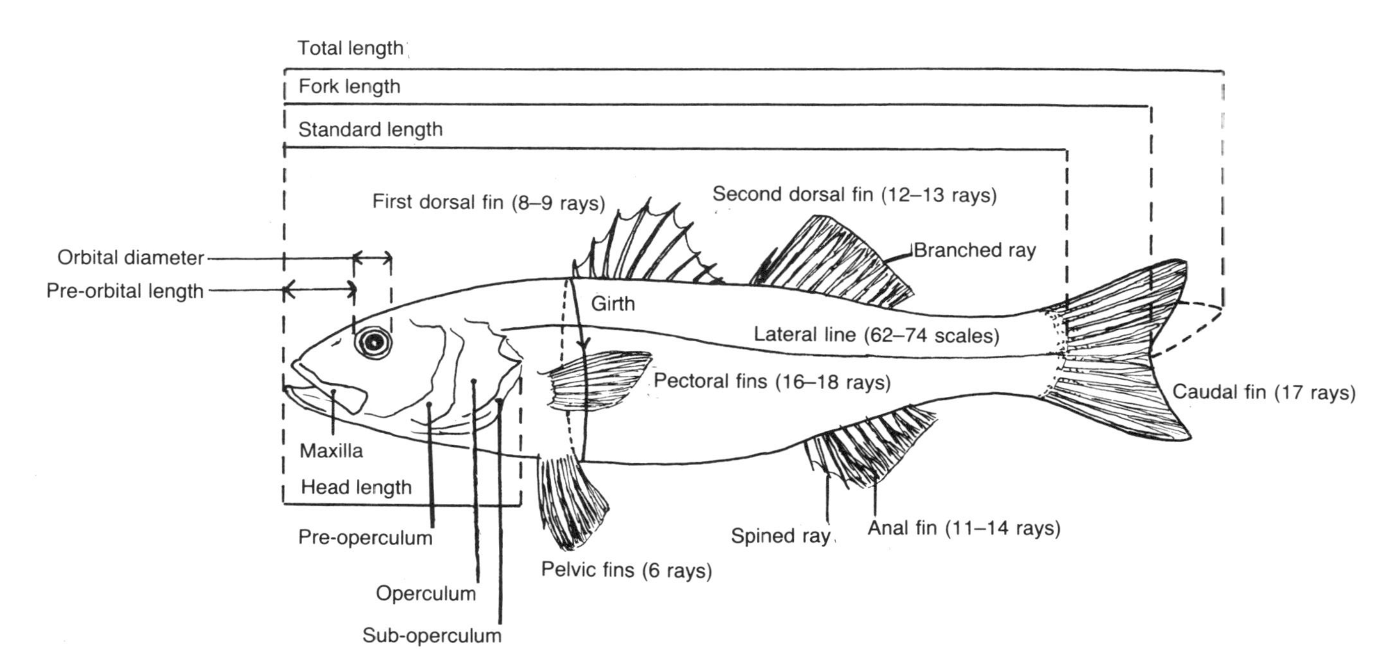

Fig. 1.2 Outline of bass with named fins and meristic counts.

labrax and *punctatus* to a separate genus was no longer supportable, and that *Morone* should take priority over *Dicentrarchus*.

For the purpose of this book, however, we retain the generic name *Dicentrarchus*, as used in Europe, but readers should note that *Dicentrarchus*, *Morone* and *Roccus* are essentially taxonomic synonyms.

1.3 COMPARATIVE MORPHOLOGY

The following key, adapted from Whitehead and Wheeler (1966), lists the basic diagnostic features of the six *Morone* species, four of which are shown in Fig. 1.3.

I. Lower border of pre-operculum (central of three bones covering the gills) has several antrorse (forward-facing) spines; dorsal fins separated by a space; (distribution: Mediterranean and eastern Atlantic; marine and estuarine),
   (a) Lateral line with 62–74 scales (mode 70); teeth on vomer (central tooth plate in roof of mouth) in crescentric band without posterior extensions. Adults without black spots on upper part of body;
   – *labrax* (Linnaeus, 1758). sea bass
   (b) Lateral line with 57–65 scales (mode 60); tooth patch on vomer anchor-shaped; adults with small black spots on upper part of body;
   – *punctatus* (Bloch, 1792). spotted bass

II. Lower border of pre-operculum with small denticulation directed downwards; (distribution: western Atlantic, eastern and southern North America).
   (a) Dorsal fins separate and spines increasingly even in length towards posterior; two sharp spines on hind border of operculum (main bone covering gills); teeth on base of tongue.
      i. Body elongate, its depth less than one third of its length; 57–67 lateral line scales; teeth at base of tongue in two parallel patches (marine and estuarine);
   – *saxatilis* (Walbaum, 1792). striped bass
      ii. Body deeper than *saxatilis*, its depth more than one third of its length; 52–58 lateral scales, teeth at base of tongue in a single row (fresh water);
   – *chrysops* (Rafinesque, 1820). white bass
   (b) Dorsal fins connected; second anal spine almost equal in length to the third spine; a single sharp spine on the hind border of the operculum; teeth present along edges of tongue, but not at base.
      i. Longest dorsal spine about half length of head; faint streaks on flanks (marine and fresh water);

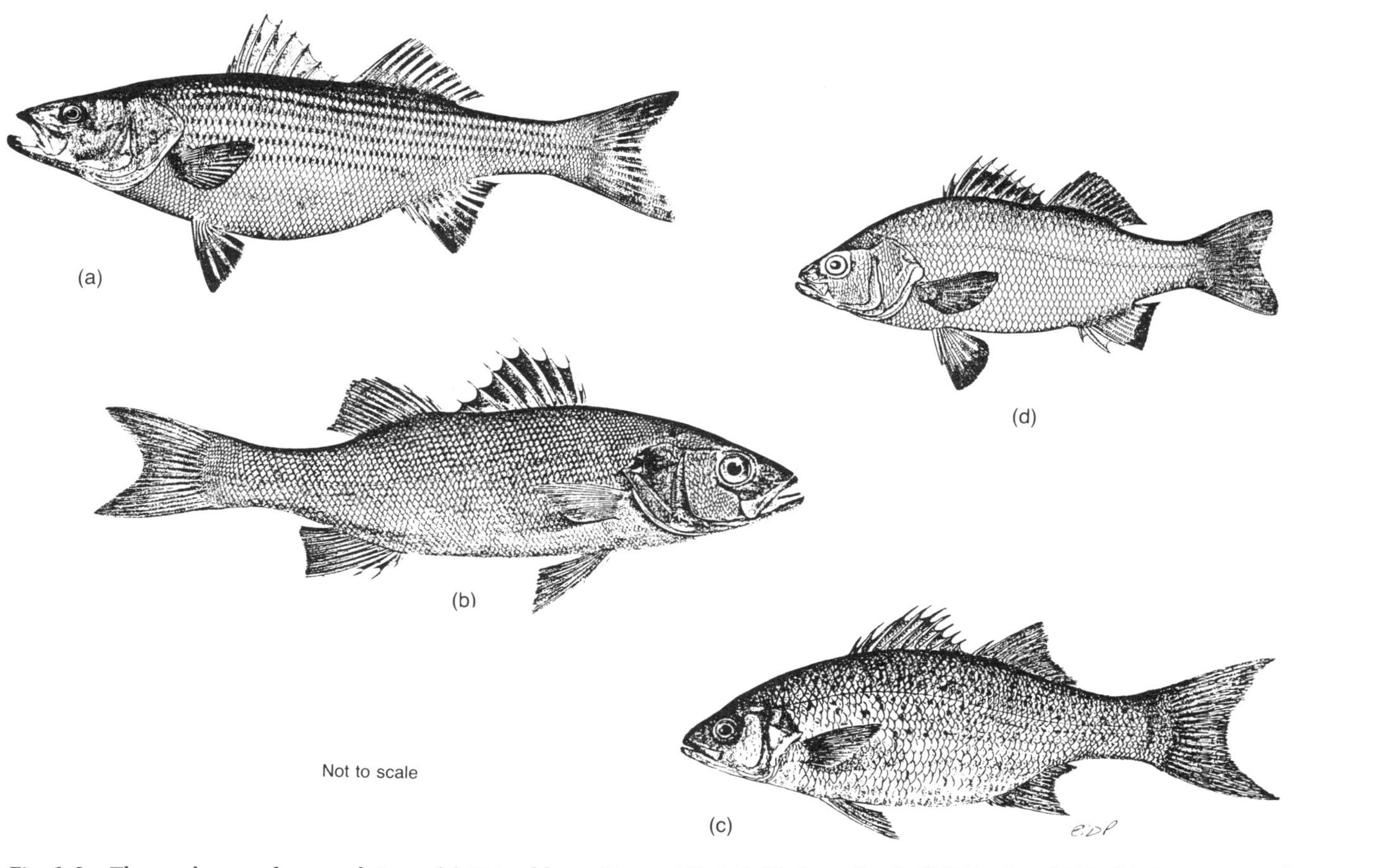

Fig. 1.3 The sea bass and near relatives. (a) Striped bass, *M. saxatilis* (adult), from Goode (1884, plate 170); (b) European sea bass, *D. labrax* (probably small juvenile), from Day (1884); (c) Afro–European spotted bass, *D. punctatus* (author's drawing); (d) white perch, *M. americana*, from Goode (1884).

– *americana* (Gmelin, 1788). white perch

ii. Longest dorsal spine greater than half length of head; seven distinct longitudinal lines on flanks, interrupted posteriorly (freshwater or lower Mississippi valley);

– *Mississippiensis* (Jordan and Eigenmann, 1887). yellow perch

The following list of meristic counts and proportional measurements for *D. labrax* (Fig. 1.2) is adapted from Boulenger (1890).

Depth of body, 22–27% of total length (TL); length of head, 27–33% of TL. Snout, 1.5–2 × diameter of eye, which is 14–20% of length of head in the adult; wide space between eyes covered with plain, circular scales; lower jaw slightly projecting; maxillary (upper jaw bone) extending to below anterior third or centre of eye, the width of its extremity being 60–75% diameter of eye; membranous pre-orbital bone (forward of eye socket) entire; four to six strong backward-facing spines on lower border of preoperculum; lower opercular spine stronger than upper. Teeth on vomer forming a crescentic group; a patch of teeth along the middle of the tongue and others on the borders. Gill rakers longer than gill fringes, numbering 16 to 18 on lower part of anterior arch. Anterior dorsal fin originating behind vertical through axil; spines rather strong, fourth and fifth longest, 40–60% length of head. Pectoral fin about half length of head. Third anal fin spine longest, 25–33% length of head. Middle caudal (tail) fin rays about 66% length of outer fin rays.

1.4 GENERAL MORPHOLOGICAL DESCRIPTION OF *D. LABRAX*

The pelvic fins of the bass (along with those of all other members of the genus) are situated well forward on the belly and slightly behind the pectoral fins, which are in turn positioned just behind the opercula. The pectoral fins each have 16–18 branched (soft) rays. The two dorsal fins are separate. The anterior dorsal fin has 8–9 spined rays and the posterior dorsal fin has one (the first) spined ray and 11–12 branched rays. The first ray of the pelvic fins, which each have 5 branched rays, and the first 3 rays of the anal fin (which has 9–11 branched rays), are also sharply spined. The caudal fin is forked and normally has 17 branched rays, but can have as few as 13, and is coloured grey, sometimes with a dark blue sheen. The other fins are whitish or grey, but sometimes the pectoral and pelvic fins have pale blue leading edges.

The body of the bass is covered by large, regular scales, and its colour varies considerably, depending on the fish's origin, ranging from dark grey, blue or green on the back to a white or pale yellow belly. The flanks are silver-blue, sometimes pale gold or bronze. Visible scale margins are often black-edged, particularly on larger specimens. The scales have tooth-like

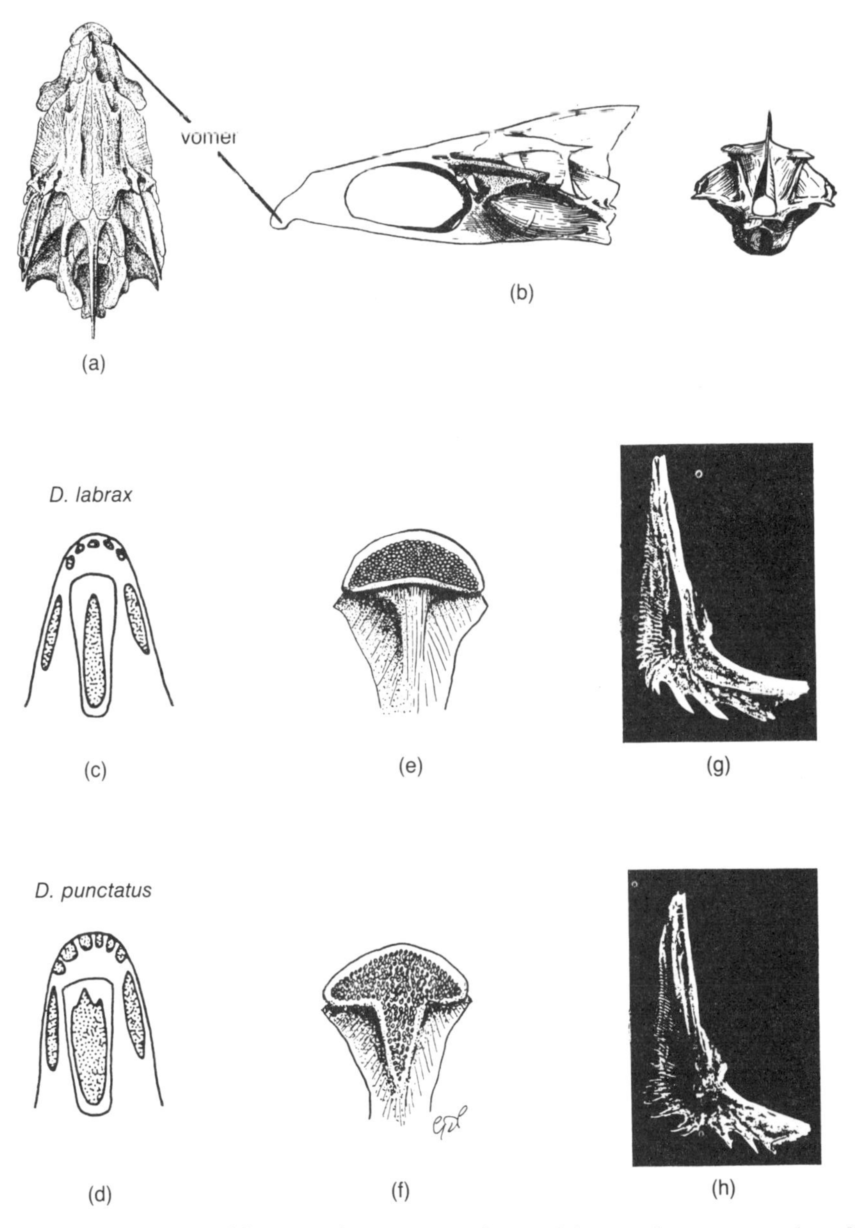

Fig. 1.4 Osteological features of European sea bass *D. labrax* and *D. punctatus*: (a and b), skull (labrax only); (c and d), tooth patch formations; (e and f), vomers showing tooth patches; (g and h), pre-opercular bone of (g) *D. labrax* and (h) *D. punctatus*. Sources: (a) from Boulenger (1890), (b) and (e) reproduced with permission from Woolcott (1957), (c) and (d) reproduced with permission from Waldman (1986), (g) and (h) reproduced with permission from Bou Ain (1977).

processes in the visible segment and those on the head are smooth with no clear segmentation, but the top of the head lacks scales. The head in young bass appears quite pointed, but becomes blunter in older fish, and shades from dark grey or black above, through metallic hues of silver, to gold and purple on the sides and opercula. The opercula each have two sharp spines on the posterior border. The eyes are relatively large for a marine fish and the iris is silver-white, shot with gold. The mouth is large with a characteristic Perciform 'Gladstone bag' shape when opened, and has prominent maxillae. The tongue has toothed margins, with tooth plates rooted in its soft tissue. There are tooth patches on the floor of the mouth, on the central tooth plate and on the palatine bones on either side of the roof of the mouth. There are also tooth patches just inside the mouth in front of the tongue. These features (along with the pre-opercular bone) can be used for diagnostic purposes (Fig. 1.4).

Bass in their first year tend to be paler in appearance than older fish, and usually have dark spots on the back and upper sides (Fig. 1.5). Normally, these spots have disappeared by the time the fish is 1 year old, although some individuals retain them well into the second year. In UK waters the occasional adult of *D. labrax* is seen with a few black spots on the back.

Fig. 1.5 One-year-old bass, some retaining dark spots.

Variants

Although there are only two recognized bass species (*D. labrax* and *D. punctatus*) in European and Mediterranean waters, it is not unusual to come across individuals or groups of fish which vary from the typical form. Barnabé (1976a), who has carried out extensive work on bass in the French Mediterranean, gives examples of four types: the 'standard' *D. labrax; D. punctatus; D. labrax* which retains black spots on the body at 1 year old; and *D. labrax* with a blunt head shape. The last he proposed as *D. labrax*, 'forme oncocephalus' which, although rare, occurs sufficiently often to be noted as a local type. The occurrence of blunt-headed individuals is well documented in striped bass (Hickey *et al.*,1977) and in largemouth bass, *Micropoterus salmoides*, (Pickett, 1979), and is thought to be the result of either a germinal defect or an oxygen deficiency associated with high organic loads and biological oxygen demand in coastal waters during the early embryonic stages of development (Mansueti, 1958, 1960). This abnormality is also seen in other fish species, notably the esocid freshwater pike, *Esox lucius* (Buller, 1981).

The shape of the head has been found to vary between individual bass in the same shoal, some fish having rather pointed heads, whereas others have blunt heads with a pronounced 'bridge' to the snout (Fig. 1.6).

There are some regional differences in morphological characteristics of bass, which may be related to the environmental conditions of their habitat. Barnabé mentions 'sand-bass' and 'rock bass' in the Mediterranean. The latter form is very dark in colouration – black on the back and head with dusky, charcoal grey on the sides – and they tend to have relatively large heads and thin bodies. They are known colloquially in the UK as 'reef bass' or 'black jacks'. Bass from sandy areas or caught in deep water are often much paler in colour. Like many fish species, bass that are kept in tanks are seen to change their colour in response to their background.

A distinct form of bass occurs occasionally in South-east England and the eastern English Channel, though its appearance may be related less to topographical conditions than to diet and growth rate. These bass are unusually deep bodied and have small, rounded heads, small caudal lobes and a pale, golden-yellow sheen on the sides. They have appeared in discrete shoals from which all fish that are caught look similar, and stayed in one location for a few days or weeks before moving on. We first saw this type of bass in November 1983 whilst tagging bass at Bradwell, Essex, when 14 where caught, ranging in size from 55 to 82 cm total length. They appeared to be present at the tagging site for only one set of high spring tides, and differed in appearance and in vigour from other bass caught in previous and subsequent weeks at the same site. They had the appearance of having ripe

Fig. 1.6 Variation in head shape of bass.

gonads, but the one fish examined internally was found to have small gonads and a large amount of mesenteric fat. Scale samples revealed very fast growth (see Chapter 6) in all the pale, fat specimens. A few months later, in the summer of 1984, similar fish appeared off Guernsey and Alderney in the Channel Islands. These fish were taken near areas where the local and characteristic 'reef' bass were being caught, mainly on longlines. They came to light when the French markets started rejecting consignments that contained pale, fat-laden bass, on the grounds that they had deteriorated too far in quality in the warm weather. In contrast, the local bass presented no such problem. Scale samples again revealed that the pale, fat fish had significantly faster growth rates than were considered normal for bass around the Channel Isles.

1.5 ANATOMY

The anatomy of bass has been rather neglected in published descriptions, but Fig. 1.7 presents a simplified anatomical diagram. For greater detail of anatomy, digestive systems, nervous systems, musculature and physiology we recommend Craig (1987), who has made a comprehensive study of the perch (*Perca fluviatilis*) which, although from a different family, shares many common features with the Serranidae.

Bou Ain (1977), working in Tunisia, studied the morphology and

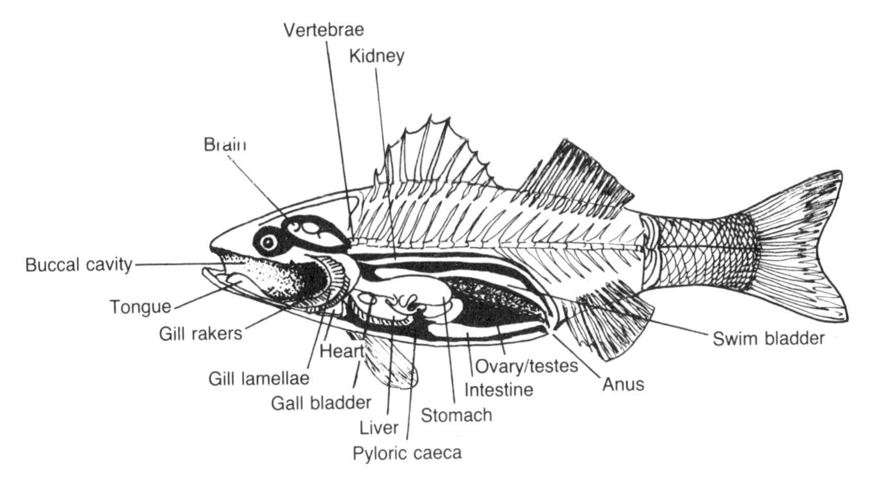

Fig. 1.7 Diagrammatic drawing of a dissected bass, showing major organs.

anatomy of *D. labrax* and *D. punctatus* and described the main skeletal features: cranium, opercular bones, otoliths, vertebrae, fin rays and scales. Specific aspects have been studied in greater detail, e.g. otoliths by Boulineau-Coatanea (1968) working in Brittany. Reference to the skeleton of the bass is often made only in connection with the counting of vertebrae for diagnostic purposes. Gravier (1961), for example, gives a count of 25 for Moroccan bass, which applies to both *D. labrax* and *D. punctatus,* although Bou Ain recorded a few *labrax* with 24 vertebrae.

In common with that of most predatory species, the digestive tract of the bass is relatively short and the intestine runs relatively straight, without loops or coils, through the body cavity. The main organs, heart, liver, stomach and spleen are situated well forward, allowing space for maturing gonads and accumulated mesenteric fat. Like all the perciforms, the bass is a physoclist, having a closed swim bladder which has no connecting duct to the oesophagus.

1.6 CONCLUSION

Despite the potential for controversy which still attends the taxonomic classification of the European sea bass, there is little doubt now that we are dealing with a single species of marine fish. What little is known about the biology of its nearest relative, the spotted bass, permits the two species to be distinguished only on the basis of morphological features and, to a lesser extent, geographic range. Even then, the variations in body shape, colour and behaviour of the sea bass, upon which previous arguments for there being more than one species of bass living around the British coast were based, probably serve more to confuse than confirm the species' identity. It should not be forgotten that the morphological similarity between the sea bass and the striped bass, though they live on opposite sides of the Atlantic, led to earlier attempts to fill gaps in our biological knowledge of the former with ideas and assumptions based on the striped bass. Much of the rest of the first part of this book seeks to establish a more credible biology for the sea bass in its own right.

Chapter two

Distribution and general behaviour

2.1 DISTRIBUTION/RANGE

Many authors give the geographical range of the bass as being in the north-east Atlantic and North Sea, from Norway and the Baltic in the north, southwards to North Africa, and including the Mediterranean and adjoining seas; but they fail to show where the bulk of the population is found. Some, quite wrongly, refer to it as a mainly Mediterranean species.

The scientific literature indicates that most research on bass has been carried out in the sea areas between the United Kingdom and Ireland to the north, and Morocco to the south, and also in the Mediterranean as far east as Israel. Research on marine fish is usually carried out where there are fishing interests, and because fisheries only exist where there are sufficient fish to make them viable, it is indicative that there have been few studies on bass in Belgium, Holland, Denmark, Sweden or Norway. The first published record of a bass in Scandinavian waters was a fish caught in August 1829 off Hven in the Sund, near Copenhagen (Smitt, 1892); subsequently single bass were recorded four times on the west Swedish coast near Gothenburg. Smitt quotes Collett as saying that the largest specimens in Scandinavia measured 35 cm fork length and that solitary bass have been recorded as far north as Tromsö.

The bass is caught in appreciable quantities only south of 53° N in the North Sea and 54° N in the Irish Sea and on the Atlantic coast of Ireland. On the coasts of England and Wales, this northern limit corresponds approximately to a line drawn between the Wash and Morecambe Bay. Occasional rod-and-line and trawl catches of bass are made on the Yorkshire coast and around the mouth of the River Tees, and small bass are known to frequent the warm-water outfall at Blyth power station on the Northumberland coast. Larger bass are also caught from time to time in

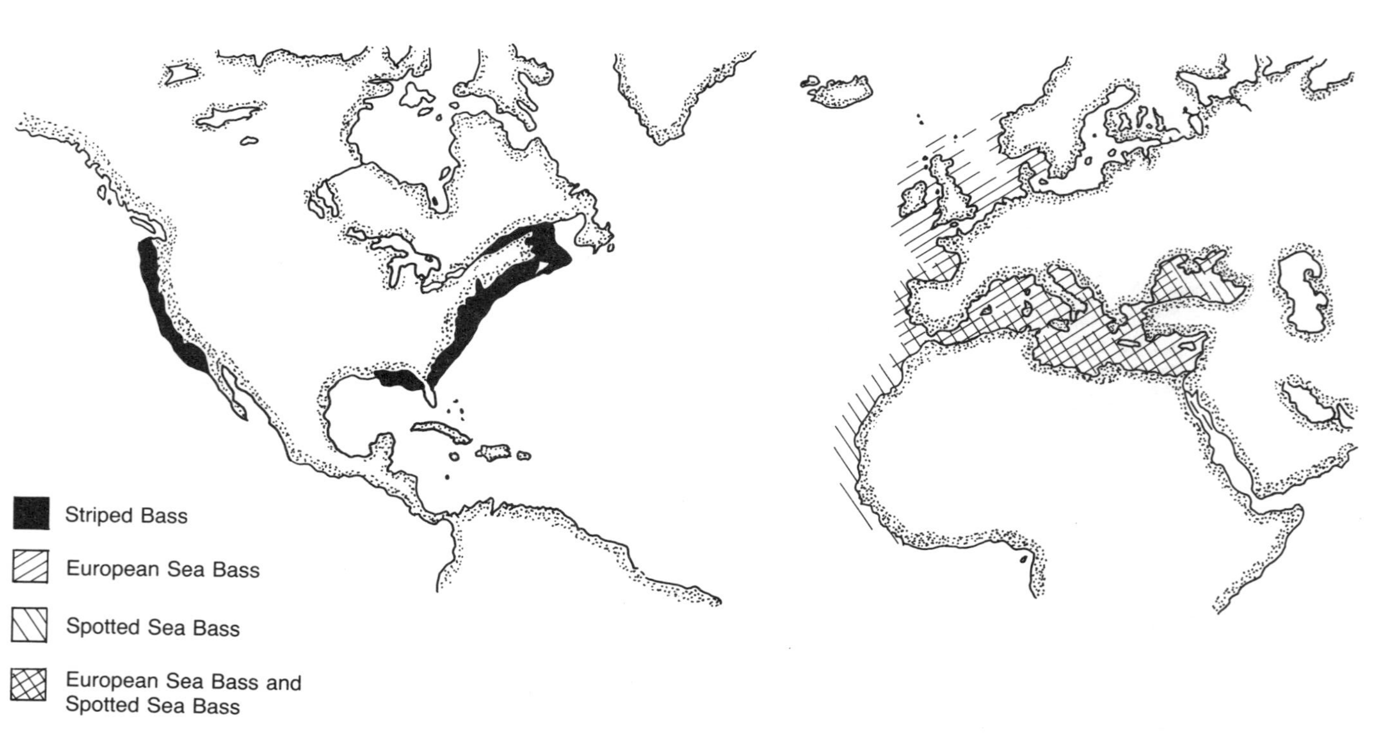

Fig. 2.1 Geographic range of sea basses.

salmon net fisheries off North-east England and East Scotland, and in fixed nets in the Solway Firth, and there are reports of bass in anglers' catches in Luce Bay in most years. Indeed, occasional bass are reported from all around the Scottish coast, including the Orkney and Shetland Isles (*Sea Angler*, Dec. 1992).

Further south, the bass is confined to coastal waters through the Bay of Biscay to Morocco, where it occurs at least as far south as Agadir (Gravier, 1961). Whilst it appears to be less common along the Mediterranean shores of Spain than on the Atlantic side of the Iberian Peninsula, the bass occurs throughout the Mediterranean and in the Adriatic and Black Seas.

The range of the bass overlaps that of its near relative, the spotted bass, which is found in the Atlantic from southern Brittany at least as far south as Senegal (Cadenat, 1935), and in parts of the Mediterranean and Gulf of Suez (Ben Tuvia, 1974), but *D. punctatus* has a much more fragmented distribution than *D. labrax* (Bou Ain, 1977). The presence of bass in faunistic lists of some of the more periferal countries is indicated below.

Russia	Svetovidov (1964)	Belgium	Poll (1947)
Turkey	Davedjank (1926)	Tunisia	Bou Ain (1977)
Western Morocco	Cadanat (1935)	Egypt	Rafail (1971)

Adult bass live inshore and in estuaries, even entering fresh water during summer, but are generally found in deeper, warmer water offshore during winter. For populations in the North Sea, English Channel and Irish Sea, there are seasonal shifts in distribution as a result of migrations (Chapter 5). Bass are rarely encountered more than 80 km seawards of main land masses, although some fish have been shown to cross the Western Approaches to the English Channel during migrations from the UK west coast to western France (Pawson *et al.*, 1987).

There is no evidence of any movement of bass across the Irish Sea between Ireland and Britain, despite extensive tagging in coastal waters of both countries. Further south, where sea temperatures are more stable, bass may be less likely to move to the north or south, or offshore in winter, although in the Golfe du Lion (Barnabé, 1976a) and in the Arcachon region (Stequert, 1972), they are known to leave shallow tidal lagoons between October and December and to return there in spring. It is not yet certain whether the recent unseasonably high sea temperatures (1988–90) will cause a permanent northward shift in the centre of distribution of the north-west European population of bass.

Local distribution and preferred habitat

Bass under 30 cm spend most of the year in or near estuaries, or along open coastline, close inshore. Young-of-the-year bass (which are termed 0-groups, because they are in their first calendar year) are found, initially as postlarvae, in the shallow waters of tidal lagoons, estuaries, harbours and marinas. They move into deeper water near by as they grow, but around England and Wales, estuaries are the main habitat of bass up to 4 years old (Kelley, 1988a).

As bass grow, they spend more time in deeper water; fully mature fish have been caught in bottom trawls fished at depths to 80 m. In recent years, commercial fishermen using rod and line have found concentrations of large bass throughout the summer around some deep-water wrecks in the English Channel. They may, however, be encountered throughout the water column, according to the location of their food, and are often seen feeding or shoaling at the sea surface under calm conditions. Shallow beaches with breaking surf and offshore sandbanks will attract bass and, in summer, they are often to be found in tide rips or races off rocky headlands, reefs or banks. These places provide feeding opportunities for small prey species, and for bass, which seem to favour turbulent water. Adult bass are seldom found over muddy bottoms, although 0-groups and 1-groups (fish in their second calendar year) regularly occur in muddy creeks cutting through marshes bordering lowland estuaries.

Influence of salinity

Bass of all ages can tolerate fresh water and will move freely into it, i.e. they are euryhaline. Whether bass are actually attracted to lower salinities is open to question, although postlarvae have been observed to congregate near the salt/freshwater interface in estuaries (Dando and Demir, 1985; Kelley, 1986; Jennings *et al.*, 1991). Lancaster (1991) suggests that the influence of temperature or the availability of food may be the main motivating factors for postlarval and older bass to enter low salinity water.

Chervinski (1974), working in Israel, reported that 0-group bass, which had been transferred from high-salinity lagoons to low-salinity ponds, survived and grew. He carried out experiments using bass of 20–34 mm in length, to determine whether transfer to fresh water could be achieved. Direct transfer from seawater to tap water (salinity 0.5‰) at around 18 °C resulted in total mortality, but all fish survived direct transfer to dilute seawater with a salinity of 3.9‰. By gradually reducing the salinity from 10 to 0.5‰ over a 24 h period, 100‰ survival of the bass fry was obtained. Freshwater adaptation is therefore quite possible under cultivation conditions; indeed the Romans are said to have kept bass in freshwater ponds where they fed freely (Columella, quoted in Couch, 1862).

In the wild state, sudden, enforced changes in salinity are unlikely to occur, and mortality of bass owing to freshwater incursion has not been

Fig. 2.2 Power station discharge channel where bass are attracted by warmed water.

observed. Kelley (1986) found that, in some West Wales creeks, 0-group bass moved towards fresh water in times of drought and away in times of flood following high rainfall inland. Bass fry seem, therefore, to distribute themselves with respect to salinity, and they are certainly more tolerant of salinity fluctuations than most marine fish.

Thermal influences

Bass are also highly eurythermic; that is, they will tolerate a wide range of temperatures. Although they occur in water which may cool to less than 5 °C in England and Wales – below which temperature, 0-groups will begin to die after a few weeks (Kelley, 1986; Lancaster, 1991) – they show a preference for conditions above about 10 °C. Temperature is likely to limit the distribution of 0-group bass to areas in which the water does not become too cold for survival. Adults can withstand water temperatures ranging from 2 °C to 32 °C (Barnabé, 1990).

The natural effect of falling sea temperatures in autumn is to cause a movement of bass into deeper, less cold water. Kelley (1988a) suggested that, for 0-group fish around the UK, these movements begin as the water temperature inshore drops below 15 °C. The larger individuals are the first to forsake the shallows, and it is possible that the smaller (slower growing or younger) fish which are left behind become torpid and may cease to feed if the water temperature stays below 10 °C (Lancaster, 1991). A sudden temperature drop in summer will cause a temporary movement of small bass to deeper water, or seawards in estuaries.

Discharges of artificially warmed water may at times attract adult bass as well as 0-groups and 1-groups. These discharges are often from industrial installations where cooling water is required: coastal power stations (Fig. 2.2), oil refineries and factories. The warmed water itself may have some benefit in terms of feeding and growth opportunities (Chapter 6) and, in severe winters, may enhance survival locally.

The presence of a warm-water effluent will tend to concentrate small bass in shallow, inshore areas where they would not normally be present in winter. Where there are several warm-water discharges in one area, such as from West Thurrock, Kingsnorth and other power stations in the Thames Estuary, the natural winter distribution of young bass may be considerably altered. Larger, adult bass, although attracted to the warmer water, tend to spend less time within the influence of the discharges, and bass which have been tagged in their vicinity have shown the normal autumn migration patterns, i.e. to the south and west, away from the summer feeding area.

2.2 BEHAVIOUR

Much of what follows is anecdotal, because there have been few scientific studies of the behaviour of bass in the wild. Some of the best available descriptions of bass behaviour come from Barnabé (1976b, 1978), who worked with larvae, juvenile and adult bass mainly held in aquaria, and many of his observations probably hold true for the behaviour of bass in the natural environment. They are supplemented by our own observations, which have not been published elsewhere. There has been little investigation into the physiological aspects of the behaviour of bass.

Shoaling

It is not known how, at an early age, bass form shoals. Eggs and larvae occur in relatively low densities in the ichthyoplankton (Thompson and Harrop, 1987; Jennings and Pawson, 1992) and metamorphosing larvae do not appear to form recognizable shoals. Several authors (Claridge and Potter, 1983; Aprahamian and Barr, 1985; Dando and Demir, 1985; Kelley,

1988a), however, have observed postlarval bass and fry congregating where physical barriers are present in upper estuaries, creeks and harbours. In these circumstances, there may be a build-up of numbers over days or weeks, possibly in pulses related to tidal cycles, until the tiny bass are able to aggregate into shoals. The shoals may vary from a few dozen individuals to many thousands, according to the strength of the year class and local conditions. Tagging returns suggest that bass may remain in distinct groups for several years at a time. To date, there have been three separate reports of two bass, which were tagged as part of a single batch several months previously, being recaptured on the same day and at the same place (Pawson *et al.*, 1987). There are many other cases of several fish from the same tagged batch being recaptured months or years later at one locality over a period of a few weeks.

In winter, large catches of bass are occasionally made by trawlers, which have encountered a shoal of fish in the 50–85 cm length range (1.5–8.0 kg). This suggests that bass may retain a shoaling habit throughout most of their lives. Anglers' observations of 'solitary' bass (Young, 1955) probably stem from the fact that when large bass are caught they are often taken in isolation from other bass. This may reflect as much on the feeding behaviour of big bass, which tend to move rapidly through an area according to the tidal cycle, as on the supposed lack of further members of a shoal.

There are many records, usually found in angling publications, of vast shoals of bass seen swimming or feeding near the sea surface under calm conditions. Before commercial fishing in South-east England began to target bass in the late 1970s, such large shoals were commonplace offshore in the Thames Estuary. Bob Cox and John Rawle, charter-boat skippers from Bradwell, Essex, took out angling parties and made some large catches of big bass in these circumstances. More recently, anglers and commercial line-fishermen have encountered large shoals of bass in the vicinity of Beachy Head (Sussex), off Portland Bill (Dorset) and near Worm's Head (Glamorgan), all sites of tidal races in which feeding bass are often to be found.

It is difficult to estimate the size of fish shoals at sea, but in late 1990, Cox (pers. comm.) reported an aggregation of small bass in the River Blackwater, Essex, which produced a signal on echosounder paper which indicated a shoal 800 m long × 150 m wide × 2 m deep. If these fish shoaled at a density 50 individuals per cubic metre (not a high density for small fish), such a shoal would contain around 10 million bass.

In the recent past, it was common practice for Cornish fishing communities to own large seine nets. These were used to encircle the large, dense shoals of thick-lipped grey mullet (*Crenimugil labrosus*) that would sometimes occur in winter near Land's End, especially off bays such as Sennen Cove. Shoal spotters, called 'huers', would watch from the cliffs for the fish to move into locations where they were catchable. Sometimes these shoals would be of bass, and catches of several tonnes of large, mature fish would be made.

Feeding – pursuit of prey

Bass are opportunistic predators throughout their life and will adopt a wide range of tactics to find and capture their prey. Generally, bass hunt as a shoal, taking whatever prey species are seasonally abundant in a particular location. If the food items are on the sea-bed, e.g. crabs, bivalve molluscs or polychaete worms, the bass shoal will spread out and will graze, head down, along the substrate. However, bass are more readily observed when they feed on pelagic prey, which they drive upwards towards the surface and attack from below at a steep angle. This feeding may become quite frenzied, and the bass will often be seen to break the water surface in pursuit of their prey. We have often observed similar feeding behaviour of bass kept in tanks at DFR's Lowestoft Laboratory, particularly when they are fed following a break of a few days (feeding is normally daily). As the feeding activity dies down, the bass become more circumspect in their approach and inspection of the food. The routine addition of food at the regular feeding time will elicit an aware response, which is indicated by rapid raising and lowering of the dorsal fins and darting movements along the side of the tank. Once one or two fish have inspected and accepted the food, all will join in the feeding, and soon the bass are seemingly oblivious to the presence of observers.

Bass in tanks will approach and engulf live brown shrimp (*Crangon crangon*) with amazing speed, returning immediately to their former position. Live prey are usually held in the buccal cavity before being swallowed. This mode of feeding is often experienced by anglers when using fish livebaits from a boat under still water conditions. There will often be a characteristic hard 'knock' on the rod top as the bass takes the prey, followed by a slack line as the bass resumes its station and prepares to swallow the bait.

Anglers are generally convinced that weather conditions play an important role in the feeding behaviour of bass. Stormy conditions, which stir up the sea-bed and reveal food items, are known to stimulate feeding; calm, clear water is claimed to have the opposite effect. Anglers' observations of the way a bass takes a hook bait vary enormously owing, in part, to the type of bait used and the manner in which it is presented. When taking a bait, the bass are expected to pull strongly at the rod and line, but at times bites can be quite delicate and the baits are often not taken convincingly, even by large fish.

In autumn and winter, bass are often found in association with pelagic shoals of mackerel (*Scomber scombrus*), sprat (*Sprattus sprattus*), pilchard (*Sardina pilchardus*) and scad (*Trachurus trachurus*) off Brittany and South-west England. Trawler skippers have reported that, when using colour echosounders to locate mackerel shoals in the western English Channel, they have sometimes seen a characteristic shape of contrasting colour appear at the bottom of the trace of the shoal. Their supposition, that this signal indicates the presence of bass, has sometimes been substantiated by fishing the trawl at the appropriate depth and catching bass.

Aggression and territorial behaviour

Although bass are known to attack prey species quite violently, their reaction to other species of their own size is not well documented. When a bass appears to be threatened by a larger animal, it will either retreat rapidly or adopt a typical defence posture. This involves raising the first dorsal fin and distending the opercula, in order, it seems, to make itself appear larger and to present as many sharp spines to the aggressor as possible. When live bass are handled they often adopt this posture, whilst flexing the spine, which makes the fish difficult to hold. This behaviour sometimes results in a rigor, possibly owing to shock, and even when the fish is returned to the water, the opercula are held rigidly open, which may cause buccal pumping to cease and lead to asphyxiation and, ultimately, death.

There is little evidence of aggressive behaviour between bass of similar size, although they may be territorial when occupying summer feeding areas (Carlisle, 1961). Barnabé (1976b, 1978) reports that the behaviour of bass in captivity is not the same as that observed in the wild, whilst diving in the littoral zone of the Golfe du Lion. It is possible that these differences in behaviour may be due to the fact that in the wild, fish can escape from a potential aggressor, whereas the movements of captive fish are much more restricted.

Swimming

Bass are strong swimmers and, as is the general rule with fish, their swimming power (and hence speed) increases with size. Bass are not inconvenienced by strong tides and turbulence, and have been caught on rod and line for tagging in the tide race off Portland Bill, Dorset, where it is estimated that the current averages 4 knots (2 m s^{-1}) on the flood tide. They seem to be particularly adept at using back-eddies and other slack water to avoid the strongest currents. Swimming power is gained from the large caudal fin and the bass propels itself forward by bursts of three or four flicks of its tail, with all other fins flattened to the body to reduce drag. Recent experiments in a flume tank at Lowestoft (Arnold, 1969) have demonstrated that bass of 24–37 cm are able to hold station in currents of up to 80 cm s^{-1}, (Moore *et al.*, 1994) (Fig. 2.3).

Fig. 2.3 Bass of 24–37 cm total length undergoing swimming speed trials in a flume tank at Lowestoft.

It was once thought that adult bass were not catchable in trawls because they swim too quickly. Today, specially designed, lightweight high-headline bottom trawls or midwater trawls have been developed, and the additional power of modern engines has enabled relatively small vessels to tow these trawls fast enough to catch bass. In surveys conducted annually by DFR in the Solent, Hampshire, using trawls designed especially to catch bass in the 25–35 cm total length range, it has been necessary to maintain a speed of 2 ms^{-1} through the water and to tow with the tide. It is rare to catch bass longer than 40 cm with this gear, although they are known to be present in the survey area.

Tagging studies suggest that bass can sustain a high average swimming speed when on migration, helped, no doubt, by the presence of darker-toned 'slow' muscle along the flanks of the fish (Pitcher and Hart, 1982). For example, a 62 cm fish released on the coast of North-west Wales in November 1972 was recaptured 18 days later near Wolf Rock off West Cornwall, a minimum distance of about 480 km (Kelley, 1979). Most migrations appear to be made along the line of the coast (Pawson *et al.*, 1987) and, because striped bass are known to use tidal flows when migrating (Kerr, 1953), it is reasonable to assume that European sea bass might also do this. Kerr also reported that in the more confined space of holding tanks and flumes, striped bass orientated into a current and rarely swam with it unless frightened or exhausted and seeking refuge. This optomotor response (Harden Jones, 1963), which enables fish to hold station against a visible background, has also been observed in bass during the flume trials at Lowestoft. When fishing for bass, most offshore trawlers tow their gear for several hours to tire the fish swimming along with the net, fish which would otherwise escape on hauling.

'Flashing'

Anyone who has held bass of any age in tanks or aquaria will have noticed the behaviour called 'flashing' (or 'grattage du flanc', Fouché, 1986). This typically occurs when the fish are resting as a shoal near the bottom. Every now and then, an individual will slowly sink towards the bottom and then suddenly move forward, turn on one side and appear to rub one flank on the substrate (Fig. 2.4).

This behaviour is readily observed in shallow, clear water, where the first indication that bass are actually present in an area is often the sudden silvery flash of a fish's flank. Although mainly a characteristic of the bass, this behaviour also occurs with grey mullets, particularly golden-grey mullet (*Liza aurata*). Two explanations for flashing have been put forward: either it is intended to disturb small crustacean food items buried in a sandy bottom, or it is an attempt to get rid of ectoparasites. Although we have observed bass flashing against a fibreglass tank base which had no overlying sediment, close examination of specimens that have flashed repeatedly has revealed no evidence of such parasites or damage to the mucous layer.

Fig. 2.4 Small bass 'flashing'.

Burying and escape mechanisms

A possible link with flashing may be found in a rather more rarely observed behaviour of 0-group bass, that of burying themselves in a soft substrate. We first observed this at Lowestoft one morning in 1983, when 10 bass appeared to be missing from a tank which had a sand/shingle bottom layer about 5 cm deep. On closer inspection, caudal fins and heads could be seen protruding above the substrate; all the bass had partly buried themselves on their sides. Later observations revealed that the movement often commenced with a couple of 'flashes', followed by a pronounced sideways dig of the snout into the sediment. Once the head was buried, the bass then shuffled into the sand, and its snout usually re-appeared before the fish became stationary. Barnabé (1976b) also observed this phenomenon, and noted that the bass lay on their sides at an angle of 15–30° and stayed buried for 30–60 s. He also reports similar behaviour of adults in intensive rearing systems.

This behaviour may be a survival mechanism in the face of perceived threat from predators. Many fishermen will have observed how bass escape an oncoming beach seine net by turning on one side and sliding under the footrope, and it is possible that bass can sometimes use this ability to avoid capture by trawls. In contrast, grey mullets will, in a similar situation, attempt to leap over the headline. In the relatively turbid waters around England and Wales there have been few scientific observations of the behaviour of wild bass in relation to fishing gear, although there are many anecdotal accounts.

As might be expected, the above adds little to ancient knowledge, as is shown by the following verse from Ovid, written some 2000 years ago (quoted by Couch, 1862):

> "In like extremity the greedy toils,
> With arts more exquisite the bass beguiles;
> Low he descends when powerful fear commands,
> And scoops with labouring fins the furrow'd sands;
> Lodged in that cave expected fate derides,
> While o'er his back the leaded foot rope slides."

This extract indicates that a seine or gill net fishery existed for bass in those days.

Couch also suggested how bass are able to escape once hooked on lines, again referring to ancient sources, in this case Oppian (*Halieutica*, 2nd century. AD):

> "Fishermen observe that they often deliver themselves from the line by cutting away the hook; and they suppose it to be done by means of the

serrated cutting edge of the gill-cover. But it is more probable that it is effected by drawing the line across the teeth; which are numerous, and capable of acting like a file or rasp."

Another verse, perhaps borrowed from Ovid, gives a less unlikely explanation of their method of escaping from hooks:

> "The crafty Bass, whene'er they conscious feel
> Deep in their jaws infix'd the barbed steel,
> Writhing with restive fury backward bound,
> The hook dismissing through the widen'd wound."

Many modern anglers are likely to have experienced bass escaping from the hook whilst in play, and again there is evidence of fishing for bass in ancient times, this time with lines and hooks.

Barnabé (1976b) gives accounts of observations, made by diving, on the behaviour of bass in relation to various fixed fishing gears.

Reaction to disturbance

There is little published information on the effects of disturbance on marine fishes, and the following account is therefore largely anecdotal. Experience with bass held in captivity suggests that bass of all sizes rapidly adapt to being kept in glass-sided aquaria, but they often remain shy when held in metal and opaque fibreglass-walled tanks and continue to show fright responses to various forms of disturbance, including noise and the presence of humans. In the wild, they are notably shy of human activity and the noise which this generates. Bass will only ignore the presence of boats and the noise of their engines when they are already feeding frenziedly. At these times, they seem to abandon caution and can often be seen striking at prey near the surface close alongside fishing boats.

Bass seem to be aware of the shadow of a boat on the sea-bed and, along with many other fish species, can detect the sound of the hull reverberating in moving water when the boat is at anchor; the vibrations are probably transmitted down the anchor rope (Cox, 1985). One of us (GP) has had the experience of watching a shoal of feeding bass moving with the tide towards an anchored angling boat. When almost within easy casting range, the shoal circumnavigated the boat, and then resumed its former course downtide of the vessel. It is this displacement effect that is the basis of the success of pair trawling, where the trawl is towed between two vessels and fish are herded and concentrated between the boats. Many other marine fish species behave in this manner, and in some circumstances, this may be the only method with which bass can be caught.

Barnabé (1976b) remarked that bass completely ignored boats in busy waterways in the south of France. This is certainly true of juvenile fish, but there is an impression that bass tend to become much more wary of such activity with increasing size and age. Kelley (pers. comm.) suggested that the increase in coastal boat traffic, moorings and general human activity during the second half of the 20th century has had an adverse effect on sport fishing for large bass. In many UK estuaries, adult bass can now only be caught occasionally, and then only in the quietness of the dark hours. It is widely accepted by bass sport anglers that 'early and late' is the motto when fishing from the shore, especially under calm conditions. It may well be that bass are avoiding the inshore areas at times when they have experienced human disturbance. Away from the shore, bass will feed at any time of day or night, according to local tidal conditions and food availability. Other activities, such as dredging for gravel and seismic surveys used in gas and oil exploration, are thought by fishermen to displace bass shoals, albeit temporarily, but attempts to assess this effect have so far proved inconclusive (MAFF, unpublished data).

2.3 THE FUTURE

Much of what we have to say about the distribution and behaviour of bass will come as no surprise to anglers and fishermen, because this is the sort of information upon which their ability to catch fish depends. The bass's attraction to warm water (in North-west Europe) and its tolerance of low salinities, are common knowledge, but the functional significance of these attributes is only now becoming apparent. As with any marine fish, the behaviour of bass has tended to be inferred from what little can be seen from above the sea's surface, but observations of fish held in aquaria are beginning to reveal the true nature of this fascinating beast. The bass is a distinctive fish in many ways, and would be an ideal subject for further behavioural research, in the laboratory and at sea.

Chapter three

Diet and feeding

3.1 INTRODUCTION

Knowledge of the diet and feeding patterns of fish is important to anglers, fishermen and scientists alike. To the angler, it is an aid to choosing the right bait and fishing at the right place and time. This also applies to the commercial fisherman who is using baited hooks, but even when trawling or gill netting, it is a considerable advantage to know when and where fish feed. The scientist may be interested in the nutrition of fish or may be more concerned with the ecological relationships between predator and prey species. Fisheries managers are becoming more aware that changes in the abundance of stocks of a particular, predatory species, sometimes effected through deliberate management action, can inadvertently alter the mortality rate and, hence, abundance of other commercial species (Brander and Bennett, 1986).

Being placed high in the food chain, the bass depends for its survival on many other species of marine animal. It is also one of the most abundant predators in some sea areas and is therefore a strong candidate for study of its place in the marine ecosystem. To date, however, work along these lines has been mainly qualitative, e.g. Chevalier (1980) and Kelley (1987). Other than in aquaculture studies (Santulli *et al.*, 1993), there have been no attempts to quantify the species' food intake, to measure digestion rates or to assess the impact of predation by bass on populations of prey species.

In its early life stages, the bass is one of the most important components of the marine fauna in coastal waters. Not only can humans physically alter the natural habitats of small bass, but in so doing, will also affect the distribution and abundance of those species upon which bass depend for food. Changes to estuarine habitats, such as drainage of salt-marsh, alteration to flow regimes, thermal effluents and chemical pollution, may all have an effect on the food chain as well as directly on the bass themselves. To evaluate the effects, actual and potential, of human activities in coastal waters, it is important to have a comprehensive knowledge of the diet of bass throughout the range of the species' distribution.

3.2 DIET

Overview of studies

Only three studies of the food of adult bass in the British Isles have been conducted over a period of more than one year. These have been undertaken in Ireland by Kennedy and Fitzmaurice (1972) and in England and Wales and the Channel Isles by Kelley (1987) and by DFR (MAFF, unpublished data). The latter included bass caught in offshore areas, which were not covered by Kelley's study. The diet of immature bass was first described in the UK by Hartley (1940) using fish caught in the River Tamar in Cornwall. The diet of 0-group bass has been studied in the Severn Estuary by Aprahamian and Barr (1985) and, more recently, in nursery areas on the South Wales coast by Lancaster (1991).

The diet of all sizes of bass has been studied in the South of France by Barnabé (1980). In Brittany, seasonal variation in stomach contents was recorded during one year, using mainly adult fish caught by nets (Boulineau-Coatanea, 1969, 1970). The diet of juvenile stages of bass in lagoons in Spain has been reported by Arias (1980), and in the Po river, Italy, by Ferrari and Chieregato (1981). Roblin and Buslé (1984) carried out further studies on fry and juvenile stages of bass in southern France and, in reviewing other work, these authors suggested that differences observed in diet could be related to the latitude at which samples were taken for the various studies.

The preferred prey types change as bass grow; they can take larger food items and as they inhabit a more diverse environment, adult bass are able to take advantage of different ranges of prey species. The types of food eaten will depend to a great extent on the distribution of the bass themselves. This aspect may have biased some of the above-mentioned studies, because the bass examined may have been unwittingly selected from certain types of site with characteristic faunas. For this reason, we have considered it useful to present the results of a cross section of studies, attempting to cover as much of the bass's local and general distribution as possible.

Changes of diet with size and age

Larvae

The diet of larval bass in the wild is not well known, but once the yolk sac has been absorbed, between 9 and 25 days after hatching, they are thought to feed on small members of the zooplankton, such as copepod nauplii. The food preferences and requirements of bass larvae reared in captivity have been the subject of several studies (Section 10.3) and these have demonstrated that larval bass require mobile planktonic food such as rotifers (*Branchionus* spp.) and brine shrimp nauplii (*Artemia* spp.).

0-Groups

Roblin and Bruslé (1984), working in the lagoons of the Golfe du Lion, catagorized all bass below 6 cm standard length as fry, and those of 6–12 cm as fingerlings. These two groups displayed distinct differences in the selection size of prey items. The range of species eaten varied between sites according to their preys' respective availability. This is a consistent theme in all published studies, irrespective of latitude, and indicates that, from a very early age, the bass is an opportunistic predator. The bulk of its diet consists of the species of crustaceans and fish which are the most readily available.

Various works show that, as bass fry grow, they switch their diet from zooplankton to larger, epibenthic prey, such as harpacticoid copepods, *Corophium* sp., gammarids and mysids (ghost shrimps). Fingerling bass will start to take larger decapods, such as brown shrimp, and small fish, such as gobies. Whilst virtually any animal-based food of a suitable size, including terrestrial insects and freshwater invertebrates, will be eaten, the diet of 0-group bass is dominated by the more abundant food organisms in the near-shore environment, particularly mysids, and in all areas, the bulk of their diet consists of small crustaceans. Table 3.1 shows the main classes of food taken in four different regions.

0-Group bass will prey on each other if there is a sufficiently wide size range to allow this, irrespective of additional feeding opportunities. In June 1990, around 500 postlarval bass of 1–3 cm were placed in an aquarium at Lowestoft and, despite a plentiful daily supply of brine shrimp and finely chopped lesser sandeel (*Ammodytes tobianus*), the larger bass soon began to eat the other fish. Within 3 months, only 25 bass of 7–12 cm remained. Barnabé (1976a) reports similar cases of cannibalism by bass kept in tanks at Sète in the South of France. Katavic *et al.* (1989), consider that cannibalism may only be significant in intensive cultivation systems, though Corps (1992) reports 1-group bass of the extremely abundant 1989 year class in southern England having eaten 0-group bass.

Juvenile diet

Owing to differences in the growth of bass in various regions (Chapter 6), it is more appropriate to compare the food of juveniles on the basis of their length than their age. Different authors have varying concepts of what is meant by

Table 3.1 Generalized diet of 0-group bass in four European studies* and shown as percentage frequency of occurrence of food items

Food items	*1 River Severn Estuary, 1980*	*2 Po River Delta, Italy 1977–78*	*3 Esteros (fish ponds), Cadiz, 197277*	*4 Etangs of Golfe du Lion, 1977–78*	
	(< 70 mm)	*(24–79 mm)*	*(21–120 mm)*	*(< 72 mm)*	*(72–144 mm)*
Annelids (unspecified)				14.3	60.0
Polychaetes		10.0			
Cladocera	3.5	5.0			
Ostracods				2.4	
Copepods (harpaticoid)	10.3			83.3	
Copepods (calanoid)	16.7–40.0	60.0	2.4		
Copepod nauplii		15.0			
Cirripede larvae		35.0			
Mysids	65.4–100.0	80.0	47.6		10.0
Idotea	11.5				
Sphaeromes				4.8	
Amphipods (unspecified)		30.0	2.9		80.0
Gammarus	19.2–33.3			52.4	
Corophium	16.7–40.0			88.1	60.0
Decapods (unspecified)			14.3		
Larvae		20.0			
Zoea larvae				4.8	20.0
Insects (unspecified)		6.0		28.6	
Coleoptera			2.4		
Diptera larvae		10.0			
Fish (including larvae)		40.0	4.7	2.0–4.0	
Debris				2.0–4.0	20.0

*Sources: 1, Aprahamian and Barr (1985) (from Table 2); 2, Ferrari and Chieregato (1981) (recalculated); 3, Arias (1980) (from Table 10); 4, Roblin and Truslé (1984) (from Table 19).

'juveniles'. Roblin and Truslé's juveniles at Canet were only 9–12 cm in length (and about 1 year old), whereas Kelley's (1987) juveniles ranged from 25 to 41 cm, and were 3–6 years old.

The numbers and types of food organisms found in the stomach contents of juvenile bass of various sizes at a range of locations are given in Table 3.2. The apparent regional differences in the food items taken are not wholly explained by the different size ranges of fish studied. It is more likely that the wide range of habitats and the resident fauna of each study area hold the key to the variability in these observations. Many of DFR's samples of bass were taken offshore, some in relatively deep water in winter, and their diet provides a marked contrast, for example, with that of similar-sized fish from the brackish lagoons off the Po river, or from the fish ponds of Cadiz. Crabs do not feature as a prey species of juvenile bass in the studies in southern France, Italy or Ireland, whereas they predominate in both of the UK investigations. Copepods, cirripede larvae and decapod larvae only appear in samples from the Po river embayment. The main difference between Kelley's data and DFR's is that small crustaceans (isopods, amphipods and mysids) do not feature in the latter, despite these bass having a lower length limit (15 cm) than Kelley's (23 cm). Kelley's samples contained small crustaceans and came mainly from the Erme Estuary in South Devon, and from estuaries in Pembrokeshire, where these food items may have been relatively abundant. DFR's samples from South-west England came mainly from the Dart, Tamar and Fal estuaries, and the remainder were obtained predominantly from South-east and southern England, particularly from the Solent.

The English and Welsh regional data on juvenile bass stomach contents are summarized in Table 3.3. As in bass from other countries, the largest component of the diet is crustaceans, although the main species vary from area to area. The diet of 1-group bass was dominated by shrimps, fish, crabs and marine worms, rather than mysids, all of which are common in the intertidal zone. Additionally, small fish were found more regularly in the diet of larger bass. It is surprising that the lesser sandeel was recorded only five times in stomachs sampled on the English south coast, considering that this is the most widely used bait in that region and many juvenile bass are caught on it. This suggests that samples obtained by angling with baited hooks are not necessarily biased towards any particular food type. Overall, crabs, particularly shore crabs (*Carcinus meanas*), were the most common food of juvenile bass in the UK, followed by the brown shrimp.

Adult diet

Fish are generally termed adult once they are ready for their first spawning and have begun to undertake regular migrations between feeding and

Table 3.2 Diet of juvenile bass in various countries*

Food items	[1]*Canet, France (18 fish, 105 ± 12 mm)*		[2]*Po River, Italy (17 fish, 230–310 mm)*	[3]*Ireland (32 fish, 105–139 mm)*		[4]*England and Wales (262 fish, 230–380 mm)*		[5]*England and Wales (467 fish, 150–400 mm)*
	Bass	*Items*	*Average per fish*	*Bass*	*Items*	*Bass*	*Items*	*Bass with item*
Coelentrates (unspecified)						2	12	
Annelids (unspecified)	13	6				9	41	
Polychaetes								13
Copepods (harpacticoid)			6.2					
Copepods (calanoid)			23.9					
Copepod nauplii			534.8					
Cirripede larvae			9.5					
Mysids	2	4	0.1	11	20	1	40	
Isopods	4	9						
Amphipods	1	6	1.7	2	6	11	627	
–Gammarids				2	3			
Decapods								
–Larvae			0.4					
–Crabs						64	99	170
–Shrimps	4	3		28	74	68	210	99
Gastropods				1	4	5	14	37
Bivalves								2
Fish	10	6	0.9	1	1	34	71	71

*Sources: 1, Roblin and Bruslé (1984); 2, Ferrari and Chieregato (1981); 3, Kennedy and Fitzmaurice (1972); 4, Kelley (1987) (England and Wales); 5, Authors (unpublished data).

Table 3.3 Stomach contents of juvenile bass in four regions in England and Wales, recorded as numbers of bass with each food (frequency of occurrence). From biological sampling carried out 1982–87 by the authors (MAFF unpublished data)

Food item	*North-west Wales*	*South-west England*	*South England*	*South-east England*	*All regions*
Annelids (unspecified)			5		5
Arenicola sp.		1	2		3
Nereis sp.			5		5
Prawns/pink shrimp		1	2		3
Brown shrimp		54	24	18	96
Crab (unspecified)		33	44	9	86
Carcinus sp.	2	16	61	5	84
Gastropods (unspecified)			7		7
Slipper limpet (*Crepidula fornicata*)			30		30
Bivalves (unspecified)			2		2
Fish (unspecified)		3	9	7	19
Sandeels		5			5
Herring/sprat	4	10	25	8	47
Total bass sampled	12	199	232	113	556
Total bass with food	6	100	189	39	334
Total bass empty	6	99	43	74	222

spawning areas. Maturity is more closely related to the length of bass than to their age (Chapter 7) and, therefore, fish with faster growth mature earlier in life. There are some slight regional differences in the size at first maturity of males and females, but over 90% of bass around the UK can be classed as adult at 42 cm.

The main studies on the food of adult bass are those of Barnabé (1976a) (based on 150 fish from the Mediterranean Sea), Boulineau-Coatanea (1970) (272 fish from Brittany), Kelley (1987) (963 fish from England and Wales) and Kennedy and Fitzmaurice (1972) (103 fish from Ireland), and we present data on 838 (UK) fish which are not published elsewhere.

Regional variation and seasonality. A common feature in all studies of the diet of adult bass is the importance of various species of crustaceans. In North-west Europe, the shore crab was the most important single species, being eaten in both the moulting and intermoult stages. In Brittany, *Carcinus* spp. provided the greatest weight of crustacean food, although they were exceeded in numbers by swimming crabs (*Macropipus* spp.). In the UK, Kelley recorded more shore crabs in the diet of bass than any other individual prey species and, in Ireland, more bass contained them than all other food items combined. In the results of DFR's study, shown in Table 3.4, shore crabs accounted for a similar frequency of occurrence to that of brown shrimp, when evaluated as a single identifiable species. There was, however, a high frequency of unidentified crab remains in some samples, which may have been taken in the soft (just moulted) form and more rapidly digested.

In this study, however, crustaceans come in second place behind fish in terms of frequency of occurrence, quite a different result from the observations made in the short-term studies in Brittany and Ireland and in Kelley's long-term UK study. This may be due to genuine regional differences in food availability and bass feeding behaviour, but it is also affected by the greater distance offshore from which many of DFR's samples were taken. The Brittany samples were all collected in one area (Rade de Brest) and Kelley's fish were taken mainly by angling from the shore or from small boats fishing close inshore.

DFR's samples include a high proportion (90%) of commercially caught bass, and some samples were taken 15–50 km offshore. They included catches taken by trawlers between December and March inclusive, from water of 50–80 m depth around South-west England, where much of the UK adult stock overwinters. Fish dominate the diet of bass from this area, even when the samples from all seasons are combined. This contrasts with Kelley's rod-and-line-caught bass from South-west England, which show higher frequencies of crustaceans than fish (Kelley, 1987). These bass were caught close to the shore and were probably feeding at the time. The explanation is, of course, very simple: bass living offshore in deep water tend

Table 3.4 Stomach contents of adult bass in four regions in England and Wales, recorded as frequency of occurrence (as in Table 3.3). from biological sampling carried out 1982–87 by the authors (MAFF unpublished data)

Food item	*North-west England and Wales*	*South-west England*	*Central southern England*	*South-east England*	*All regions*
Annelids (unspecified)				11	11
Aphrodite (sea mouse)				1	1
Arenicola sp.	5		5	1	11
Nereis sp.	1	1		3	5
Prawns/pink shrimps				3	3
Brown shrimps	4	3	6	49	62
Crab (unspecified)		17	10	44	71
Carcinus sp.	5	6	12	34	57
Other crustaceans	1	5	4	40	50
Gastropods (unidentified)			1	1	2
Slipper limpet			3	1	4
Cephalopods (squid/cuttle)		3	1	4	8
Fish (unspecified)		15	10	22	47
Herring		1		7	8
Sprat		2		6	8
Pilchard		6			6
Clupeid remains		2	10	7	19
Mackerel		43			43
Gadoids		8	5	9	22
Sandeels		9		8	17
Gobies spp.		1	2	1	4
Total bass with food	13	237	56	223	529
Total bass empty	2	199	19	89	309
Total bass sampled	15	436	75	312	838

Table 3.5 Diet of adult bass (frequency of occurrence of food items) in England and Wales (all regions combined) by month, and proportion of adult and juvenile bass with empty stomachs. From biological sampling carried out 1982–87 by the authors (MAFF unpublished data)

Food item						*Month*						
	Jan.	*Feb.*	*Mar.*	*Apr.*	*May*	*June*	*July*	*Aug.*	*Sep.*	*Oct.*	*Nov.*	*Dec.*
Annelids (unspecified)					11							
Arenicola Sp.					1	1			1		5	
Nereis sp.				1	4							
Prawns/pink shrimps									2			
Brown shrimps					10	4	5	2	38	4		
Crab (unspecified)			1		22	4	8	7	27	2	2	
Carcinus sp.				1	17	17	12	2	7	1	3	
Other crustaceans			1	4	8		26	1	4		4	
Gastropods (unidentified)						1	1					
Slipper limpet					3	1						
Cephalopods (squid/cuttle)	1					1	2				1	
Fish (unspecified)	39		4	10	9	11	4	20	21	28	20	13
Herring	1			4		2			1			
Sprat	1				2	1	1		1		1	1
Pilchard	1									1		1
Clupeid remains					1	1	15		1			
Mackerel	27			1						6	9	
Gadoids			2				3			4	12	1
Sandeels				5	2	1	4	5				
Gobies spp.	1				1				2			
Number of adult bass sampled	75	4	29	44	89	77	79	93	106	50	73	47
% empty	43	100	76	39	40	36	43	35	29	26	32	57
Number of juvenile bass sampled	42	8	41	15	65	43	110	21	196	15	0	0
% empty	86	100	39	13	20	35	30	29	46	33	–	–

to encounter large shoals of pelagic fish such as mackerel and pilchard. Bass living inshore, whether in summer or winter, are also able to find more shallow-water types of food, including crustaceans and annelid worms.

The seeming regional variation in the type of prey species may be the consequence of erratic sampling or related to seasonality, as already discussed. There are, however, a few examples of genuine regional differences, and a knowledge of the distribution of certain prey species supports this. The most obvious example (Table 3.5) is the brown shrimp, which was recorded in six months of the year and in all regions, but generally at a fairly low level. In the Thames Estuary in South-east England, however, it was the most common item in bass stomachs, particularly in September. Shrimp are particularly abundant on the sand banks and in the tidal creeks in this area, and probably represent the most easily obtainable crustacean component of the bass' diet, at a time of year when they are still feeding in warm, shallow water.

There are other seasonal trends revealed by these data. Small gadoid fish, mainly poor cod (*Trisopterus minutus*) and whiting pout (*T. luscus*), are mainly taken in the autumn, October to December. Polychaete worms, particularly ragworms (*Nereis* spp.), are eaten in spring. At this time, they leave their burrows to form spawning aggregations in shallow water, and bass may feed on them exclusively whilst easy pickings are to be had. In April 1966, Donovan Kelley (pers. comm.) recorded one bass of 3.5 kg which contained 70 individuals of *Nereis* spp. Similarly, shore crabs are particularly vulnerable at the moult stage, which usually occurs between May and September.

Concerning evidence for cannibalism, Kelley (1987) records only three bass, all adult, out of 1225 specimens which had their stomach contents examined, as having eaten other bass (one each). Similarly, Boulineau-Coatanea (1969, 1970) also records three prey bass in three adult bass out of 272 samples. Of some 2000 bass stomachs that we have now examined since 1982, not one has contained identifiable bass remains.

Comparison of the diet of juvenile and adult bass

The difference in the occurrence of major food items in the diet of the broad groupings juveniles and adults around the UK (1982–87) is shown in Table 3.6. This presentation is somewhat artificial, as the change of items taken is not sudden and varies with predator size within both categories.

The most important feature to note is the overall increase in the proportion of fish in the adult diet, although juvenile herring and sprat feature more strongly in the juveniles. The decline in the proportions of shrimps and crabs with the size of bass is also noticeable. Various species of cephalopod (e.g. *Loligo* sp., *Sepia* sp.) were present at a low level in these

Table 3.6 Comparison of the diets of juvenile and adult bass in England and Wales, shown as percentage frequencies of total bass sampled* and total bass containing food. From biological sampling carried out 1982–87 by the authors

	Fish			*Crustaceans*			
	Unspecified	*Sandeels*	*Sprat, herring*	*Unspecified*	Carcinus	*Prawns and pink shrimp*	*Brown shrimp*
Juveniles (15–42 cm TL)							
F_f	5.8	1.5	14.2	26.1	25.5	0.9	29.2
F_a	3.4	0.9	8.7	15.5	15.1	0.6	17.3
Adults (42–86 cm TL)							
F_f	51.0	3.2	5.1	22.9	9.3	0.6	11.7
F_a	35.2	2.2	3.5	15.8	7.4	0.4	8.1

Table 3.6 continued

	Molluscs			*Annelids*			*Other*
	Unspecified	Crepidula	*Bivalves, Squids*	*Unspecified*	Arenicola	Nereis	
Juveniles (15–42 cm TL)							
F_f	2.4	9.1	0.6	1.5	0.9	1.5	0.3
F_a	1.4	5.4	0.4	0.9	0.5	0.9	0.2
Adults (42–86 cm TL)							
F_f	0.4	0.8	1.5	2.1	2.1	0.9	0.2
F_a	0.3	0.5	1	1.4	1.4	0.7	0.1

*Total number of bass sampled, 556 juveniles and 776 adults.
F_f, % frequency of bass sampled with food in stomach.
F_a, % frequency of all bass sampled.

samples, and they are thought to be taken mainly by large bass, which is why some specialist anglers use them as bait. Although the proportion of crabs and shrimps declines in adults, this does not necessarily mean that the total quantity of these species eaten by an individual bass decreases as it grows. Though it is still likely that the most readily available food items of a suitable size will usually be taken, there are indications that certain groups of bass feed selectively on a limited range of foods. Bass that spend most of the year in offshore locations, such as around deep-water wrecks, are likely to be mainly fish eaters, simply because of the availability of such prey.

3.3 FEEDING BEHAVIOUR

Feeding periods

Feeding studies, as opposed to those on diet, are concerned with when, why and how much rather than what food is taken. Lack of feeding may be due to low temperature, satiation, spawning activity or time of day, but these effects are not easily determined under natural conditions. Many anglers have observed and firmly believe that bass feed more intensively at night (or dusk and dawn) rather than during the day. Numerous articles in angling magazines over the years attest to this phenomenon. A lack of bites on fishing tackle may not necessarily indicate that the bass are not feeding: they may be disturbed, or out of range of the bait, and cannot be captured until the appropriate feeding conditions at a particular location, daylight, tide, etc., prevail.

In order to know how often bass feed, it is desirable to know how quickly they digest food and evacuate the gut. For 0-group bass, this subject has been investigated by Lancaster (1991), but little is known for adult bass in the wild. Under artificial conditions (cultivation), such knowledge is important in order to determine economically viable feeding regimes and conditions for efficient growth. So, for cultured bass taking prepared foods, feeding rates and conversion ratios (the increase in fish weight relative to the weight of food eaten) at different temperatures are well known (Barnabé, 1990; Santulli *et al.*, 1993). Feeding patterns are unlikely to be the same in the wild, where more energy is spent chasing food and migrating, and temperature and light periodicity vary considerably.

Seasonal variation in vacuity

A clue to the frequency of feeding and the conditions that might control feeding may be obtained from an analysis of vacuity, the proportion of fish examined that are found to have empty stomachs. The interpretation of the outcome of such an analysis needs to be considered carefully, according to the way in which the sampled fish were caught. Unlike salmon in rivers and overwintering mackerel, bass caught on rod and line are most probably feeding at the time, and one might reasonably expect to find that a high proportion of these fish would contain some recently taken food. Netted samples may give a truer representation of the feeding activity of the population, because they will also include fish that were not feeding. Empty stomachs do not always signify non-feeding fish, as regurgitation may sometimes occur during capture. This is particularly likely to happen when fish are caught in deep water and the decrease in pressure as they are brought to the surface allows the swim bladder to expand, owing to their closed swim bladder system. This forces the stomach, and its contents, out of the fish's mouth, thereby increasing the apparent level of vacuity in subsequent sampling. Usually, however, the presence of the everted stomach in the fish's mouth will enable mistakes to be avoided.

Mean monthly values of vacuity for adult bass sampled by DFR over the years 1982–87 throughout England and Wales are given in Table 3.5. Like the adults, juveniles appear not to feed in February, and the second-highest values of vacuity are in January for juveniles and March for adults. This result supports the hypothesis that because juvenile bass remain in cooler water than adults during winter (Pawson *et al.*, 1987), they might be expected to feed less avidly then. Three-quarters of the fish in samples from South-east England, which included the only bass examined in January and February, had empty stomachs, and in March, adult bass are usually spawning in deep water off southern England (Jennings and Pawson, 1992), at which time they might be less inclined to feed. It is probable, therefore, that both low temperature and spawning activity may limit food intake. In captivity, adult bass continue to feed in winter at water temperatures of 9–11 °C. As few adult bass are caught at temperatures below this, we do not know the threshold below which feeding ceases, although Barnabé (1990) suggests 7 °C. In experimental conditions at Lowestoft, the feeding rate of juveniles dropped dramatically at water temperatures below 10 °C, and they may cease to feed at around 7 °C (Lancaster, 1991; MAFF, unpublished data).

Seasonal patterns of feeding levels

The seasonal pattern of feeding has been investigated further in work aimed primarily at investigating maturity cycles (Chapter 7), using a data-set which included details of stomach fullness and contents, and total stomach weight. Around 1500 fish were sampled for stomach contents, and approximately 65% of these contained food. Almost half the values of a feeding index (the weight of the stomach's contents divided by the fish's somatic weight, given as a percentage, Le Cren, 1951) for these fish, were

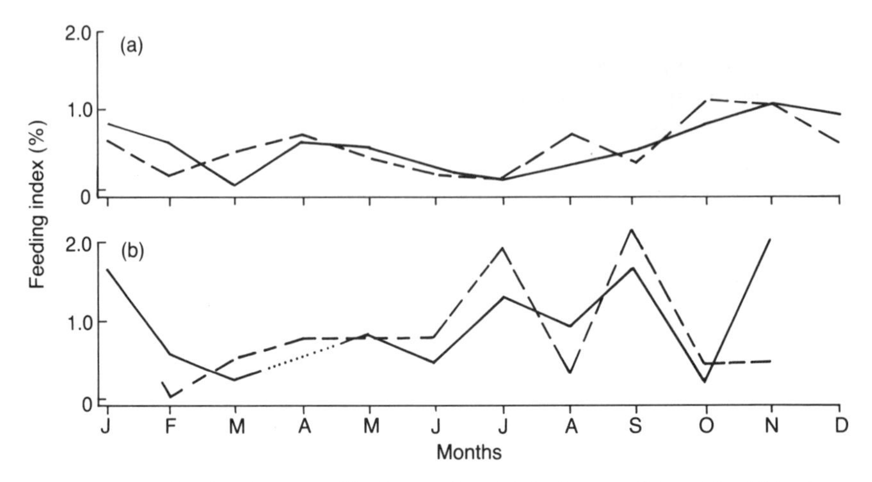

Fig. 3.1 Monthly distribution of feeding index, (a) for bass >42 cm TL, (b) for bass 32–42 cm TL. Solid curves, males; broken curves, females; dotted segments, no data (MAFF unpublished data).

less than 1%. For fish over 42 cm, the mean feeding index appeared to be highest from October to January, and to reach an annual low in June and July (Fig. 3.1a).

During the period November to January, the feeding levels of the few fish (mainly females) taken towards the northern limit of the distribution of bass around the UK were lower than those of both sexes from the South-west of England. Between June and September the situation was reversed; the more northerly fish then had the fuller stomachs. There were no overall or seasonal differences in feeding index between the sexes.

For bass in the 32–42 cm range, mean and maximum values of the feeding levels of both males and females appeared to be generally higher and with a less clear seasonal pattern than were those for larger fish (Fig. 3.1b). The main contrast occurred in July, when significantly higher values were recorded for bass under 42 cm, principally owing to fish caught in the northern region at that time. Apart from in January, when all sizes of bass had similar values, feeding levels of 32–42 cm fish in the southern region appeared to be much lower than those in fish caught further north. Bass of less than 32 cm tended to have empty stomachs between October and March, inclusive, with the highest mean feeding index (at around 4%) in April.

Whilst there was no significant difference in the feeding indices for either sex between fish with resting or maturing gonads, a decrease in feeding by ripe males was apparent; ripening and spent fish, both females and males, had the highest levels of food in their stomachs.

3.4 SUMMARY

It can be concluded from the foregoing that bass are opportunists which take advantage of almost any available food item, of an appropriate size, that moves on the sea-bed or anywhere in the water column. It is not a specialist feeder, though crustaceans and fish provide the bulk of the bass's diet because they account for the greatest biomass of available, prey-sized organisms. Prey size is governed chiefly by the need for bass to engulf and swallow its food whole, though how these fish can extract slipper limpets (*Crepidula fornicata*) from their shells, remains a mystery. The bass is an active predator, which pursues fish in the open water or seeks out crabs and shrimps among rocks and seaweed. Bass will also position themselves in slacker water near piers, overflows and sandbanks, lying in wait for smaller fish to be swept past them by the current.

It seems that the most important factor controlling feeding in bass is water temperature, and, because bass around the UK feed much less avidly in water below 10 °C, it might be expected that their food intake is much reduced between the months of November and May. Digestion rates also decrease as temperatures fall, so a bass with its stomach half full in January is not feeding at the same rate as a similarly endowed fish in July. Consequently, little is known about how much food is consumed by bass in the wild, or how much of this is converted to body weight (the conversion ratio). However, recent advances in marine aquaculture have led to a much fuller understanding of the nutritional biology of the bass (Chapter 10).

Chapter four

Early life history and causes of mortality and disease

4.1 INTRODUCTION

Adult bass spawn in the seas around southern England and Wales from February to July. Bass eggs are planktonic and hatch between 4 and 9 days after fertilization, depending on sea temperature. During the following 2–3 months, the growing larvae drift from the open sea inshore towards the coast, and eventually into creeks, backwaters and estuaries. These sheltered habitats are used by the juvenile bass for the next 4–5 years, before they mature and adopt the migratory movements of adults (Fig. 4.1). The recruitment, feeding, growth and population dynamics of juvenile bass has been the subject of two PhD theses (Jennings, 1990; Lancaster, 1991), from which much of the information in this chapter has been taken.

4.2 SPAWNING AREAS

In early February, bass begin to spawn offshore in the western English Channel and Celtic Sea. Temperature probably provides an important cue for the initiation and location of spawning, because bass eggs are rarely found where the water is colder than 8.5–9.0 °C (Thompson and Harrop, 1987). As with many other marine fish species, bass spawn in mid-water, and the eggs are widely distributed in the open ocean and may be found throughout the water column. As a consequence, their spawning areas are not so clearly defined as those of demersal spawners such as herring (*Clupea harengus*), which deposit eggs directly onto the sea-bed. The results of a series

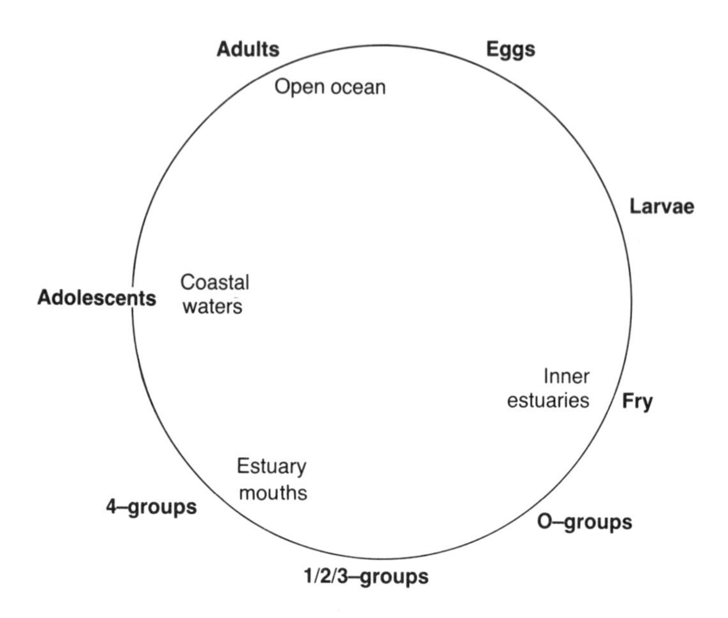

Fig. 4.1 The life cycle of sea bass.

of plankton surveys in the Celtic Sea in 1989 and 1990 have indicated that, off the west coast at least, there is one favoured spawning area, some 20–50 km to the north-west of Trevose Head, Cornwall (Fig. 4.2) (Jennings and Pawson, 1992). The main spawning area in the English Channel is less distinct, and spawning bass appear to move progressively further east as the season advances (Fig. 4.3) (Thompson and Harrop, 1987). Bass may continue spawning until late June, and again temperature may provide the cue for cessation of spawning; bass eggs have rarely been found in the sea around Britain at water temperatures above 15 °C (Thompson and Harrop, 1987).

4.3 EGG AND LARVAL DEVELOPMENT

During the spawning season, each mature female bass may produce between a quarter and half a million eggs per kg of her total body weight (Mayer, 1987). The eggs are about 1.3 mm in diameter and contain low-density oil droplets, which provide the developing larvae with nutrition and also ensure that the eggs remain close to neutral buoyancy and stay within the

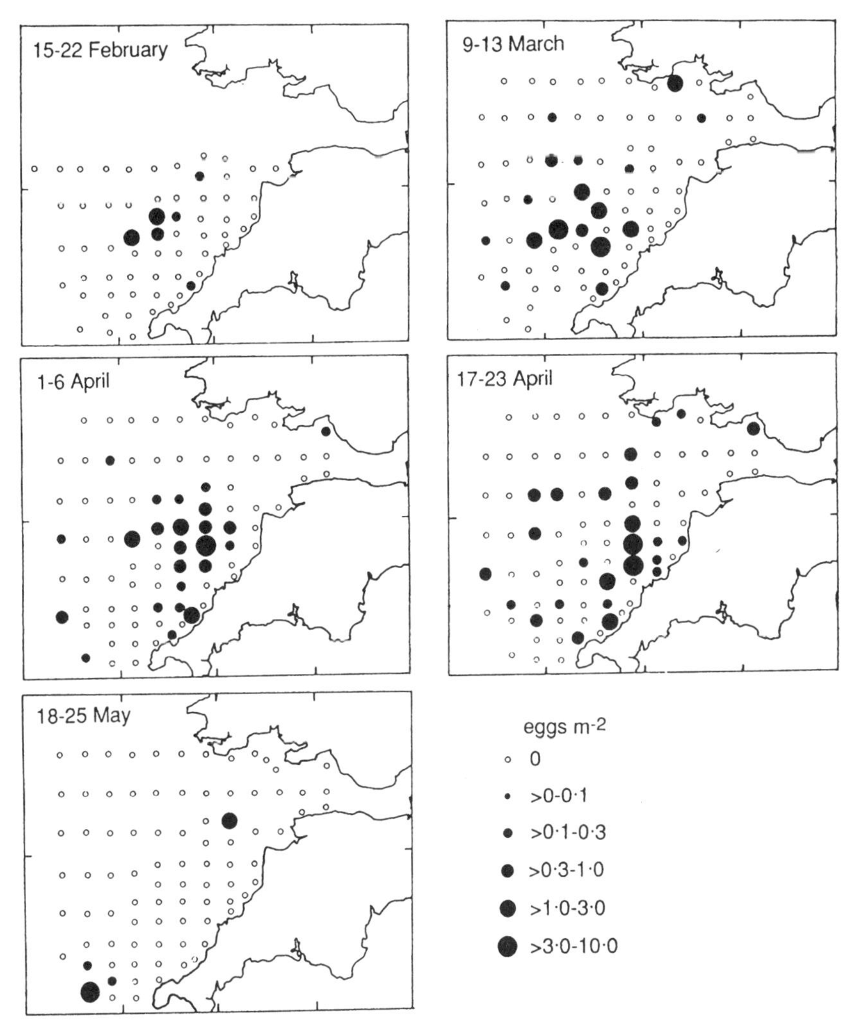

Fig. 4.2 Spawning areas of bass in the Celtic Sea, February–May 1990. Key depicts number of eggs per m^2. From Jennings and Pawson, 1992.

plankton. Initially, they are transparent, but 1–2 hours after fertilization the first cell divisions are visible (Fig. 4.4) (Jennings, 1990).

Egg development in bass and several other marine fish species has been divided into a series of clearly identifiable stages. The relationships between temperature and the time taken for bass eggs to reach each development stage are shown in Fig. 4.5. This allows eggs to be aged and, therefore, it is possible to estimate the day on which the eggs were spawned.

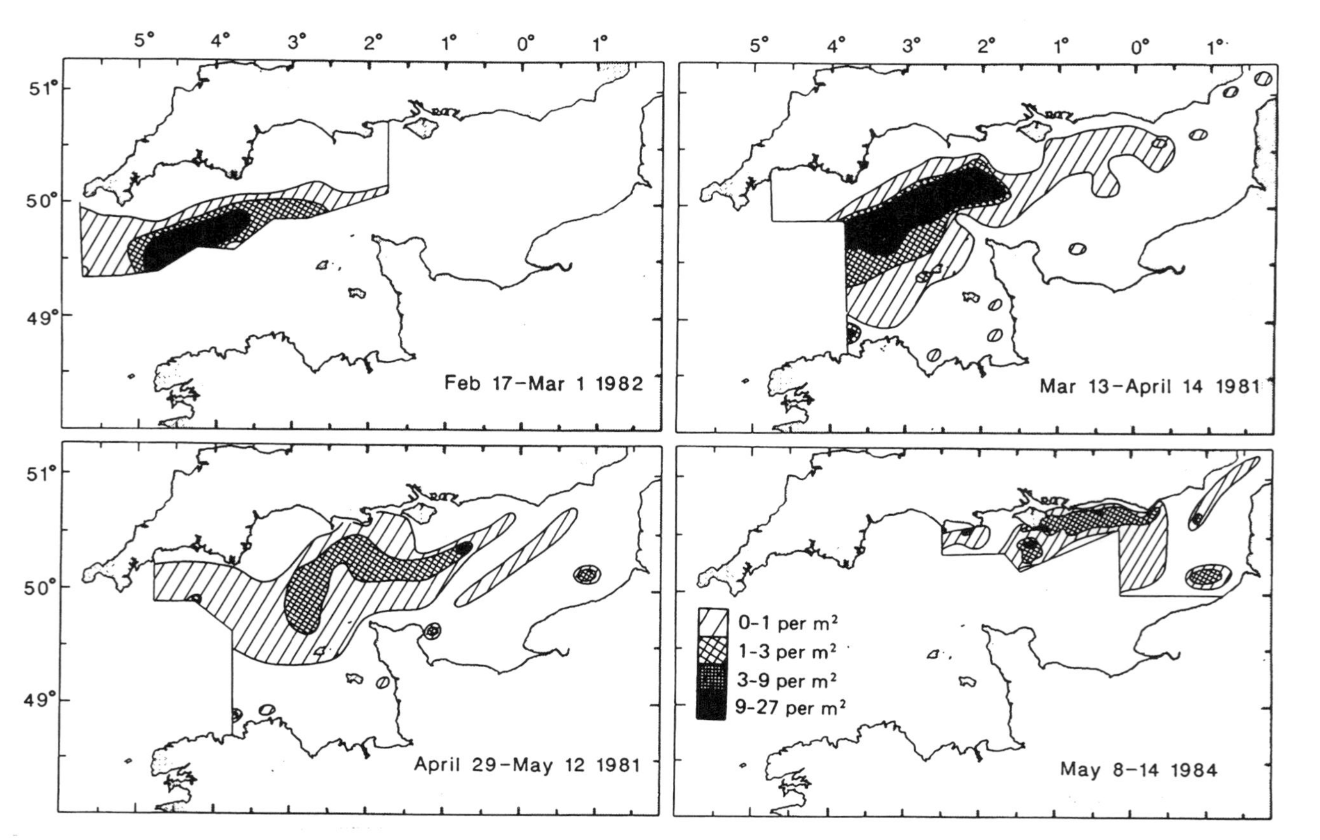

Fig. 4.3 Spawning areas of bass in the English Channel. Key depicts number of eggs per m^2. Reproduced with permission from Thompson and Harrop, 1987.

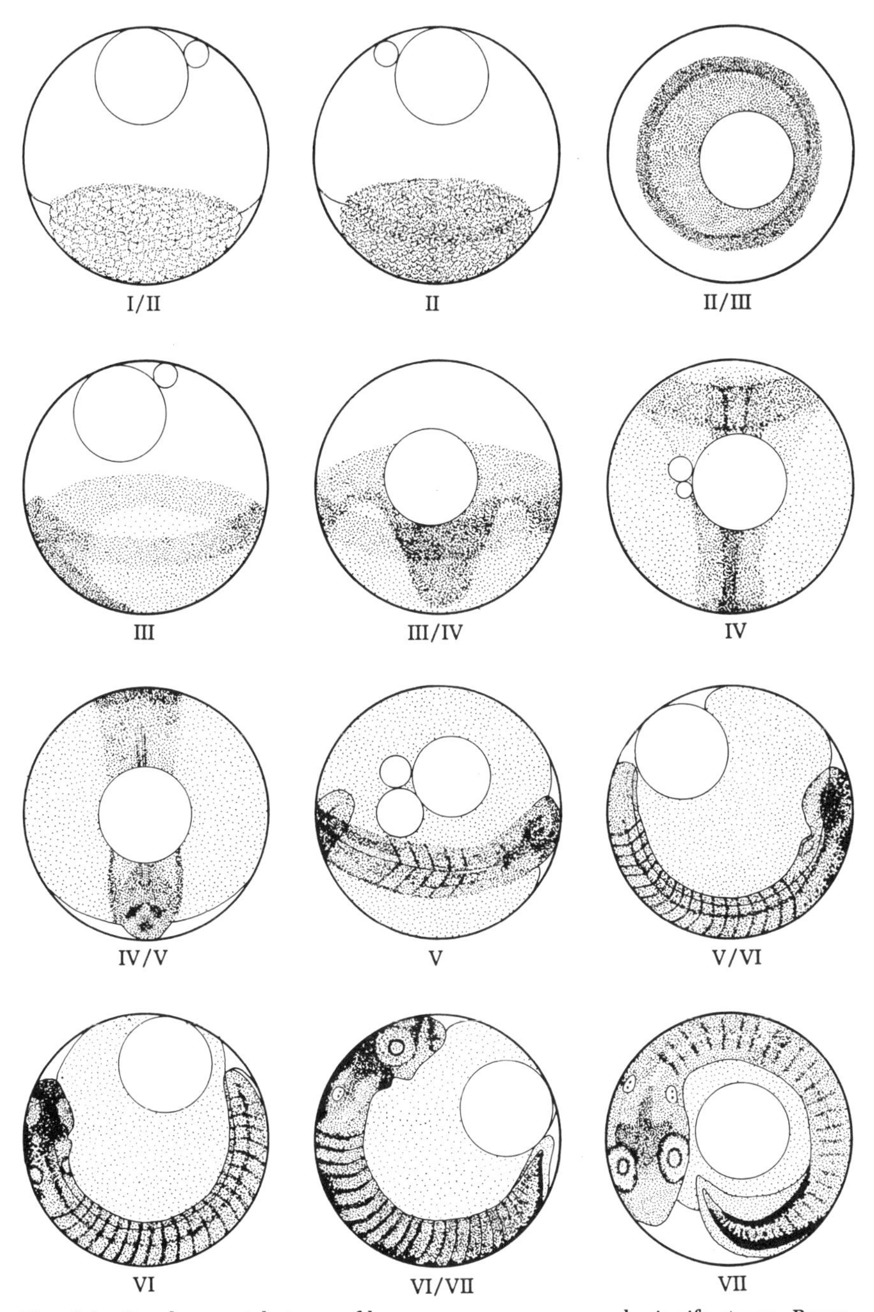

Fig. 4.4 Developmental stages of bass eggs; roman numerals signify stages. Reproduced with permission from Jennings, 1990.

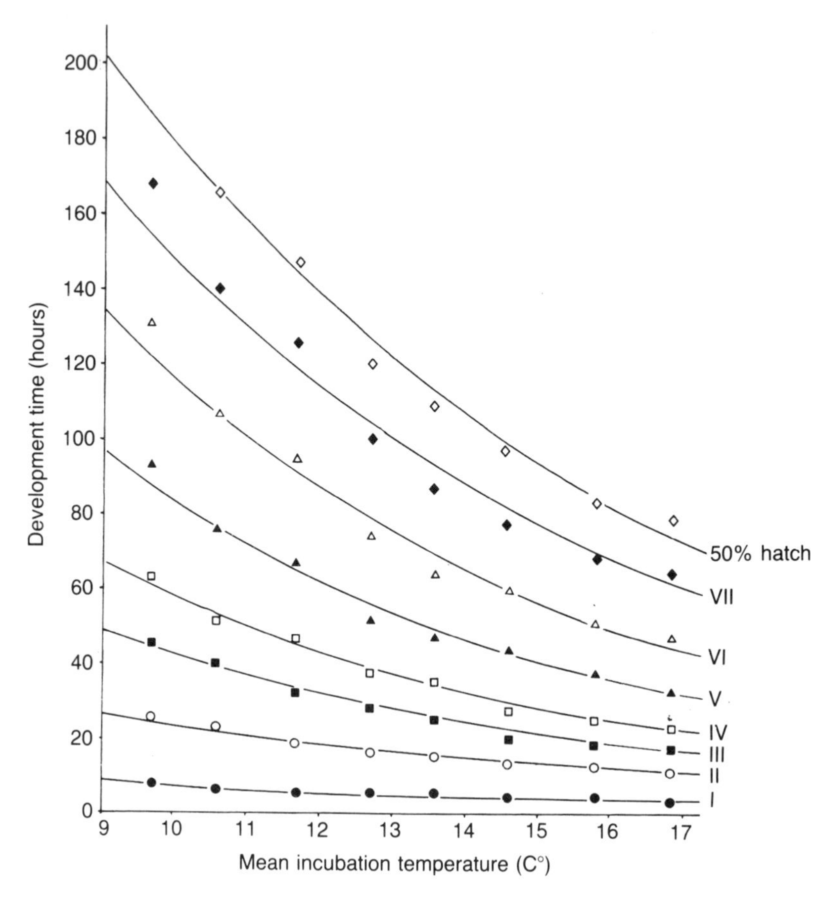

Fig. 4.5 Development rate of bass eggs in relation to temperature: Stages I – 50% hatch. Curves fitted using regression equations. From Jennings and Pawson, 1991.

At the beginning of the spawning season of bass around southern Britain, when eggs are found in water with temperatures of around 9 °C, they will hatch after about 9 days. This time will decrease to a minimum of 4 days as the water temperature rises to 15 °C towards the end of the spawning season, in June. In trials using eggs spawned naturally by captive bass in marine aquaria, in which the temperature regime was controlled to be similar to that experienced off Plymouth, the eggs developed abnormally and failed to hatch when incubated at temperatures below 8.7 °C or above 17.7 °C (Jennings and Pawson, 1991).

Bass larvae are 4.0–4.5 mm long at hatching. Observations in marine aquaria suggest that in calm water they would initially float upside down

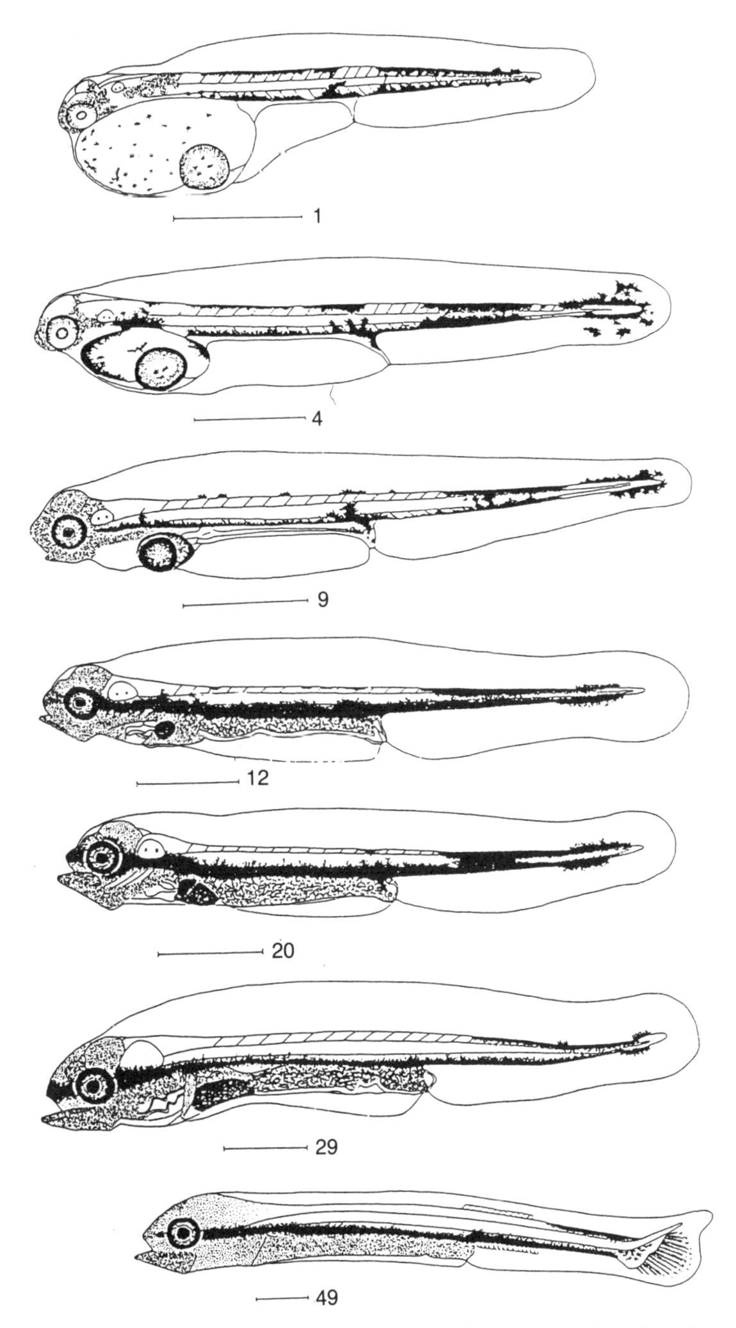

Fig. 4.6 Development of bass larvae at 12 °C (numbers are age from hatch in days). Scale bars, 1 mm. Adapted from Jennings, 1990.

close to the sea surface. The larvae have small areas of dark pigment around the notochord (which develops to form the vertebral column), but otherwise they are almost transparent. Larval development, like that of the eggs, accelerates at higher temperatures.

At 12 °C, the larvae first open their mouths and begin to feed after 9 days (Fig. 4.6). At 9 °C this stage is reached in 14 days, and at 15 °C, in 5 days. Around the time of mouth opening, bass larvae have a continuous line of dark pigment on their ventral surface, a characteristic which clearly distinguishes them from other fish larvae in the sea and, incidentally, enables them to be easily identified in plankton samples. In addition, the swim bladders are then visible as reflective ovoid bubbles just below the midline of the larvae. These allow the larvae to maintain neutral buoyancy without swimming, now that their low-density oil droplets have been absorbed, and thus provide energetic benefits.

4.4 MOVEMENT TOWARDS NURSERY HABITATS

In comparison with many other coastal marine fish, such as herring, plaice or cod, few wild bass larvae have been found. However, on the basis of their known geographical distribution, size and age, it is assumed that they move to within 10 km of the coast within one month of hatching (Fig. 4.7) (Jennings and Pawson, 1992).

The mechanism by which the larvae move inshore is not fully understood. During plankton surveys in the Bristol Channel in 1989 and 1990, sea surface drifters were released close to the main bass spawning area in the eastern Celtic Sea, to monitor the rate and direction of wind-driven surface water movements. Many of them did travel towards inshore bass nursery habitats, but at a rate which was too slow to account for the inferred movements of larvae, most of which were found inshore at an age of 20–40 days. Alternative evidence for the passive transport of larvae was provided by data from current meters positioned 15 m below the water surface. At this depth, which coincides with the depth at which bass larvae have been captured, the residual currents were faster than those at the sea surface. It is possible, therefore, that these currents could have been used selectively by the larvae to move into inshore waters.

Once inshore, bass larvae feed and grow until, at age 2–3 months (Fig. 4.8), they begin to migrate actively into juvenile nursery habitats in estuaries, harbours, backwaters, creeks and shallow bays. The cue for movement of postlarval bass into these habitats appears to be both environmental and developmental. They first arrive in UK nurseries in June, attracted, possibly, by an environmental stimulus such as higher water temperature or reduced salinity. However, as the summer

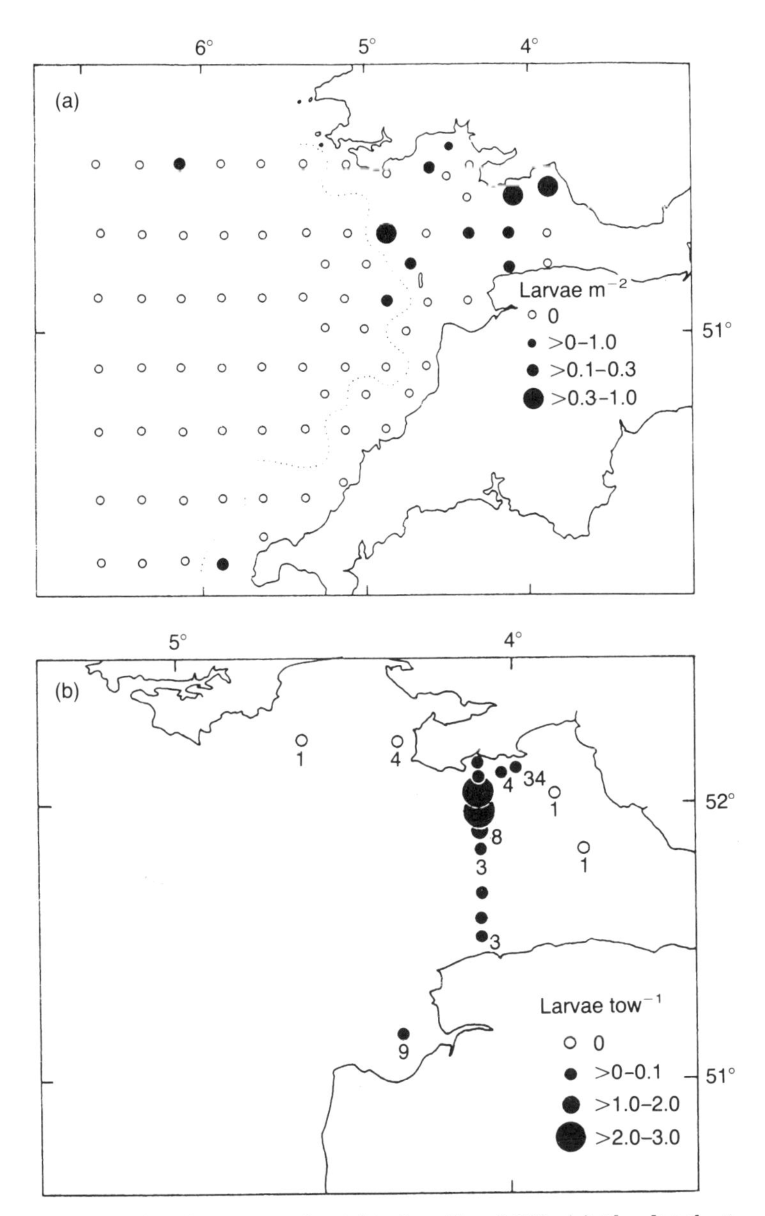

Fig. 4.7 Larval distributions in the Celtic Sea, May 1989. (a) The distribution and abundance (number per m^2 of sea surface) of bass larvae observed during plankton survey, 19–25 May 1989. The broken line shows the eastern boundary of water with a stratification value of less than 10 joules m^{-3}. (b) The distribution and relative abundance (mean no. of larvae per tow) of bass larvae captured by ring net, 25–27 May 1989. Each station was sampled twice except where a subscript indicates the number of tows. From Jennings and Pawson, 1992.

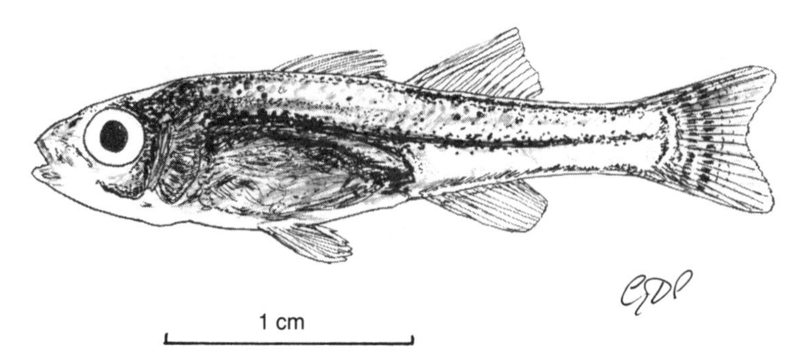

Fig. 4.8 Bass postlarva aged 2–3 months.

progresses, there is no decrease in the size of postlarval bass that continue to enter nursery habitats. It appears, therefore, that bass will not migrate into these habitats, even when environmental conditions are suitable, unless they have reached a particular developmental stage. Sampling of 0-group bass in nursery habitats in autumn sometimes reveals a bimodal length distribution (D.F. Kelley, pers. comm.), which suggests that there may have been two peaks of recruitment of postlarvae.

The majority of 0-group bass are found within brackish waters. When they first arrive inshore, the water temperature of these nurseries is usually higher than that of the open sea, and this may explain why bass are so abundant there (Jennings *et al.*, 1991). Other benefits of estuarine habitats are that they may provide better feeding conditions and contain relatively few predators such as mackerel and scad, which are common on the open coast at this time of year.

4.5 DISTRIBUTION OF YOUNG BASS

0-Group bass are found in most southern English and Welsh estuaries during the summer months. They are frequently the most abundant species present, with the exception of gobies, a fact that was only recognized recently (Kelley and Reay, 1988). Estuaries are also used by the juveniles of other species of fish, in particular grey mullet, flounder (*Platichthys flesus*), sprat and whiting (*Merlangius merlangus*). As Table 4.1 demonstrates, fine-meshed trawl catches in the Solent and associated harbours tend to contain more bass up to 5 years old than any other species.

Bass from small shallow estuaries probably move to larger estuaries at the end of their first summer, though estuaries with deeper channels, which retain water at low tide, may be used from the first summer through to an age of 4 or 5 years. Otherwise, it appears that there is limited

Table 4.1 Species composition of standard trawl catches in Solent Survey, September 1989 (45 stations) (MAFF unpublished data)

Species	*Numbers*	*No. of stations present*
Bass (mainly 2–4 groups)	1344	44
Plaice	174	33
Flounder	97	12
Pout	96	9
Mullet (golden)	89	14
Scad	78	11
Herring	77	11
Sole	62	9
Eels	51	14
Black bream	45	15
Dab	39	5
Whiting	27	5
Mackerel	13	4
Thornback ray	13	5
Wrasse	12	8
Smelt	12	2
Brill	9	6
Pollack	3	3
Smooth hound	2	2
Gurnard	1	1
Pogge	1	1
John Dory	1	1
Garfish	1	1
Total other than bass	903	

movement of young bass between estuaries. When a pollution incident killed bass in the Ogmore River in South Wales during July 1988 (Jennings, 1990), there was no sign of young bass reappearing there that summer, although mullet, flounder, sprat and shad (*Alosa* spp.) had returned within 2–3 weeks, and there were other large populations of juvenile bass within 20 km of the Ogmore River. This suggests that, if the environment remains suitable, young bass tend to remain within the nursery estuary to which they first recruit. A corollary is that there is probably little interchange of juvenile bass between nursery estuaries and that immigration from other nearby stocks does not necessarily occur.

4.6 TEMPERATURE EFFECTS ON GROWTH AND SURVIVAL

The growth and movement of 0-group bass in nursery habitats appears to be linked with temperature. In the UK, the highest natural summer water temperatures seldom exceed the 22–24 °C optimum for the growth of 0-group bass (Barnabé, 1990), therefore bass tend to grow faster in warmer regions and during warmer summers (Alliot *et al.*, 1983). This can be seen in the annual growth increments on scales (Chapter 6), even in older fish (Kelley, 1988b). Generally, 0-group bass grow fastest around the south and south-east coasts of England, whereas growth is slower in West and North Wales. There is evidence (Lancaster, 1991) to suggest that food availability in the wild does not usually limit the growth of bass, and differences in growth rate have been attributed mainly to temperature effects. In warm summers, growth of 0-groups at a given site may be twice as fast as during cool summers.

Postlarvae and 0-group bass are normally found in the warmest waters within estuaries, and this is important for achieving their potential for growth. By actively seeking eddies, and remaining close to the river bed, young bass can maintain position in estuarine channels even when the main current is flowing in excess of 0.4 m s^{-1}. Using this ability, they can stop themselves from being swept downtide, and they are often found in warm marsh pools when the tide has receded. Further evidence for the propensity of young bass to seek and select warmer water is provided by observations of their movements into and out of tidal creeks. Bass will be present in creeks when the tide has risen over sun-heated sand and mud, which may raise the water temperature by 2–3 °C. However, when the tide rises in the morning over creek beds that have been chilled on a cool night, and the water temperature there falls 2–3 °C, small bass are usually absent.

The older and larger 0-groups appear to leave the nursery habitats first as the winter approaches, followed by smaller and younger fish. Some small bass may remain close inshore throughout mild winters, and although they may feed at this time, growth is slow or negligible. Most juvenile bass are thought to overwinter in the deeper channels of estuaries, or just offshore, where they can find slightly warmer water than that which is being cooled by night frosts or cold fresh water entering the estuaries. Nevertheless, mortality of young bass may be high if they experience unusually low winter temperatures. Young bass can tolerate this situation for several weeks, but they ultimately lose condition and become weaker with passing time. They become lethargic and less mobile at lower temperatures, and are therefore more vulnerable to predation. In early May 1986, D. Kelley (pers. comm.) caught 157 1-group bass in a seine net in the Taw Estuary in North Cornwall, after prolonged winter frosts had reduced water temperatures to below 5 °C. These fish were in poor condition and few survived after being put into tanks. Other bass of the same size and age, netted from the R. Tamar in South Cornwall where the sea temperatures had dropped only to around 8–9 °C, were in good condition and most survived after being put in tanks. Kelley (1986) notes that severe winters may have devastating effects on the broods of the previous summers, and cites the almost complete absence of the 1962 year class around England and Wales following the hard 1962/63 winter. It appears that, even in some south-western estuaries, which are normally used by bass all the year round, nearly all overwintering 0-groups died during the exceptionally cold weather early in 1963. Overwintering mortality may be size related, with fish larger than 60 mm at the end of their first growing season (Kelley, pers. comm.) being less likely to die. Bass that grow faster during the summer are also generally fatter (Lancaster, 1991) and, as a result, are likely to have greater energy reserves. Once temperatures fall below 10 °C, bass feed less and less and have to rely on stored energy reserves. Thus, the faster-growing fish are probably better prepared to face the inevitable period of starvation that a UK winter brings.

Young bass are more likely to survive their second winter; they are larger than 0-groups and normally move into deeper water where the effects of cold weather are less. At the time of writing (1992), the 1989 and 1990 year classes are abundant around England and Wales, following two warmer-than-average years (summer and winter 1989 and 1990). The winter of 1990/91 was much colder and, by February, the sea temperatures in the Solent had been below 5 °C for four consecutive weeks (Corps, pers. comm.). Both 1-group (1990 year class) and 2-group (1989 year class) bass caught in trawls at this time appeared to be in very poor condition, unlike similar fish caught at other times of the year. On transfer to tanks, all the catch of one trawl haul (56 fish) died within 48 h. It is assumed that whilst they had been stressed by low temperature, it was the shock of capture that led to death. This phenomenon may be a cause for concern in areas where large numbers of small bass are caught and returned to the sea in winter, in small-mesh fisheries for shrimp, sprat or herring, such as occur in eastern and southern England. It may be particularly appropriate to consider the indirect effects of winter fishing (for other species) in nursery areas that sustain a year-round population of young bass.

The relative abundance of 0-group bass can vary considerably, by as much as two orders of magnitude (100 ×) between years (Chapter 14). It is likely that the strength of year classes is to a large extent determined during the egg and larval stage, although, as described above, over-wintering mortality at the end of the first year may also be a significant factor. Strong year classes of bass around the UK are generally associated with warm, settled weather in spring and summer, and it has been suggested that prolonged periods of offshore winds may reduce the numbers of juvenile fish recruiting to the inshore nursery areas (Holden and Williams, 1974). It has been demonstrated, however, that the climatic conditions associated with warm sea temperatures are likely to lead to better-than-average survival of juvenile bass (Pawson, 1992). Settled conditions may lead to stratification of the water column, and concentration of suitable food for the bass larvae, at a time when mortality by starvation could otherwise occur. Studies with northern anchovy (*Engraulis mordax*) off California (Methot, 1981) and sprat in the Irish Sea (Coombs *et al.*, 1992) have shown that food densities which are sufficient to support the larvae are more likely to occur when the water column is vertically stratified, because the prey organisms collect at high density in a relatively narrow layer. When storms disturb this stratification, the layer of food organisms is dispersed, and the resulting decrease in the mean density of food in the water column might then be insufficient to maintain the larvae.

From March onward, juveniles in their second year of life begin to return to summer habitats. Larger estuaries are still used, but the fish may also be abundant in shallow bays on the open coast near their original nursery areas. At the end of their second year, the juvenile fish are usually 12–16 cm total length. For the next 2–3 years of their life, juvenile bass continue to migrate to deeper water each autumn and return inshore the following spring. Their growth in UK waters throughout this period is slow compared with that of the young of many other commercial marine species: bass attain a length of about 25 cm at the end of their third year and 33 cm at the end of their fourth. When the juvenile bass are aged 4 or 5 and close to maturity, their movements become increasingly wide ranging until they adopt the adult migrational patterns.

4.7 CAUSES OF MORTALITY AND DISEASE

Predators

The smaller the bass, the more it risks being eaten by predators. When young bass are swimming near the sea surface or in shallow water, they may be attacked by birds (e.g. gulls, *Larus* spp; gannets, *Sula bassana*; cormorants, *Phalocrocorax carbo*, or herons, *Ardea cinerea*), although the dorsal spines of the bass may provide some deterrent. During routine sampling, one specimen of 37 cm was encountered which appeared to have suffered damage from a bird's beak earlier in life, resulting in a natural colostomy (Fig. 4.9) (Pickett *et al.*, 1988). However, such findings in bass may be rare compared with the frequency with which trout in reservoirs are known to suffer non-lethal attacks from cormorants (Russell, pers. comm.).

0-Group bass are probably most at risk from other piscivorous fish species. Occasionally, large shoals of pollack (*Pollachius pollachius*) or whiting occur in bass nursery areas. When local fishermen reported that pollack were eating young bass in the Solent (Hampshire) in 1987, an examination of the stomachs of pollack in the 30–40 cm range revealed each to contain

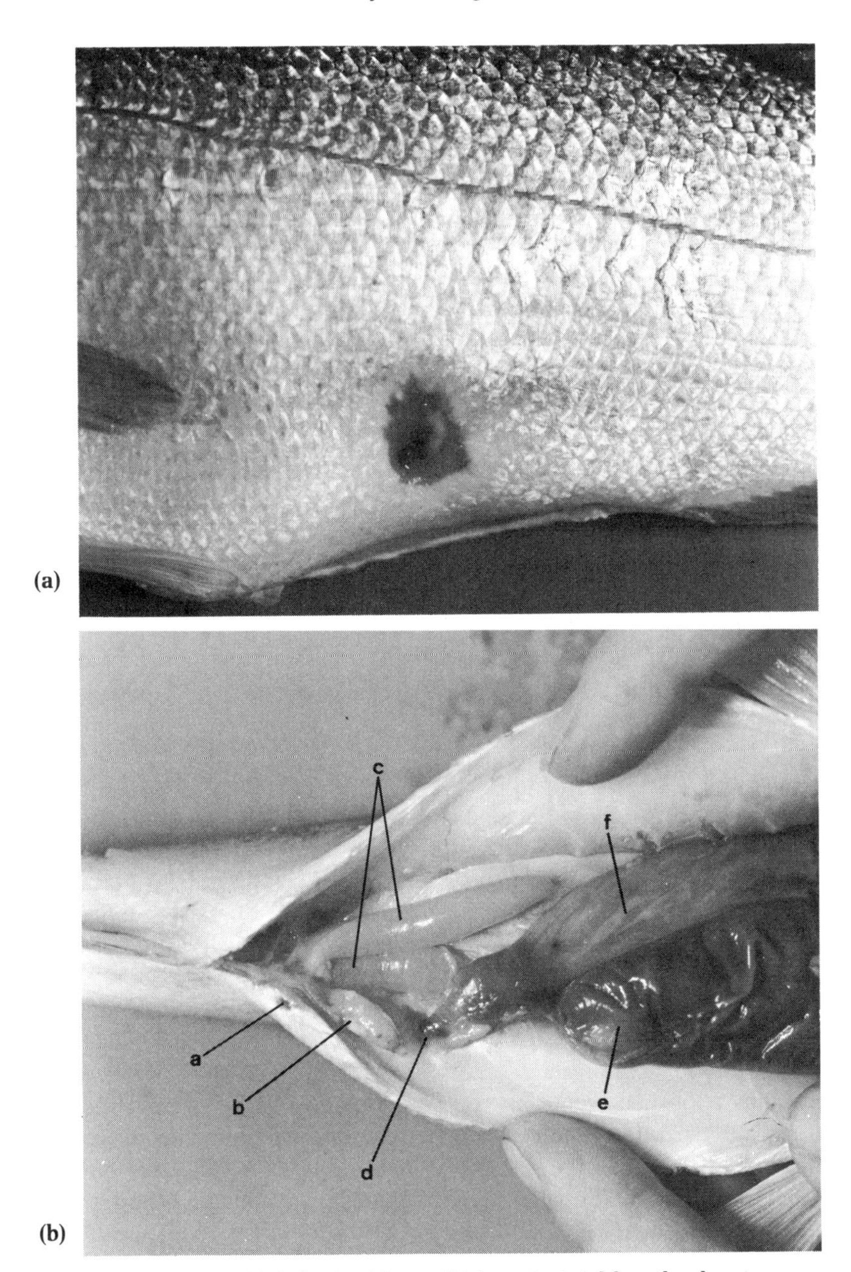

Fig. 4.9 (a) View of left flank of bass, 37.4 cm in total length, showing a wound midway between the pectoral and anal fins. (b) Ventral view of the bass' body cavity showing: (a) the urino-genital pore; (b) the redundant rectum; (c) the ovaries; (d) fusion between the anterior intestine and the body wall, new 'anus'; (e) stomach; (f) the duodenum. From Pickett *et al.*, 1988.

4–6 0-group bass of 5 to 8 cm (MAFF, unpublished data). Similarly, in 1989, whiting were found to be feeding on 0-group bass at Bradwell in Essex (Cox, pers. comm.).

Any bass surviving to be an adult might be expected to have a long life; they spend much of the time offshore in shoals, and they are fast swimmers and thus able to avoid most predators. Occasionally, specimens have been found with injuries which may have resulted from attacks by other animals. On the Welsh coast, grey seals (*Halichoenus grypus*) are suspected of preying on juvenile bass, and D. Kelley (pers. comm.) has reported large numbers of seals near the mouths of estuaries where juvenile bass were shoaling. However, an examination of the contents of seals' faeces, carried out in South Wales by the Sea Mammals Research Unit (D. Kelley, pers. comm.), failed to find evidence of bass in the seal's diet. Even if bass remains were found, such data would not necessarily confirm or deny that predation on bass is part of the seals' normal behaviour; they are well-known pilferers of fish caught in gill nets (Rae and Shearer, 1965). Fishermen using longlines for bass near the Isle of Wight, on the English south coast, have sometimes recovered only the head of a hooked bass on hauling (Baldacchino, pers. comm.). They claim that this is the work of the tope (*Eugaleus galeus*), which are also taken on the lines. Again, the hooked bass are easy prey for other opportunistic predators.

Parasites

Although the bass is high in the food chain, and might therefore be expected to act as host to a number of parasites which infect its prey, macroscopic examination of several thousand bass has revealed few parasites, internal or external. There have been few purposeful studies on bass parasites, however, compared with those on the striped bass, for which a wide range of both internal and external infesting organisms has been reported (Bonn *et al.*, 1976). Many of these observations, however, relate to fish under cultivation conditions, where parasites and diseases are perhaps more likely to occur.

The main published works on the parasitology of bass around the UK relate to gill parasites. Paling (1966) described the attachment of the monogenean (trematode worm) *Diplectanum aequous* to the gills of bass taken near Salcombe, South Devon. This parasite is small, less than 1 mm in length, and it may easily be missed by casual observers, but by using a stereomicroscope, 30–250 parasites per fish were recorded.

Davey (1980), working in the Lynher Estuary near Plymouth, Devon, found that the copepod *Lemathropes kroyeri* was common on 2–3-year-old bass. Up to seven parasites were found on each infested fish, with a higher infestation rate in older, larger fish. Another gill parasite, specific to bass, is

Atrispinum labracis, which Winch (1983) reported occurs mainly on bass from the open sea, but is seldom found on fish living in estuaries. Infestation was rarely more than two individual parasites per bass. Nematode infestation of the abdominal wall has occasionally been recorded in adult bass from South-west England (D. Kelley, pers. comm.). It is, perhaps, surprising that, with its varied diet and, for a marine fish, relatively great longevity, the bass appears to be so lightly parasitized.

Diseases

Even less is known about the diseases of bass than about their parasites. Although the striped bass has been observed to be infected by some diseases found in freshwater fish, such as the fin-rot diseases, *Pasteurellaosis*, *Columnaris* (which is often responsible for mould-like growths on skin and fins) and *Epitheliocystis*, we have found no reports of similar cases in wild bass. At Lowestoft, we had an outbreak of both fin-rot and subcutaneous lesions in bass, which had probably suffered handling damage before being placed in tanks. Fin-rot can be caused by a number of organisms including *Vibrio*, which could also cause the skin lesions. Several sick specimens were successfully treated by applications of the antibiotic Cycatrin (Wellcome Labs) in powder form. Three years later, these fish had entire fins and all their scales were regenerated; they subsequently spawned naturally in the laboratory. Bass in intensive culture at 16–18 °C are particularly susceptible to bacteria of the *Vibrio* family, and a liquid vaccine, which can be administered by injection or immersion, is widely used by fish farmers.

Reaction to pollutants

There have been few studies so far of the effects on bass of contamination by chemical substances. Jennings (1990) reported a kill of bass and other fish species in the Ogmore River (South Wales) in 1989, which was due to the deoxygenating effects of an accidental sewage discharge. The bass involved were mainly 0-groups and, afterwards, none remained in the estuary. Later the same year, young grey mullet, sprat and gobies (*Gobius* spp.) re-colonized the Ogmore, but bass did not. In drought conditions, 0-group bass have been observed to move out of estuarine tributaries in which sewage effluents were less diluted than usual (Kelley, 1988a).

Physiological changes have been noted in bass on exposure to heavy metals under experimental conditions. Establier *et al.* (1978) monitored the uptake of inorganic mercury in bass exposed to 0.1 ppm in seawater for up to 62 days. Rapid changes in blood cells, depression in haematocrit and haemoglobin levels, and morphological erythrocyte changes were noted. Accumulation of cadmium by bass from seawater containing 25 ppm Cd for

96 h, was studied by Establier and Gutiérrez (1980). Histological alterations were noted in some organs, including kidney, intestine, gall bladder and pancreas.

4.8 FURTHER RESEARCH

Knowledge of the biology of juvenile bass was very limited prior to the early 1980s and, despite being one of the most abundant fish species in many estuaries, their presence was rarely recognized. Ten years later, we have achieved some knowledge of the major bass spawning areas around England and Wales, and the feeding, growth and movements of juvenile bass in nursery areas have also been studied. However, several fascinating questions about the early life history of bass remain unanswered. Little is known about the cause and relative magnitude of mortality during the egg and larval stages, because these stages appear to be widely dispersed at low densities, and are not amenable to study without recourse to expensive research-vessel time. The stimulus or cue that draws postlarval bass to their estuarine nursery habitats is still not understood, although several candidates have been put forward. Fortunately, bass can now be reared in captivity, and laboratory experiments in which developing larvae are exposed to water flow in which temperature, salinity and other chemical components are manipulated, could provide an insight to this phenomenon. Such studies would provide an experimental basis for subsequent fieldwork, and help us to understand how the sub-lethal effects of chemical contamination might influence the role of polluted estuaries as bass nurseries. Recent research suggests that the few young bass that are found outside the estuarine nursery environments grow perfectly well, and are thus not dependent upon them (Jennings *et al.*, 1991; Lancaster, 1991). Yet the postlarvae appear to seek estuaries and 0-groups are vastly more abundant there than elsewhere. It is important to know which particular features of the estuarine environment provide the most favourable habitats for young bass and, possibly, to estimate what area of nursery habitat must be protected from development, reclamation and pollution to ensure that bass populations are sustained, biologically and as an exploitable resource (Jennings, 1992).

Chapter five

Migrations, movements and stock identity

5.1 INTRODUCTION

Bass have been tagged in several studies around the British Isles, and the information obtained from the subsequent recaptures has been used to provide a description of the migrations and seasonal distributions of bass and of its fishery, and an indication of the population structure around the British coast. Kennedy and Fitzmaurice (1972) recorded a limited distribution of recoveries of bass released along the coasts of South-west and South-east Ireland, and suggested that bass populations were essentially local, a conclusion which was supported by Holden and Williams (1974) from the results of taggings carried out by anglers around the south-west coast of England. Both of these investigations took place in situations which were perceived, at the time, to be near the centre of the distribution of bass around Ireland and Britain, respectively. It was Holden and Williams, however, who suggested that the low return rates could be due to large bass emigrating from release sites to areas where fishing effort likely to catch them was low. Since 1975, bass tagging exercises have been carried out along the west coast of Britain, in the Thames Estuary and off Hampshire and Dorset (Kelley, 1979; Pawson *et al.*, 1987). As a consequence, bass have now been tagged in most areas around the coasts of England and Wales in which the species is caught.

5.2 TAGGING

Organization of bass tagging studies

Almost all bass tagging in Britain has been supported by funds provided either by the Natural Environment Research Council (NERC) or MAFF. Since 1971,

Donovon Kelley has co-ordinated the efforts of a small group of experienced personnel, to catch fish for tagging by means of rod and line or, in some localities, by beach seine. In 1970 and 1971, a MAFF exercise was run in conjunction with the National Federation of Sea Anglers and the National Anglers' Council (Holden and Williams, 1974), and relied mainly upon the members of selected angling clubs to catch and tag the fish. Since 1981, DFR scientists have tagged bass caught by commercial fishermen using a variety of gears: trawls, longlines, handlines, rod and line and gill nets.

Prior to 1978, tagging took place in areas where angling was the main fishing activity for bass and where there was a good chance that tagged bass would be recaptured and reported, i.e. in South-west England, southern Ireland and Wales. More recently, an effort was made to tag bass in those regions where the commercial fishery for bass has been expanding, e.g. the Solent and Thames Estuary, at locations for which little information was available, e.g. Portland and Runnel Stone, and in some of the bass nursery conservation areas designated in 1990 (Great Britain–Parliament, 1990).

The choice of specific tagging sites in each area has been determined

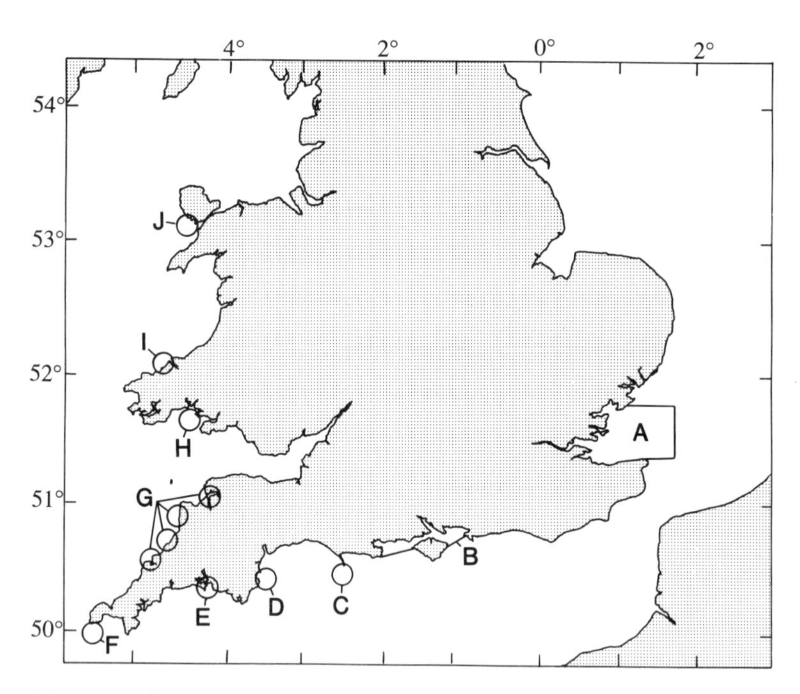

Fig. 5.1 General areas of release of tagged bass around the UK: A, Thames Estuary; B. Solent Harbours and Poole Bay; C, Portland; D, East Devon; E, Plymouth; F, Land's End; G, North Cornwall and North Devon; H, South Wales; I, West Wales; J, Anglesey. From Pawson *et al.*, 1987.

chiefly by the pattern of the fishery and by the potential catch rates of bass taken by fishing methods that produce fish suitable for tagging, i.e. rod and line, beach seines and trawls. In the Thames Estuary, most bass were tagged as the opportunity arose on angling charter boats, otherwise, bass have been tagged during short visits to the chosen site, timed to coincide with tides that offered the best chance of fish being caught.

The distribution of releases of bass in the main tagging areas around England and Wales (Fig. 5.1) is, therefore, biased seasonally, according to the fishing conditions and the need to ensure that a sufficient quantity of fish might be caught, and geographically, in view of the need to tag throughout the species' range and for ease of access by shore or boat to a fishing area. As a consequence, most of the tagging has been carried out in the summer in areas of high bass-fishing activity. Unfortunately, bass have not been tagged near the northern extremities of the species' usual range around Britain (Cumbria–Solway Firth on the west coast, and Suffolk–Yorkshire on the east) or outside the coastal waters. Neither have attempts been made to tag bass in the winter fishery, which takes place mainly offshore in the south-west. The relative severity of winters affects the distribution of this fishery and its duration varies considerably, and hence it is difficult to arrange visits with good prospects of tagging bass.

Tags are normally attached singly, and the tagging technique, subject to slight variations with tag-type, has been described by Kelley (1979). It is possible to reduce damage and stress to the fish by measuring them, tagging and taking a few scales for age determination all in one short operation; the quicker the fish can be returned to the sea, the better. Lively specimens can be subdued with a damp cloth over their heads, and two operators greatly speed up the procedure. Usually, the catch rate of bass is sufficiently slow to allow fish to be dealt with as they are caught, but larger catches can be held in tanks on boats or in a water-filled inflatable dinghy (when using beach seines), both preferably having a flow-through of seawater.

Details of the tag number, fish length and condition, and tagging location and date, are recorded for each tagged bass on the paper envelopes used to store scale samples, and a record is entered into a field logbook. All tags have carried a message in English, or English and French, asking for their return, either to DFR or to the Marine Biological Association at Plymouth, or alternatively, to local fishery officers. There is an internationally agreed award (£3 in 1993) paid for the return of any tag being used in a government-run scheme, plus the value of the fish if its body is surrendered.

Factors influencing recapture

Angling has been the main method reported as being used to recapture bass tagged around Britain, accounting for between one-third and all of returns from each release, depending on the fishing area and the development of the commercial bass fishery with time. The relative proportion of tags returned by anglers and by fishermen using other catching methods does not, however, reflect the relative levels of sport and commercial fishing activity, because many commercial fishermen use rod and line to catch bass.

In view of the low overall recapture rates (usually a 10–20% return of tagged fish released), it is evident that the majority of tagged bass are lost to science. Returns of tags following a particular release have tended to decline sharply after 2 years, and it is thought that a significant proportion of those bass that are recaptured, go unrecorded. Non-reporting may occur unintentionally, because tagged fish are unnoticed in large catches made by commercial fishermen, or when small fish are returned alive to the sea. Some tags will be withheld deliberately to avoid attracting attention to the captor's fishing activities, or by anglers who are reluctant to surrender or kill fish bearing a tag asking for return of tag and fish. Many tags are lost simply through ignorance or apathy, despite extensive publicity by leaflets and posters (Fig. 5.2) distributed to fishery officers and angling clubs in all north-western European countries, to fishing tackle outlets and the local press in the release areas, and also through articles submitted to angling periodicals and the fishing trade press.

In addition, bass under the minimum landing size have not been reported as frequently as larger fish, despite attempts to explain that it is not illegal to land and report any fish involved in a scientific experiment. The effectiveness of various gears in catching bass, and their distribution within the fishery, determines the probability of recapture in relation to the movement patterns of released fish, and this often results in high recapture rates at or near tagging sites. There is also evidence of regional variations in the susceptibility to recapture of bass of different sizes.

Recovery rates by type of tag

A factor in the success of any tagging scheme, and which is under the direct control of the tagger, is the choice of tags used (Fig. 5.3).

A comparison of the recovery rates of tagged fish will show the effectiveness of different types of tag, but this is complicated by a number of other factors, including those mentioned above. Nevertheless, some tags are clearly more prone than others to being lost from the fish (shedding) or to being overlooked by the captor of a tagged fish. Shedding occurs with all types of tag at a rate determined by:

1. the method and site of fixing the tag on the fish's body;
2. removal before recapture – e.g. by becoming snagged in gill nets or being bitten or pulled by other fish or crustaceans;

£3 REWARD

For return of TAG, details of CAPTURE, and FISH SIZE, and 5-10 SCALES

TO

MAFF FISHERIES LABORATORY, LOWESTOFT, ENGLAND

OR

BODY of FISH to LOCAL MAFF OFFICE
£3 REWARD and MARKET PRICE of FISH

THANKYOU

Fig. 5.2 Tagging poster.

Fig. 5.3 Types of tag used for tagging bass in UK waters, showing the usual attachment positions: (a) Petersen disc; (b) Carlin-Ritchie pennant; (c) Lowestoft flag; (d) mini-Howitt; (e) Floy tag; (f) Hallprint 'T'-bar.

3. mechanical failure of the tag bridle; and
4. the skill of the person performing the tagging operations.

These effects can be reduced by good tag design and method of application, and through the correct choice of tag with respect to the target species and the size range to be tagged. An attempt to estimate the degree to which this influenced overall recovery rates of bass has been made using bass held in a marine aquarium for 3 years (Pawson *et al.*, 1987). This showed that

Petersen discs (Fig. 5.3(a)) attached firmly below the first dorsal fin were not lost, though placement at a point on the fish's body that is more liable to flexure, e.g. below the rear of the bass' second dorsal fin, resulted in the tagging wound becoming sore and the tag being shed within 6 months. Bass with Petersen discs that were deliberately attached loosely below the first dorsal fin, to allow for growth, also showed soreness owing to drag on the tags, which were eventually shed during the second year of the trial. A tendency with Carlin–Ritchie pennants (Fig. 5.3(b)) for the bridles to break was remedied by using stouter wire (26 SWG) than is usual, and fish tagged either below the first dorsal fin or above the anal fin retained their tags for 3 years with no adverse effects.

Trials with bass tagged with the Lowestoft flag (Fig.5.3(c)) were inconclusive, though there were indications that these tags were shed more readily than either of the foregoing types. Recent (1990–91) tests with Hallprint internal-anchor 'T' bar tags (Fig. 5.3(f)), using 1-group (14–18 cm) bass held in aquaria over 12 months, resulted in a 90% retention rate.

A double-tagging field trial, using both Petersen discs and mini-Howitts (the main tag used since 1981 by MAFF; Fig. 5.3(d)) on bass within the length range 25–48 cm, produced a recovery rate of 19.5%, which is much higher than was found with mini-Howitts used alone. Of these recoveries, twice as many fish lost their mini-Howitt tag as those which had the Petersen disc missing. It appeared, however, that the Petersen discs were only more resilient than the mini-Howitts during the first 100 days of the fishes' liberty, and that they were susceptible to snagging in gill nets.

An empirical comparison of the effectiveness of tag-types can be obtained from releases in which bass are fitted with single tags of different types. Although the efficacy of each type of tag can be compared only with others used in the same release, it is apparent that the Petersen disc (attached below the dorsal fin), the Howitt and the Carlin–Ritchie pennant are, in general, equally effective. The choice between them depends upon the size of bass being tagged and the purpose of the exercise. The need to attach Petersen discs tightly – and not to allow space for growth – suggests that these tags are more suited to use on large bass. The majority of bass recovered after more than 3 years at liberty bore Petersen discs.

The advantages in tagging speed obtained using Floy 'T-bar' tags (known in the USA as Dennison tags; Fig. 5.3(e)) on small bass – which are usually caught in higher numbers than adults and are more susceptible to handling stress – were negated by this tag's poor recovery levels compared with Howitts, both in the lower return rate (tag conspicuousness) and in the shorter duration over which recaptures were reported (tag loss). Similar problems have been encountered using Dennison tags on striped bass in the USA. Waldman *et al.* (1990)

recorded a high loss rate (58%) over a period of one year with this type of tag, (Floy type, FD-68B) compared with 2% for internal-anchor tags (Floy type FTF-69).

Size and maturity phase of tagged bass

In most releases of tagged bass, there has been a tendency for the majority of fish to be either smaller or larger than 40 cm (Table 5.1). This was usually influenced by the choice of tagging site and the age structure of the local bass population, and by the selectivity of the catching gear, which was often directed towards the most abundant size group. In general, inshore and estuarine sites produced a preponderance of small bass and offshore or deepwater sites produced the larger fish.

In the earlier investigations (Holden and Williams, 1974; Kelley, 1979), it became evident that the extent of movement and the geographical range of bass change with increasing size, and it is also likely that the seasonal distribution and movements of bass are influenced by the sex of the fish (Chapter 7). Attempts to sex bass externally have been largely unsuccessful (Kelley, 1979), and the generally poor response to requests for the body of recaptured fish to accompany tag returns (~5% of recaptures on the south and east coasts, ~10% on the west coast) has not permitted a satisfactory analysis of distributions of recaptured tagged bass by sex. Examination of the gonads of more than 2000 bass caught around the coasts of England and Wales during 1982 to 1990 has indicated approximate length limits for three maturity categories in the combined sexes as follows:

1. juveniles: fish of less than 32 cm, which have immature gonads;
2. adolescents: fish between 32 and 42 cm, of which many males, but very few females, are found to have maturing gonads during the winter and spring;
3. adults: fish over 42 cm which have fully mature or spent gonads during or after the spawning season, February to July.

In the following account of the combined results of all bass tagging exercises around England and Wales, bass tagged as juveniles and recaptured as adolescents or adults have been classed as adolescents for analytical purposes; otherwise these categories refer to the length of bass at the time of tagging. All tagging sites are clearly identifiable and distances from the release position to the recapture position have been measured along the shortest sea route.

Table 5.1 Length distributions of tagged bass released around England and Wales, 1970–84. From Pawson *et al.*, 1987

Total length	*Thames Estuary*	*Solent*	*Portland Race*	*S. Devon*	*Plymouth*		*W. Cornwall*	*N. Cornwall*		*Carmarthen Bay*	*N. Pembroke*	*Anglesey*	
(cm)	*1979–84*	*1981–84*	*1983*	*1970–71*	*1970–71*	*1983–84*	*1982*	*1975–78*	*1981–84*	*1982*	*1979–81*	*1971–75*	*1984*
Not known	10	3	0	0	0	0	4	0	0	3	0	3	0
Under 24	10	90	0	43	21	1	0	4	0	0	0	2	0
(24–26)	12	158	0	7	37	0	0	4	12	0	0	10	1
(26–28)	9	287	0	23	23	8	0	7	26	1	11	31	4
(28–30)	6	461	0	25	5	1	0	11	39	0	31	21	1
(30–31)	16	281	0	30	6	4	0	27	39	0	18	48	1
(32–34)	25	261	0	39	4	6	0	36	34	5	55	42	1
(34–36)	38	186	0	34	1	2	0	23	41	25	36	111	1
(36–38)	43	95	0	55	3	4	0	11	25	51	17	61	0
(38–40)	63	52	0	73	1	0	0	13	6	33	7	59	4
(40–42)	79	54	0	43	0	0	2	10	0	34	6	48	0
(42–44)	91	32	0	50	0	0	14	1	2	24	7	50	3
(44–46)	79	19	0	27	0	0	15	3	0	6	3	28	1
(46–48)	78	13	1	24	0	0	13	6	0	4	7	33	6
(48–50)	61	8	0	18	0	0	28	5	0	2	2	29	6
(50–52)	51	4	5	11	2	0	24	3	1	2	3	38	4
(52–54)	34	3	8	17	0	0	25	1	0	0	6	33	2
(54–56)	23	1	8	9	2	0	19	3	0	1	3	51	9
(56–58)	22	2	21	14	0	0	22	1	0	0	1	44	5
(58–60)	18	1	16	8	0	0	15	0	0	0	2	56	4
Over 60	74	2	89	15	0	0	10	9	0	0	2	114	9
Total	842	2013	148	565	105	26	191	178	225	191	217	912	62

5.3 ADULT MIGRATIONS

Southern England and English Channel

Thirty-four of the adult bass released in the Thames Estuary were recaptured within the general release area in successive summers (May–Oct.), and only three fish were taken in summer at a greater distance; two off the Kent coast, north of the Straits of Dover, and one fish which was reported from Raz de Sein, West Brittany, after 4 years at liberty. Apart from two fish which were taken at the warm-water outflow from Bradwell Power Station in November and December after 1 year at liberty, and 2 caught in April further offshore in the southern North Sea, no adults were recovered in winter (Nov.–April) at a distance of less than 400 km from the tagging site. All these recaptures were made offshore, outside the UK 12-mile zone, with the majority being taken by French pelagic pair trawlers fishing in mid-Channel between South Devon and Brittany. The most distant fish was recovered in the Bay of Biscay, 50 km west of La Rochelle, France.

All recaptures of adult bass tagged in the Solent were taken in the release area between March and October. Of 14 recaptures of large bass tagged in the summer at Portland Bill, all but three were retaken in the release area between July and October. The two other summer recaptures were on the central South English coast, which indicates that the summer movements of adult bass there may be quite limited. The single winter recovery was caught near Santander, northern Spain in the second year after tagging. This fish represents the most southerly recapture of a bass tagged in British waters, at a minimum distance of 850 km from the tagging site. Its capture raises the possibility that some adult bass, which are found in summer along the South English coast, might migrate there from the Bay of Biscay.

Very few of the adult bass which were released near Berry Head, South Devon, in 1970–71 were recovered in summer more than a few km from the tagging site. In winter, most fish were recaptured along the Cornish coasts, and one was taken in mid-Channel, 125 km south-west of the tagging site.

West coast of England and Wales

As with adult bass tagged in the Thames Estuary, there was a clear distinction between the distribution of summer and winter recaptures of adult bass released in summer at Anglesey. During the summer, all reported recaptures (29 fish) were from the general release area and within 33 km of the tagging site. In fact, half of them were retaken at the tagging site itself. Two fish were taken near the release area early in November, but all the other winter recaptures were recorded at more than 400 km to the south, off the South Cornish coast. One fish was taken off the

Ile d'Yeu on the west coast of France, and is the only record of a 'west coast' bass being recovered nearer to the Continental land mass than to Britain.

This southerly movement is unlikely to be emigration, because 21 adult bass were recaptured within 16 km of the Anglesey tagging site at the same time of year as they were tagged (May–Nov.), but after spending at least one winter at liberty after tagging, during which time some of these fish had probably travelled as far as Cornwall. It is not known whether this extremely limited distribution of recaptures is chiefly a reflection of very localized fishing effort along a migration route, but it does demonstrate that adult bass will revisit the same local habitat in at least two successive summers.

Nine of the adult bass tagged at Anglesey between mid-October and the end of December were recovered between November and March inclusive, most being reported from within 92 km of the tagging site. The others were recaptured off North and West Cornwall, in locations similar to those of adult bass tagged at Anglesey earlier in the year. Two-thirds of the summer recaptures of fish tagged in early winter were taken near the release area, with the remainder being reported to the north, off the coasts of Cumbria and Lancashire. The combined results of the summer and winter taggings indicate that adult bass that were caught late in the year at Anglesey had spent the summer to the north, and that many of them probably did not go as far south in winter as those fish caught near Anglesey in summer. Of the adult bass tagged in North Pembrokeshire, those which were reported as being recaptured in summer were in the release area; again, winter recaptures were all made to the south, around the Cornish Peninsula. Similarly, most adult bass tagged in summer in the Camel Estuary, North Cornwall, were recaptured in summer in the release area, and only one winter recovery was made, at the Eddystone 200 km from the tagging site. These few recoveries suggest a limited range of movement of adult bass in summer off North Cornwall.

Apart from three recaptures reported from within 16 km of the Land's End tagging site a few months after they were tagged in October 1982, the other nine adult bass which were recovered were taken in summer in an area extending from the north coast of Cornwall to the Duddon Estuary in Cumbria. This confirmed the speculation that some adult bass will travel through the whole distribution range of the population tagged at Anglesey, which appears to extend at least from Cumbria to the waters around Cornwall. Most of these recaptures, however, were made between June and August in the Carmarthen Bay–Gower area. This suggests that South Wales is probably the northerly limit of movement for many bass taken near Land's End in October, and that these fish represent only a part of the whole west coast population.

Recapture patterns of adult bass around England and Wales are summarized in Fig. 5.4.

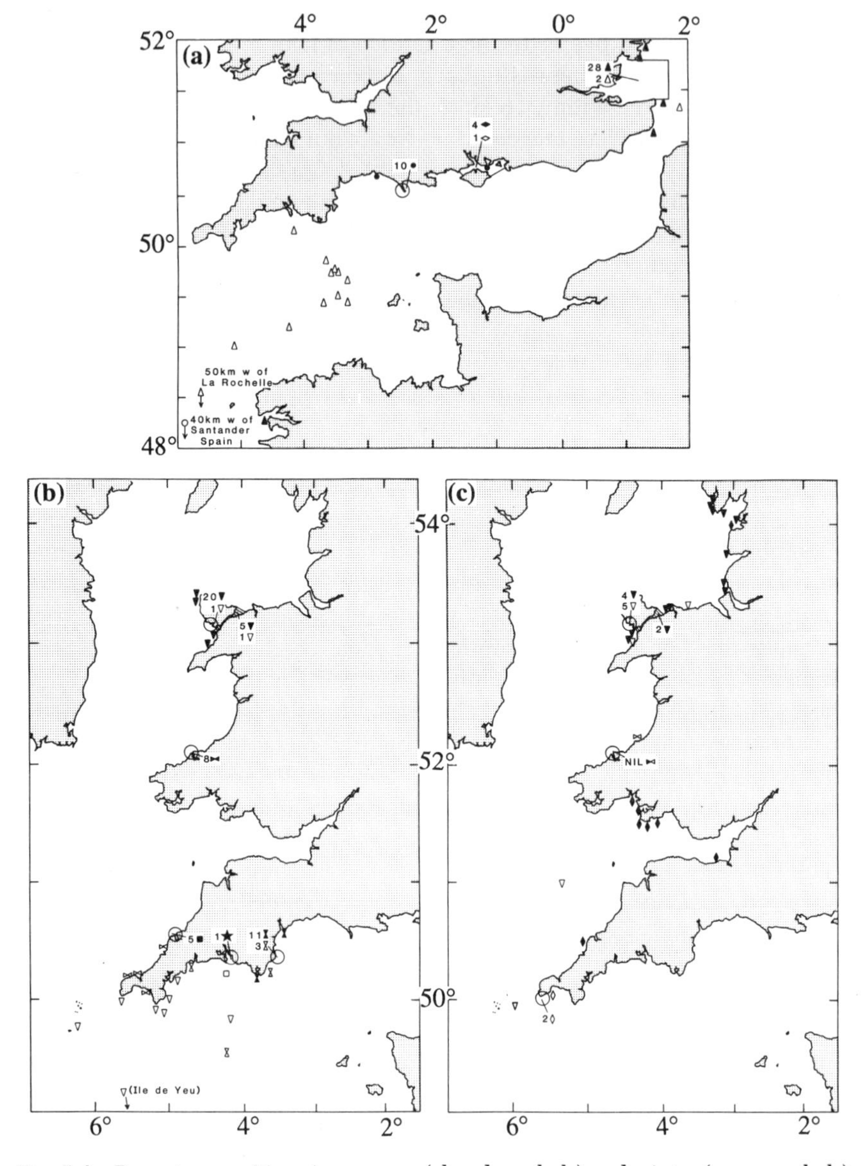

Fig. 5.4 Recapture positions in summer (closed symbols) and winter (open symbols) of adult bass tagged in England and Wales between 1970 and 1984. (a) From releases in the Thames Estuary, Dorset and Hampshire, (b) from releases in summer in south-west England and Wales. (c) from releases in winter in south-west England and Wales. ▲, Thames Estuary; ◆, Solent; ●, Portland Bill; ⧗, South Devon; ★, Plymouth; ◆, Runnel Stone; ■, North Devon and Cornwall; ◖, Carmarthen Bay; ⋈, North Pembrokeshire; ▼, Anglesey. Reproduced from Pawson *et al.*, 1987.

5.4 DISPERSION OF JUVENILES AND ADOLESCENTS

Recapture patterns of juveniles by position and season are shown in Fig. 5.5, and the distribution patterns of bass classified as adolescents which were recaptured from all releases are combined in Fig. 5.6.

The usual range of movement of juvenile bass was well within 80 km of the release area, and fewer than 3% of bass tagged and recaptured as juveniles have been observed to move outside the normal summer distribution of each of the tagged populations. The exceptions tended to be west coast fish taken in summer to the north and in winter to the south of their respective ranges. Juvenile bass tagged around the coasts of Devon and Cornwall did not, in general, appear to move far; nearly all those recaptured as juveniles being reported from within their respective release areas. Juveniles tagged further to the north on the Welsh coast and to the east in the Solent and Thames Estuary – i.e. in waters with greater seasonal temperature fluctuations and, incidentally, where estuaries become progressively shallower in comparison with the rias of South-west England – appeared to move more extensively. The small proportion of recoveries taken in winter, however, came from within a similar distance of each release area, indicating that the seasonal movements of juvenile bass are restricted in relation to the geographical position of their nursery area and are not necessarily influenced by absolute temperature regimes or their position in the overall geographical range of the species.

The pattern of dispersion of adolescent bass appears to be similar to that of juveniles tagged at the same site, though some adolescents were recaptured well away from their release areas, often having moved into areas where neither juvenile nor adult fish from the original populations were recovered. This was particularly apparent in the Thames Estuary releases; some adolescent fish were reported in summer from around the coasts of the southern North Sea outside the Thames Estuary (and in one case from the west coast of Brittany) and, in winter, several turned up on the English coast of the central English Channel. Four adolescent bass tagged on the south-west and west coasts were recovered along the English south coast, well outside the corresponding juvenile and adult ranges, having attained a size commensurate with their being adults at the time of recapture. Although juvenile and adolescent bass, tagged along the southern coast of England and in the Thames Estuary, exhibited similar patterns of movement to those of west coast fish, relatively more adolescent bass caught off South-east England in winter were recaptured over 5 km offshore. These regional variations are probably a result of local topography and hydrographic conditions influencing the distribution and pattern of the bass fishery, as well as the distribution of the bass themselves.

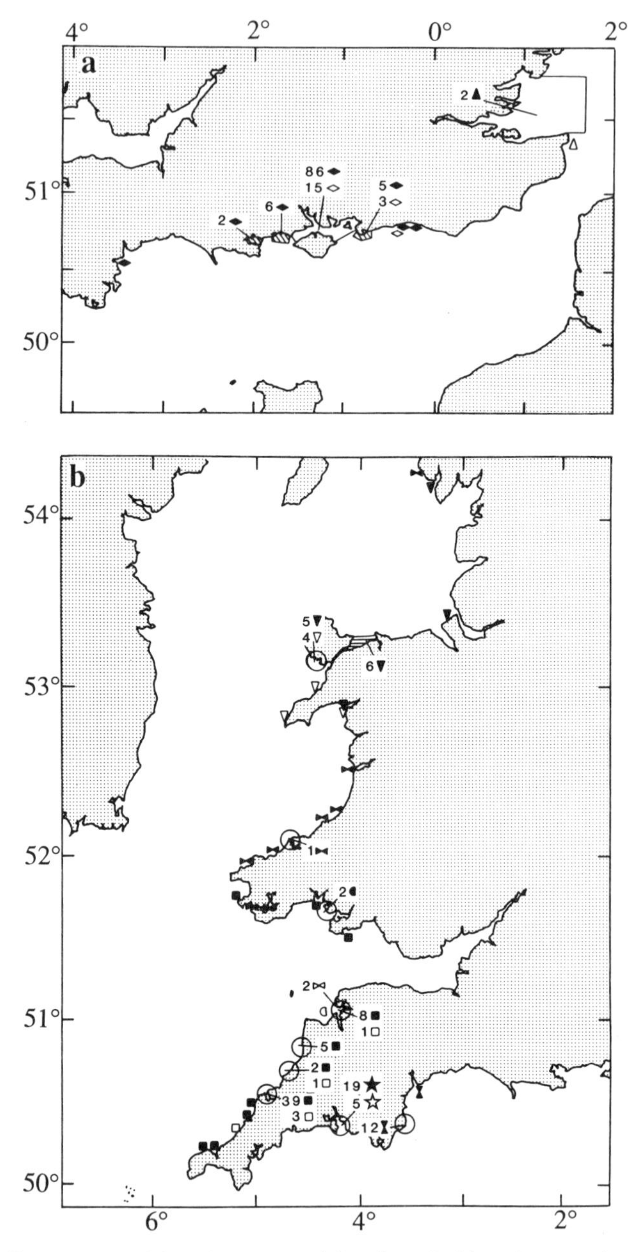

Fig. 5.5 Recapture positions in summer (closed symbols) and winter (open symbols) of immature bass tagged in England and Wales between 1970 and 1984. (A) As juveniles released in the Thames Estuary and Solent. (B) As juveniles released in south-west England and Wales. ▲, Thames Estuary; ◆, Solent; ●, Portland Bill; ⧗, South Devon; ▲, Plymouth; ◆, Runnel Stone; ■, North Devon and Cornwall; ◖, Carmarthen Bay; ⋈, North Pembrokeshire; ▼, Anglesey. Reproduced from Pawson *et al.*, 1987.

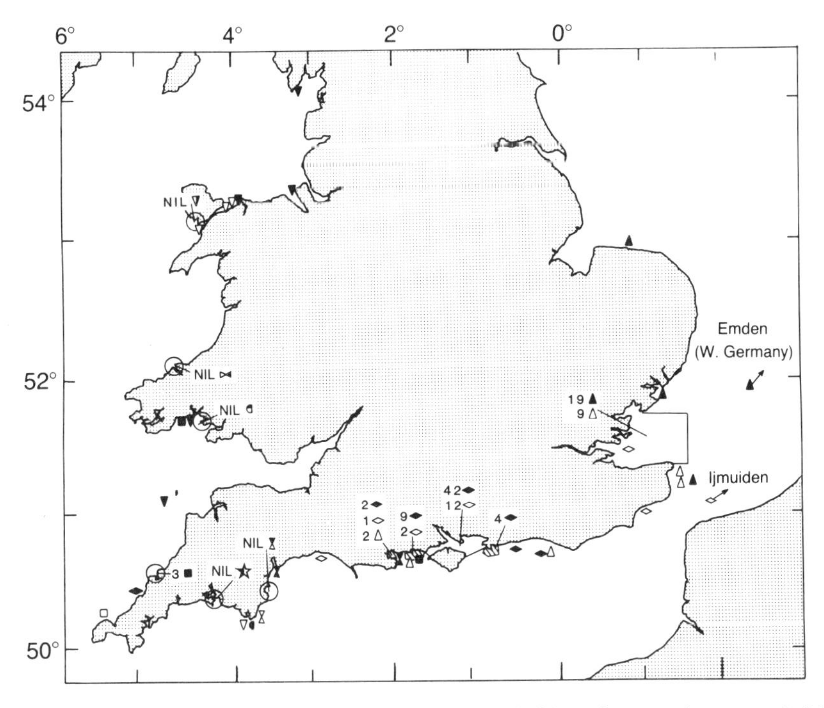

Fig. 5.6 Recapture positions in summer (closed symbols) and winter (open symbols) of adolescent bass tagged in England and Wales between 1970 and 1984. ▲, Thames Estuary; ◆, Solent; ●, Portland Bill; ⧗, South Devon; ★, Plymouth; ◆, Runnel Stone; ■, North Devon and Cornwall; ◖, Carmarthen Bay; ⋈, North Pembrokeshire; ▼, Anglesey. Reproduced from Pawson *et al.*, 1987.

5.5 RECOVERY PATTERNS OF TAGGED BASS

A summary of the dispersion of bass from tagging sites around England and Wales is given in Table 5.2, as a breakdown of recaptures taken within the ranges 0–16, 16–80 and over 80 km from the release position, by fish size category and season of recapture. The seasonal differences in these distribution patterns are shown in Table 5.3, where these data are presented as summer (May–Oct.) and winter (Nov.–Apr.) recaptures, and are also grouped within the 3, 3–12 and outside the 12 mile coastal zones to show the extent of offshore movement.

Nearly half of the recaptures, irrespective of location, were recorded within 4 months of the fish being tagged. A quarter of recaptures were reported during the summer following tagging. Up to February 1993, only 26 of the total of 979 tagged bass recaptured and reported since 1970 had been at liberty for more than 3 years: 14 between 3 and 4 years, 8 between

Table 5.2 Distribution of recaptures of bass in summer (May–October) and winter (November–April) in England and Wales in terms of distance from tagging site and time after release. From Pawson *et al.*, 1987

Size category	*Distance from tagging site (km)*												*Number recaptured*
	0–16	*16–80*	*80+*	*0–16*	*16–80*	*80+*	*0–16*	*16–80*	*80+*	*0–16*	*16–80*	*80+*	
Summer taggings													
	Same summer			Following winter			Following summer			Subsequently			
Juveniles	101	24	5	19	5	2	34	5	4	15	10	4	228
Adolescents	32	21	2	12	7	9	19	6	3	7	5	4	127
Adults	37	7	1	3	4	21	32	7	1	10	8	11	142
Winter taggings													
	Same winter			Following summer			Following winter			Subsequently			
Juveniles	9	2	4	18	1	0	2	1	0	0	1	0	38
Adolesents	0	0	0	3	1	2	1	1	0	1	2	0	11
Adults	6	3	2	5	5	14	3	0	0	1	1	1	41

Table 5.3 Seasonality of recaptures of bass tagged in England and Wales, 1970–84, distance from tagging site and distance offshore

	Recapture period						Recapture period						
	May–Oct.			*Nov.–Apr.*			*May–Oct.*			*Nov.–Apr.*			
	Distance from tagging site (km)						*Distance offshore (miles)*						
Tagging category	*0–16*	*16–18*	*80+*	*0–16*	*16–80*	*80+*	*0–3*	*3–12*	*12+*	*0–3*	*3–12*	*12+*	*Total recaptures*
Summer-tagged													
Juveniles	143	37	10	26	8	4	189	0	0	39	0	0	228
Adolescents	51	30	8	19	9	10	71	16	1	35	2	2	127
Adults	78	21	3	4	5	31	7	0	1	14	42	0	38
Winter-tagged													
Adolescents	3	3	2	2	1	0	7	0	1	3	0	0	11
Juveniles	18	2	0	11	3	4	19	1	0	18	0	0	142
Adults	6	6	15	9	3	2	27	0	0	12	0	2	41

4 and 5 years and 2 after 5 years. Recaptures from all releases were most numerous in summer, whilst the level of recoveries in winter was higher for adults than for younger fish.

The majority of recaptures of all bass have been made within 16 km of the tagging site, the proportion being higher for the smaller fish. The proportion of adolescent bass caught between 16 and 80 km distant along the coast from the tagging site is higher than for either juveniles or adults, suggesting that bass tend to have a lower affinity for a particular area during the transition period between juvenile and adult. The proportion of recaptures taken at over 80 km from the tagging site increased with the size of the fish, indicating that bass move greater distances as they mature. Similarly, recaptures outside the 3 mile zone increased in the same size order, and proportionately more adults were taken outside the 12 mile zone in winter than in summer. Recaptures of juvenile and adolescent bass tagged on the west coast of Britain in summer tended to be made at greater distances from the release area in successive years following tagging, whereas the distribution patterns of other juveniles were similar at all periods following release. Adolescent and adult bass recaptured in the opposite season to that in which they were tagged were more likely to be reported well away from the release area than those recaptured at the same time of year as that in which they were tagged; one-third of adolescent and two-thirds of adult returns in the opposite season were reported at over 80 km from the tagging site. It seems, therefore, that a considerable proportion of all size groups within the bass population remain in or return to well-defined inshore areas, particularly in summer, and that the fish also extend their range of movement both along the coast and offshore as they grow.

5.6 CONCLUSIONS

Despite the rather fragmented distribution of the fishery taking bass, and the unwillingness of some fishermen and anglers to report recaptures, these tagging data can be used to give a comprehensive picture of the dispersion and distribution of bass around England and Wales. When tagged bass move into areas with little or no fishing activity of a type likely to catch them, their presence there goes unrecorded (Holden and Williams, 1974). The lack of a winter fishery capable of taking bass in deep water to the south of Ireland, which would correspond to that around Cornwall and offshore in the western English Channel, for example, could account for the impression of a much more local distribution of bass around Ireland (Kennedy and Fitzmaurice, 1972). In the 1980s, however, French trawlers caught many bass in winter in an area about 100 km south of Tuskar Rock, off the south-east coast of Ireland (M. Kennedy, pers. comm.).

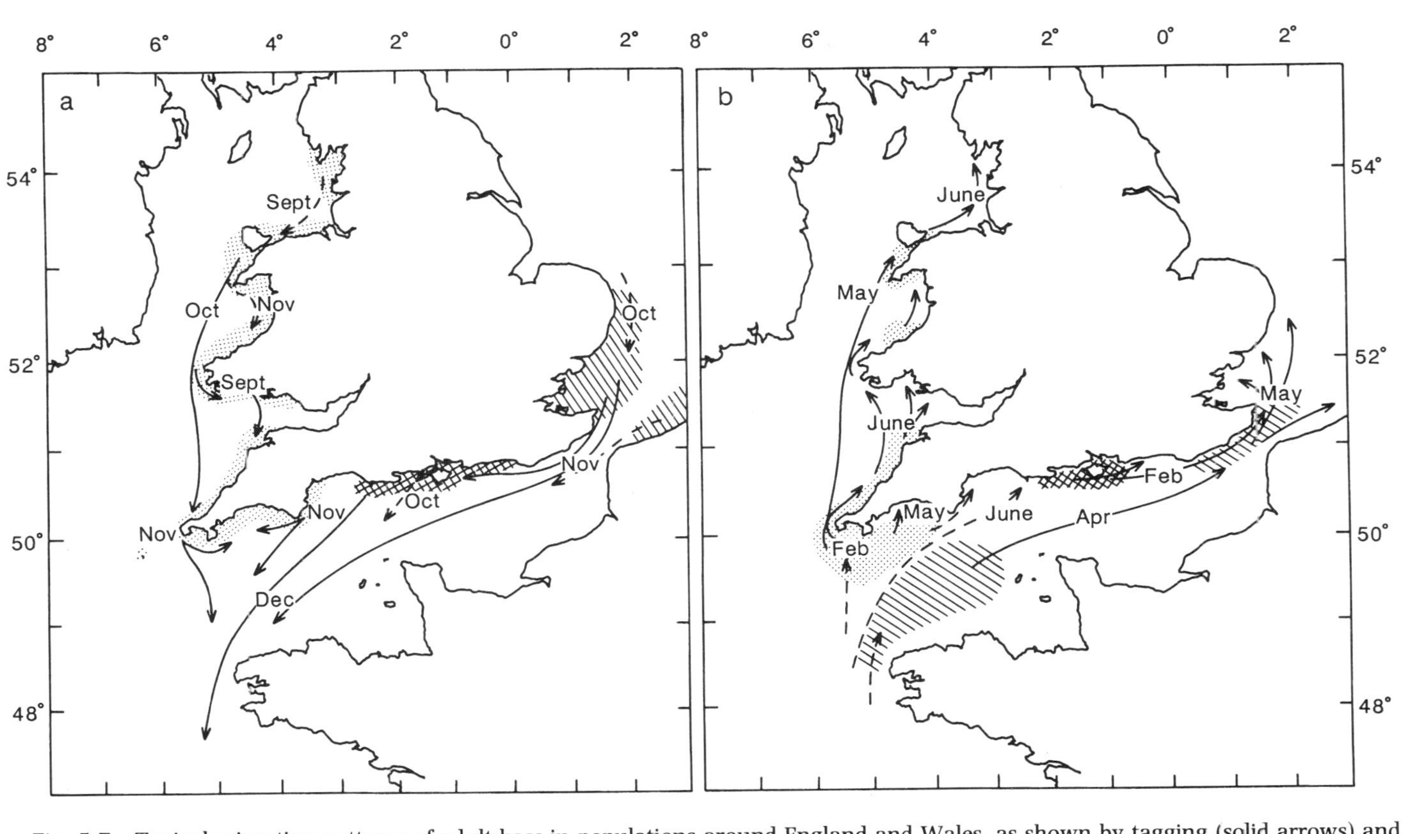

Fig. 5.7 Typical migration patterns of adult bass in populations around England and Wales, as shown by tagging (solid arrows) and postulated movements (broken arrows) in (a) autumn and (b) spring. Redrawn from Pawson *et al.*, 1987.

Our interpretation of the results of these tagging exercises in terms of the distribution, movements and migration patterns of the bass population around the shores of England and Wales is summarized in Fig. 5.7.

The marked increase in the extent of movement of bass as they approach the size of first maturity suggests that they may anticipate the movements of adults, which occur in the autumn between summer feeding areas and winter pre-spawning areas and in the spring back to the summer areas. The distances involved may be less than 100 km for adult bass living around Cornwall and South Devon, but are more typically of 400–500 km for fish that spend the summer around North Wales or in the Thames Estuary, and winter around Cornwall or offshore in the western English Channel, respectively. Adult bass from the south coast of the UK may also move into the Bay of Biscay in winter, at distances occasionally exceeding 800 km from their summer feeding area.

The movements of adult bass appear to be much more definite than those of adolescent fish. There was no case where an adult bass, tagged in summer and recovered in the same or any subsequent summer, was recaptured in an area different from that in which it was tagged. There may therefore be behavioural separation of groups of adults, which segregate to specific summer feeding grounds even though they may share common migration routes during spring and autumn and may possibly mix during spawning.

5.7 STOCK IDENTITY

Tagging studies along the North-east Atlantic coast of the USA have revealed seasonal movements of striped bass, which are similar to those we have described here for *D. labrax*. Unlike the European bass, however, the striped bass spawns in fresh water. Clark (1968) suggested that different 'contingents' of striped bass can be recognized by their seasonal distribution patterns, and that each contingent uses a particular river or group of rivers for spawning. Genetic races of striped bass have been identified in samples containing both juveniles and spawning adults from some of the main rivers entering Chesapeake Bay (Morgan *et al.*, 1973) although, unlike the Atlantic salmon, homing of adult striped bass to their natal streams has not been demonstrated.

Even if adult European bass used spawning grounds from which larvae eventually recruited to their parental population, the observation that some adolescent bass have moved between populations on the south-east, south coast and west coast of the UK, suggests that sufficient interchange occurs to minimize the chances of 'stocks' evolving and remaining separate. Nevertheless, electrophoretic analysis has revealed some evidence for genetically distinct stocks of bass on the west coast of England and Wales and on the south and east coast of England (Child, 1992). The pattern of tag recoveries described above shows that bass, particularly as adults, have a strong affinity for summer feeding areas, which may or may not be close to their nursery grounds. This suggests that, even if local populations are not genetically distinct, they can have characteristic distribution ranges. On this basis, separate contingents could be recognized for management purposes. For the same reasons, it seems likely that it is adolescent rather than adult bass which repopulate areas in which the bass population has been depleted through poor local recruitment or overfishing.

Evidence of the integrity of local stocks of juvenile bass is lacking for many areas, but this is an important aspect of management of the bass fishery, as will be seen (in Chapter 17) when the concept of nursery areas is discussed. For this reason, tagging of juvenile and adolescent bass has continued in a number of nursery areas around England and Wales since 1984 (Kelley) and 1989 (DFR). The aims have been:

1. to check the validity of the assumption that there is no significant movement of 1- and 2-group bass away from, or between, nursery areas, and
2. to assess the proportion of 3- and 4-group bass recruiting to the locally exploited population, and the size and age at which bass leave nursery areas.

Although data are still being collected, it is apparent that there is a very low recovery rate of these small bass, despite the increased use of a tag-type particularly suited to small bass (Hallprint internal-anchor 'T'-bar). This could indicate a high tag loss rate, poor tag visibility, non-reporting, or a low exploitation rate on small fish. Of course, because all the bass tagged in nursery areas were under the legal minimum landing size (MLS) of 36 cm, there may have been a reluctance on the part of captors to report them. Despite this, all recaptures of 1- and 2-group bass so far reported have been from within the original nursery areas.

Most recaptures of 3- and 4-group bass made in summer have been in or near their original nursery areas, though some fish recaptured at above 36 cm have been taken well away from the tagging sites, in adult wintering areas. It is tentatively concluded that bass below about 35–37 cm can be assumed to be native to the nursery area in which they are caught. More returns of bass tagged as juveniles or adolescents and recaptured as adults in summer are needed to determine the integrity of local stocks as a whole.

Chapter six

Growth and age

6.1 INTRODUCTION

There are three main reasons for studying growth and age in bass: (a) to contribute to our understanding of the species' biology and, especially, which factors influence growth, (b) to enable the effect of fishing and exploitation on the age and size structure of the population and on yields to the fishery, to be assessed, so that advice can be given on its management, and (c) to attempt to characterize populations on the basis of growth patterns, so that stocks can be identified for fishery management purposes. Most biological studies of bass have involved measurements of size and determination of age, and information on growth is now available for most of the species' geographical range. The geographical location and authors of these studies, which includes those referred to in this chapter are shown in Fig. 6.1.

Although the basic methods for measuring and describing fish growth are well documented, (e.g. Weatherley, 1972; Weatherley and Gill, 1987) it is useful to review them before describing and commenting on the growth patterns of bass observed around England, Wales and continental Europe.

6.2 GROWTH CHARACTERISTICS

Growth can be defined as a permanent increase in size with time. It is distinguished, therefore, from any temporary (usually seasonal) gain in weight that might be due to gonad development or the accumulation of fat deposits when feeding conditions are good. When recording size, it is necessary to choose the most suitable dimension for the species being studied. To an angler, size generally means 'mass', how big a fish appears, either in the imagination or on the scales, and it is usually measured in terms of weight at the time of capture. Weight is also used as the unit upon which payment is based when fish are sold. Marine fishery scientists

Fig. 6.1 Geographical distribution of growth studies on bass. **A**, Kennedy and Fitzmaurice, 1972; **B**, Kelly, 1988; **C**, Parsons, 1982; **D**, Sabriye, 1983; **E**, Pickett and Pawson, this volume; **F**, Boulineau Coatanea, 1969; **G**, Arias, 1980; **H**, Le Masson, 1981; **I**, Muyard, 1978; **J**, Bertignac, 1987; **K**, Lam Hoai Thong, 1970; **L**, Do Chi and Lam Hoai Thong, 1971; **M**, Stequert, 1972; **N**, Barnabé, 1976; **O**, Gravier, 1961; **P**, Bou Ain, 1977; **Q**, Rafail, 1971.

generally prefer to use length to record the size of fish because, until the development of motion-compensated scales, accurate weighing at sea has been impossible. Moreover, measurement of skeletal length is probably the most consistent and accurate means of showing how much a fish has grown.

There are various length measurements which can be used (Fig. 1.2). These are: **total length** (TL), the measurement from the tip of the snout to the tip of the straightened caudal fin; **fork length** (FL), from the tip of the snout to the cleft or fork of the caudal fin; and **standard length** (SL), from the tip of the snout to the end of the fleshy base of the central caudal fin ray. TL is most commonly used by fisheries scientists and in fisheries legislation in

Europe, though American researchers appear to prefer FL. Throughout this book we have used TL for bass, and where FL has been the dimension used by other authors, this has been converted to TL, using a relationship obtained by Kelley (1988b), ourselves and others: TL = 1.07 FL.

SL is used mainly for postlarvae or 0-group fish, being the length of the skull plus that of the backbone, and is probably a more precise, but relatively time-consuming, measurement than either FL or TL, which can be affected by worn or split caudal fins.

Generally, lengths are measured to the 1 cm below, using a flat-faced measuring board rather than a tape, which is inclined to curve over the fish's body. It is easy to see how TL has become the most used variable when large numbers of fish have to be measured in a limited time.

To the above variables can be added various girth measurements (Fig. 1.2), which are useful when considering the size selectivity of fish by nets. The girth of a fish is not particularly useful for describing growth, because it too closely reflects seasonal changes in gonad size, feeding activity and bodily condition.

The other dimension shown in Fig. 1.2, i.e. head length, is not normally used in growth determination, but Barnabé (1976b) reports a tendency for the head/total length ratio to be different for females and males; it certainly varies with the growth rate of many species of fish.

6.3 MEASURING LINEAR GROWTH

In captive fish, growth may be recorded simply by measuring (and perhaps weighing) them at time intervals. Fish can be given distinctive marks so that the growth of individuals can be followed. It is possible to measure growth in this way in wild populations of fish by using mark-and-recapture procedures (tagging live fish), though there are problems with this approach:

1. the fish that are tagged can seldom be considered to be a truly representative sample of the whole population; particularly because,
2. stress or injury suffered in the initial capture and marking procedures may cause the growth of marked fish to be slow relative to that of others in the wild population; and
3. adequate recaptures cannot be guaranteed, either because fishing effort is low or because few fish survive to be recaptured after a sufficiently long (2–3 year) time period.

Most growth studies of bass have had to rely on measurements obtained from samples of dead fish landed by the commercial fishery or anglers. The problem of attempting to examine growth in the population as a whole is that fishing gears will often be selective of small or large fish, and there can be a bias towards slow or fast growers of particular age groups. In order to describe growth within a single season, rather than from year to year, samples from a known population of fish need to be taken regularly, say at monthly intervals. Again, these fish must be representative of the population in terms of age class and size. Owing to the difficulties in obtaining adequate samples, only one study (Kelley, 1988b) has described growth of a wild population of bass within a season. Most other studies have relied on the determination of annual growth increments and have required an accurate determinant of the age of the fish.

In temperate waters, it is assumed that fish do not grow at a constant rate throughout the year, and this will normally be reflected in the composition of skeletal structures. Sagittal otoliths or opercular bones of bass can be used to determine age, but the most convenient material is undoubtedly the scales, which, in bass, are easily dislodged, quite large, and have growth patterns that are relatively easy to interpret. The use of scales has another advantage over other skeletal materials, in that the fish do not have to be killed or unduly damaged to obtain them. Tagged fish can be released unharmed (though minus a few scales), and there is no need to buy fish or pay compensation for disfiguring them, when sampling on commercial markets, where bass are sold whole. Scales are taken from the sampled fish using blunt-nosed forceps and are put in small paper packets, numbered and dated and provided with details of place of capture, length and, if available, other biological information such as sex and gonad maturity stage.

6.4 AGEING BASS BY SCALIMETRY

Ageing bass sampled from the commercial catch is a fundamental part of the assessment of the fishery (Chapter 12) and, for this purpose, DFR ages several thousands of fish each year. With the minimum of cleaning and preparation, the growth patterns on bass scales can easily be read using a hand lens. Our preference is for a projecting microscope with a magnification of × 10 to × 20. Other workers have used photographic slide projectors, placing the scale between two glass slides and projecting its image onto a screen or wall. Good-quality optics and a range of lighting effects and choices of magnification are the main requirements for age reading, plus, of course, patience and experience. Some workers have stained the scales to enhance their pattern's definition and, if necessary, surface debris, dried mucus and similar superfluous matter can be removed by cleaning the scales in hydrogen peroxide solution. With large scales, which have a tendency to curl up when dry, dampening will help them to lay flat between the glass slides, thereby producing a sharper image over the whole scale surface. It is generally accepted that, in bass, the most satisfactory scales are obtained from either flank, near the tip of the pectoral fin, below the lateral line. However it is viewed, the image of a scale with back light usually shows well-defined concentric rings (Fig. 6.2).

Bass scales have tooth-like processes in the posterior (exposed) scale segment. The anterior portion of the scale, which is embedded in the skin, is less corrugated and consequently provides the clearest rings. The many dark rings, which are called circulae, are often discontinuous, and their

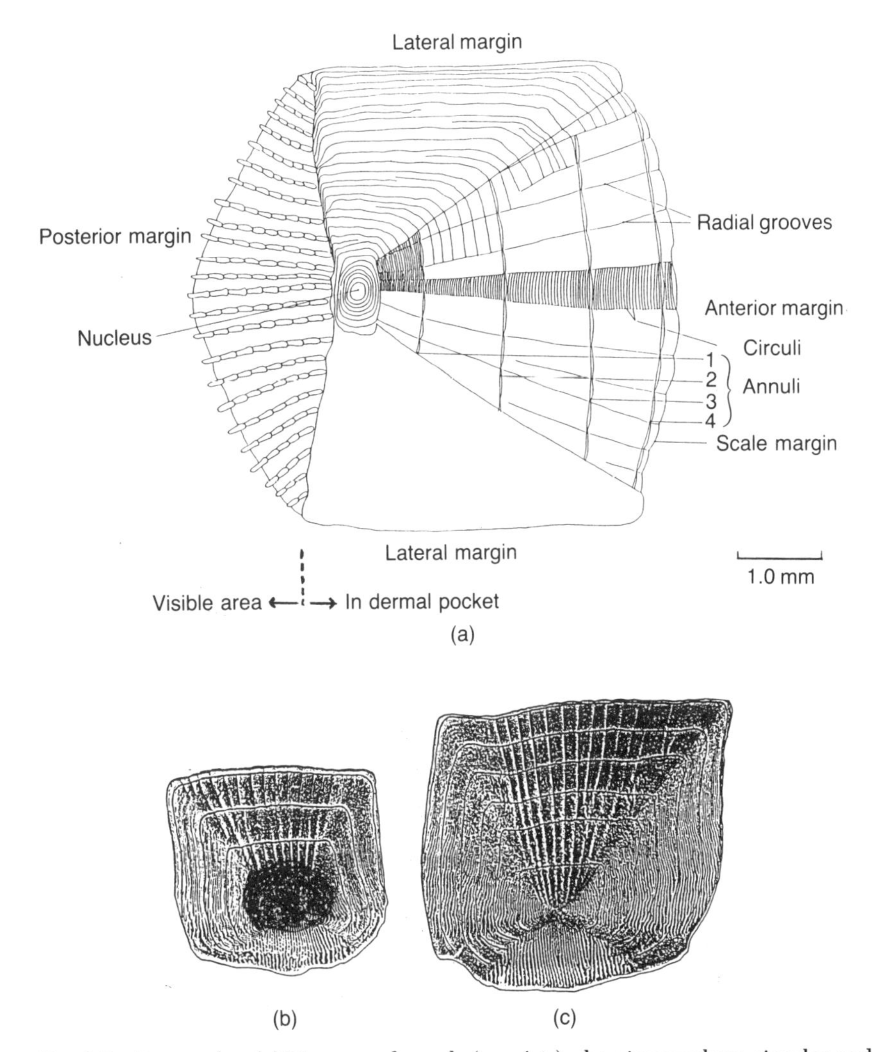

Fig. 6.2 Bass scales. (a) Diagram of a scale (age 4 +), showing nucleus, circulae and annulae (reproduced with permission from Parsons, 1982); (b) replacement scale and (c) original scale (annuli have been marked in black).

frequency varies from scale to scale on the same fish. They probably have little significance for ageing, although their spacing, close or wide, may be related to periods of slow or fast growth. Breaks in the pattern of circulae on the dorsal and ventral scale segments often correspond to the 'white' or clear bands, called annuli, which appear in the anterior scale segment. These are sometimes described as 'winter rings', but they are actually formed when a new season's growth begins in May or June, and a growing edge begins to appear around the scale margins. The annuli are sometimes emphasized by a difference in the density of circulae, which become closer spaced towards the end of the growing season.

Bass scales sometimes lack a discernible first annulus (i.e. the first clear ring moving out from the nucleus); this is particularly likely in scales from old fish owing to thickening of the scale with increasing age. The first annulus is commonly obscure, or missing altogether, on scales of juvenile bass caught in the vicinity of warm-water outfalls. This suggests that these fish may have continued to grow without a check throughout the winter and spring. As a consequence, ageing of these fish can be difficult. In many cases, however, examination of otoliths or opercular bones taken from the same fish will enable a good estimate of its age to be made. From the second to around the eighth or tenth annulus, bass scales are usually easy to read. At greater ages, the annuli tend to lie closer together, and their discrimination becomes more difficult.

Another problem associated with ageing bass is the occurrence of replacement scales. Bass may lose scales throughout their life and, as each one is lost, a new scale is regenerated to replace it. Although it may eventually acquire the size of adjacent original scales, the replacement will lack the record of the fish's growth history up to the time of its own formation. Fortunately, replacement scales are easily distinguished from originals (Fig. 6.2). Generally, the older the bass is, the more it is likely to have replacement scales; it is not unusual to have to take 20 or more scales from a large, old (20+ years) bass to yield perhaps only one or two originals, which could then be used to age the fish. However, samples of 4–5 scales usually provide enough original scales for the age of most bass to be determined with confidence.

Growth can be interrupted if sudden changes in environmental conditions affect feeding or, with tagged fish, owing to the shock of being caught and released, and this may result in false rings or 'checks' appearing on bass scales. Kennedy and Fitzmaurice (1972) found that these were uncommon in Irish bass and could usually be recognized as such, whereas Kelley (1988b) regards false rings as being quite common on scales from UK west-coast fish, and sometimes difficult to distinguish from true annuli. Experience of reading characteristic scale samples from particular bass populations is the best aid to distinguishing false rings from true annuli.

The usual method of assigning age to scales is to count the annuli outwards from the nucleus. The outer edge of the scale is counted as an extra year only if the scale was obtained from a fish caught between the first of January and the commencement of new growth in late spring or early summer. This is because, in northern Europe, it is conventional to assign marine fish a 1 January birthday, and thus to agc thcm to calender year classes. This causes no problems with bass because, under normal conditions, growth ceases during the winter. Once new growth has formed outside an annulus, the age of the fish is given as the number of annuli, or complete annual bands of growth, followed by a plus sign, e.g. 1+, 6+, until the end of the year. From 1 January onwards, a year is added and the plus sign omitted. In either case, fish that have entered their second, third or fourth calendar year are termed 1-, 2- or 3-groups, respectively. Fish with an undefined number of annuli beyond those actually counted are sometimes given a double plus, e.g. 20+ +.

Providing good scales are available, ageing bass is a relatively speedy process. It takes only a few days to train most people to read them, although experience in ageing scale samples from bass belonging to different year classes and coming from a variety of areas is of great value when interpreting difficult samples.

6.5 OTOLITHS AND OPERCULAR BONES

Otoliths (sagittae) can also be used to determine the age of fish, and for many marine species they are an important material for fisheries assessment.

Boulineau-Coatanea (1968) described bass otoliths in detail and Fig. 6.3 is reproduced from her work. The position of the otolith in the head of the bass is to the posterior when compared with that of many other marine roundfish species. The otoliths are found in the sacs of the labyrinth at the anterior end of the vertebral column. They can be reached by making a vertical incision at the back of the skull (from above) with a knife, but this can often result in damage to the otoliths. It is perhaps more satisfactory, with large bass, to remove the gills and to locate the otoliths from below. Bass otoliths are less easily prepared for ageing than are scales and, in our opinion, give less reliable results for fish over 2 or 3 years old. They are useful, however, in ageing bass of 1–2 years old in situations where scale-rings are not easy to interpret, e.g. where bass live around warm-water outfalls and seasonal growth patterns are not as apparent on scales. In these circumstances, it may be useful to investigate the relationship between otolith weight and age of fish, because at a given length, older fish are expected to have the heavier otoliths (Pawson, 1990).

The otoliths of bass larvae or fry are visible just posterior to the head, and

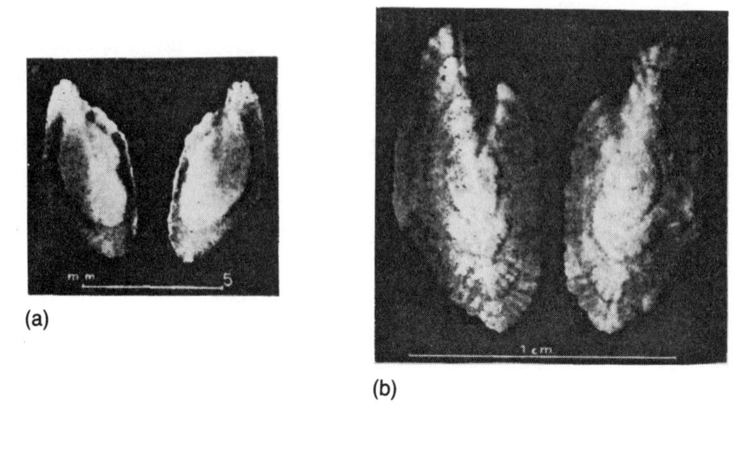

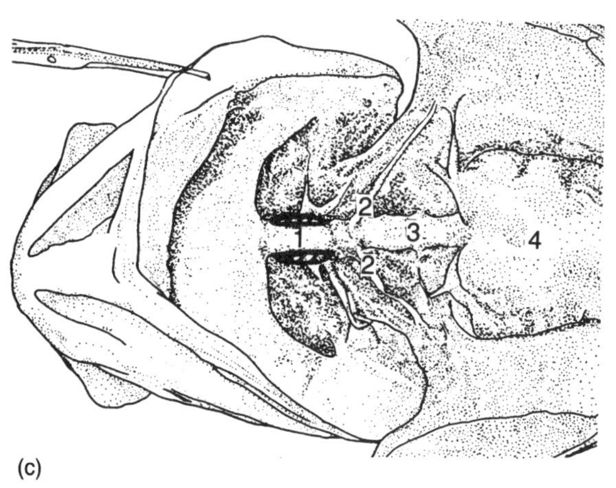

Fig. 6.3 Bass otoliths; (a) 2 years old and (b) 4 years old; (c) location of the sacs containing the otoliths (shaded). 1, sacs; 2, 10th nerves; 3, vertebral column; 4, swim bladder. Reproduced with permission from Boulineau-Coatanea (1968).

a vertical cut behind the pre-operculum will reveal the otoliths, which can be teased free with fine probes (Jennings, 1990). Mounted in resin, sectioned, and examined microscopically, a layered structure is visible (Fig. 6.4).

The first ring is deposited at around the time when larvae first open their mouths and begin to feed, and subsequent rings are deposited at a rate not significantly different from one per day (Morales-Nin, 1985; Ré *et al.*, 1986). Consequently, counting these rings provides a means of ageing bass during the first few months of their lives. Unfortunately, once bass are larger than

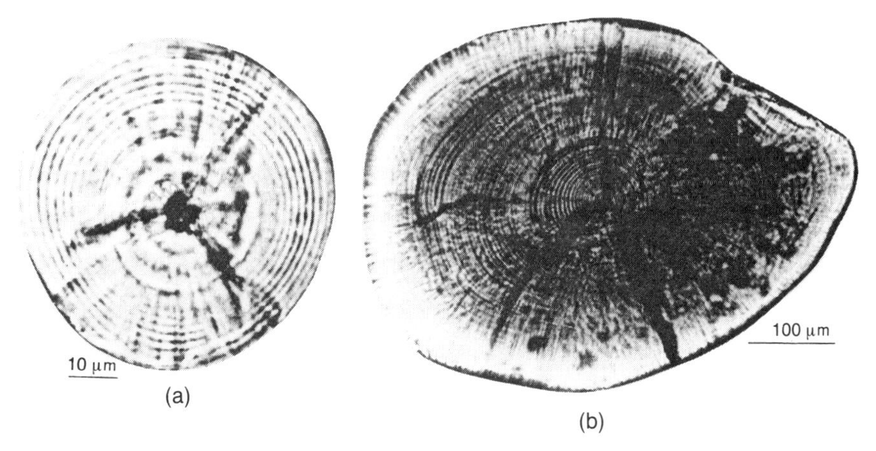

Fig. 6.4 Bass larval otolith sections, (a) 16 increments, (b) 63 increments. Reproduced with permission from Jennings, 1990.

about 35 mm in length, the daily rings on otoliths become very difficult to distinguish from each other and ageing by this means is not really practical.

The opercular bones of the bass are composed of three main parts (Fig. 1.2), the main operculum, the pre-operculum, a smaller component forward of the main bone, and the sub-operculum, which lies below the main operculum. These bones are covered with skin and some scales. Only the main operculum is used for age-reading (although the pre-operculum can be used for diagnostic purposes, e.g. Bou Ain, 1977), as shown in Fig. 6.5. To obtain a clean opercular bone, it is necessary to remove the whole gill cover

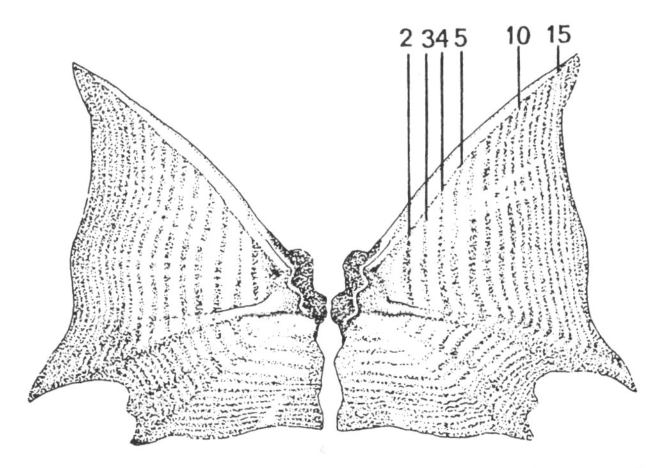

Fig. 6.5 Opercular bones of sea bass, showing interpretation of age (numbers denote years). Drawn by G. Pearce and reproduced with artist's permission.

Table 6.1 Example of an age–length key for bass, using 252 scale samples in 1 cm length groups (MAFF unpublished data)

Data specification for DEMERSAL BIOLOGICAL SAMPLES
Stock: BASS *Sex: Males + Females*
Year: 1990 *Period: All*
Gear: ALL TRAWL *Area: All*
AGE–LENGTH KEY

Length group start (cm)	*Total no.s of fish*	*Age (years)* 3	4	5	6	7	8	9	10	11	12	13	14	15	16	17	18	19	20	21
36	1	3																		
37	3	2	1																	
38	3	3																		
39	2			1	1															
40	2			1	1															
41	3			1	2															
42	4		1	1	2															
43	8				5	3														
44	6				3	2	1													
45	6				3	2	1													
46	9				2	3	4													
47	18		1			13	4													
48	20				1	11	7	1												
49	15				4	9	2													
50	12			1	1	6	3		1											
51	15					3	6	3												
52	16					6	9	1												
53	13					6	6	1												
54	14				1	6	3	3	1											
55	6					1	4	1												

56	8					1		2	3	1		1								
57	7						4	2				1								
58	6					1	1	3				1								
59	6						3	2		1										
60	5					1		2				1	1							
61	6						1		1				3	1						
62	10						3					1	5		1					
63	4									1			3							
64	3												2		1					
65	4							1					2		1					
66	4						1					2			1					
67	6												5						1	
68	2												2							
69	0																			
70	2												2							
71	2											1	1							
72	0																			
73	2														2					
74	1												1							
75	0																			
76	0																			
77	0																			
78	0																			
79	0																			
80	0																			
81	0																			
82	1																			1
Total	252	6	3	5	26	74	63	22	6	3	0	5	30	1	6	0	0	C	1	1

from its cartilaginous hinge, using a scalpel blade. Gentle boiling with a solution of hydrogen peroxide will remove tissue and skin. The bones are then washed in clean water and dried for storage. The clarity and readability of opercular bones is usually good, though, with fish of 5 years and older, thickening may obscure the first annulus.

6.6 AGE–LENGTH DISTRIBUTION

Ideally, to estimate growth of fish in samples from the landed catch, all fish should be both measured and aged, but subsamples of scales by length groups can provide adequate ageing material. These samples are usually stratified on a quarterly or annual basis by region and catching gear group, and are often obtained over several months and from many different catches. Provided length measurements are taken to determine the length-frequency distribution of the catch that is being sampled, scale samples from a total of 5–10 individual fish, selected at random from each 5 cm length group, are usually adequate. One disadvantage of using data from market samples for growth studies is that bass are sold whole, and therefore cannot be sexed. Growth data for each sex can only be obtained when internal examination, which usually involves the purchase of the whole fish, is carried out.

The unraised array of length versus age data, obtained by ageing scale samples, is known as an age–length key (ALK) (Gulland, 1965): an example for bass is shown in Table 6.1.

When the catch length-frequency distribution is used to raise the ALK to an age–length distribution (ALDs), Table 6.2, this should then represent the fishery's landed catch. The ALD can be used to calculate the mean length in each age class in the catch. This is the usual way of determining growth in a fished stock, but it has several disadvantages. First, the length measurements may be biased owing to selectivity by the fishing gear, which may result in a restricted range of size groups of fish being included in the sample. A comparison of their structure in one fishery area will demonstrate the effects of selectivity, and shows why separate ALDs are required for each main gear group or fleet of vessels. To more closely represent a population in one region, gear-related ALDs can be combined to smooth out any sampling bias. Second, if the ALD is constructed from fish aged over a whole year, it does not compensate for seasonal growth which, in bass, usually occurs between May and November. As ALDs refer only to observed lengths at the time of sampling, the true annual length increment between ages (i.e. growth) may be biased, according to the contribution of fish which have plus-growth in the second half of the year, When stock assessments are being carried out on the more important, quota-managed species, such as

Table 6.2 Example of fully raised age–length distribution, showing estimated numbers of fish caught at each age and mean lengths and weights

Length group (cm)	*Total fish landed*	*Total otoliths read*	*Numbers at age*												
			5	*6*	*7*	*8*	*9*	*10*	*11*	*12*	*13*	*14*	*15*	*16*	*17*
35	89	1	89												
36	148	3	99	49											
37	415	6	208	208											
38	590	3		590											
39	845	9	94	751											
40	682	4		682											
41	993	11		813	181										
42	1967	11		1610	358										
43	2424	14		1212	866	173	173								
44	2776	12		1157	1388	0	231								
45	1183	9	131	657	394										
46	2165	13		999	999	0	167								
47	1472	13		453	566	0	340	113							
48	1047	8			655	262	0	0	131						
49	325	3			108	108	0	0	108						
50	281	4			141	0	70	0	0	70					
51	119	3			40	0	79								
52	30	1							30						
53	59	1			59										
59	59	1									59				
60	89	2								44	0	0	0	44	
64	59	1											59		
66	59	1													59
Totals	17 879	134	621	9181	5754	543	1060	113	269	115	59	0	59	44	59
Total otoliths read			8	64	39	4	9	1	3	2	1	0	1	1	1
Mean length (cm)			39.1	42.9	45.7	47.1	46.5	47.5	49.3	54.4	59.5	0.0	64.5	60.5	66.5
Mean weight (kg)			0.641	0.839	1.009	1.107	1.067	1.126	1.264	1.730	2.213	–	2.819	2.326	3.089

plaice, sufficient age data are collected that ALKs can be constructed quarterly, which will reduce this effect.

Provided that sampling is adequate, ALDs for successive winters should give the best estimate of annual length at age of bass in the samples, and for the fished population as a whole. This is the basis of the 'direct' method of determining growth, proposed originally by Petersen (1891). The data can be conveniently presented using age–length frequency histograms, of which an example is provided in Fig. 6.6.

When ALDs are prepared, several age groups and year classes are sampled simultaneously, and a growth rate which is based on observed lengths for each age group may be truly representative of the population at that particular time. This may be distorted if growth rates differ between successive year classes, because a combination of year classes is being used to produce a population growth estimate. For example, there may have been selection of the faster-growing individuals from more recent (younger) age groups, which have experienced a change in climatic conditions favourable to growth. A similar effect often occurs when samples are taken from commercial catches which contain fish near to the legal minimum landing size; again, faster-growing fish of the younger age groups are being selected. Such data should be used with care when determining growth for analytical or predictive purposes, because sampling of the younger age groups is clearly not representative of the

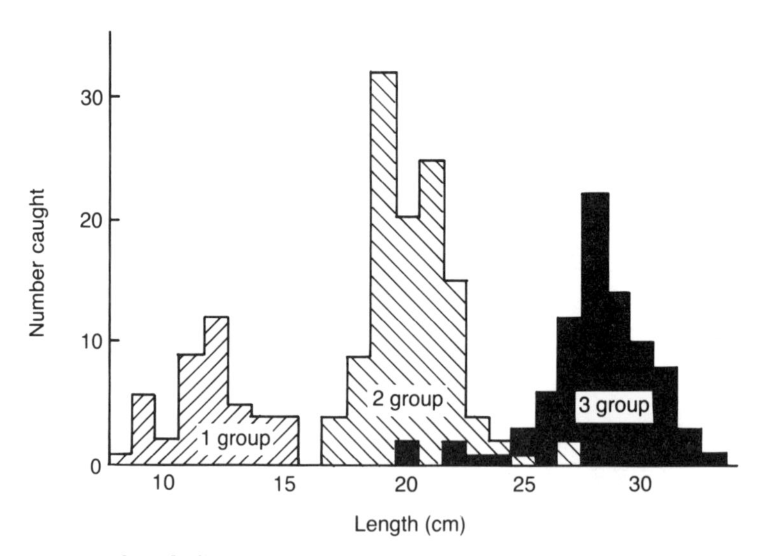

Fig. 6.6 Age–length frequency distribution of bass caught during the Solent pre-recruit surveys, May 1990. Mean lengths: 1 group = 12.3 cm (n = 43), 2 group = 20.8 cm (n = 114) and 3 group = 28.6 cm (n = 85).

population. To obtain more reliable growth information for age groups of bass not selected by the fishery, ALDs can be based on samples taken during directed population surveys using, for example, small-meshed trawls.

6.7 BACK-CALCULATION OF LENGTH AT AGE

Instead of using scales merely to determine the age a fish has reached when it was sampled, they can also be used to estimate the length attained at previous ages using a process known as back-calculation. Einar Lea (1910) showed that the size of a herring scale, and therefore the relative positions of the annuli along a radius moving out from the nucleus, was proportional to the fish's length at any particular time. Using Lea's formula, it is theoretically possible to calculate the length (L_n) of the fish at the time that each annulus appeared, using:

$$L_n = \frac{R_n}{R} L \qquad (6.1)$$

where: L_n is measured length of fish at time of capture, R is scale radius at time of capture, and R_n is radius of scale to annulus n.

This formula does not make any allowance for the size which the fish had attained before its scales first formed. In bass, this occurs at around 25 mm (Kennedy and Fitzmaurice, 1972) to 31 mm (Barnabé, 1976b). Back-calculated lengths are therefore under-estimated to this extent, though this error tends to decrease proportionately at each successive older annulus/length calculation. Even when this factor is accounted for, back-calculation can lead to an underestimate of the lengths of old fish in their early years, and an overestimate for later life. This is known as the Rosa Lee effect and was originally noticed in herring (Lee, 1920); some workers claim to have observed it in bass (Ottaway and Simkiss, 1979). Kelley (1988b), however, could find no evidence of the phenomenon in over 6000 UK bass taken between 1946 and 1985, despite using the basic Lea formula in back-calculations. It has been suggested that one cause of the Rosa Lee effect, is that the older fish, and therefore the better survivors, include the slower growers; faster-growing fish of any one species tend to have shorter life spans and are the first of any year class to recruit to the fishery and be caught (Bagenal and Tesch, 1978).

Modifications of the basic Lea formula have been used to compensate for potential sources of error in back-calculation. For example, Tesch (1968) used $L_n - c = (R_n/R)(L - c)$, in which the constant c is obtained from the regression of fish length against scale radius. If an accurate description of growth is required, it is necessary to include such a correction factor (Carlander, 1982). Theoretically, however, the chosen value can have a considerable range and, in practice, a high degree of accuracy is not usually important. For bass, we have relied solely on the original Lea formula, because:

1. errors in estimating length at age will be small and reasonably consistent with age;
2. initial errors in length measurements may affect the back-calculated lengths at age to a greater degree;
3. if all samples are treated in the same way, any error or bias will be commonly applied. The main aim of this work is to compare regional and seasonal growth, so precision is more important than accuracy in estimates of length at age;
4. the results will be compatible with those from earlier UK investigations, e.g. Kelley, 1988b).

When samples are measured to the nearest whole centimetre below the actual total length, as is usual in large-scale sampling, it is necessary to correct calculated mean length values by adding 5 mm.

Automatic back-calculations

Lengths at age can be back-calculated from scales using a semiautomatic system in which a macroprojector (e.g. Projectina Ltd, Switzerland) is interfaced with a microcomputer for data processing and storage. When a pointer, mounted on a rack and pinion, is moved across the image of the scale magnified on a ground-glass screen, the electrical potential recorded on a digital voltmeter changes in proportion to the scale's radius. On the assumption that the scale radius is proportional to the length of the bass at any age, readings between the scale's nucleus and the scale edge can then be taken at the position of each annulus to give proportional lengths. The operator sets the pointer to the scale nucleus and records zero length, and then moves it to the anterior scale margin and inputs total length on the microcomputer keyboard. A fairly small program, which includes the Lea formula, transforms the voltmeter readings at each annulus to lengths at age for each fish, and stores the data on file for later analysis. With this method it is possible to process 40 or more scale samples per hour.

Back-calculation has many advantages over the direct methods of estimating growth in fish. One benefit is that actual growth of fish in the year of capture (a problem with ALKs determined in summer) can be measured at the scale margin, and taken into account. Another gain is that growth in preceding years (or during the years that an adult fish spent as a juvenile) can be estimated for various age classes of fish. As mentioned earlier, however, there may be aspects of gear selection and growth-related survival which need to be taken into account when estimating the average length at age of fish in the population.

6.8 GROWTH PATTERNS IN UK BASS

Growth patterns are normally described using mean length at age (growth curves) or annual length increments for the studied population. Under normal climatic and environmental conditions, bass show a characteristic pattern of seasonal growth in which the largest monthly increments of length and weight occur during the warmer months, usually June to October in the UK. In autumn and through the winter, growth slows or (more likely) stops, although the period during which there is no growth probably depends on the sea temperatures and the state of maturity of the fish. The age of the fish is also relevant; marginal scale growth apparently starts later in the year with each successive season in the fish's early life and the percentage increments become smaller. The seasonal growth pattern as shown by observed lengths at age for bass aged 3–7 years is shown in Fig. 6.7.

Kelley (1988b), who has sampled UK bass for length and age for over 30 years, described regional differences in size at age, but these were not necessarily consistent from year to year; growth rate may vary between year classes, and climatic conditions appear to have a large influence on growth in bass. It may therefore be necessary to choose a 'national' growth pattern for comparison with growth rates of bass in other countries. Growth curves of male and female bass, incorporating data from all regions sampled throughout the year by DFR between 1982 and 1988, are shown in Fig. 6.8 (MAFF, unpublished data).

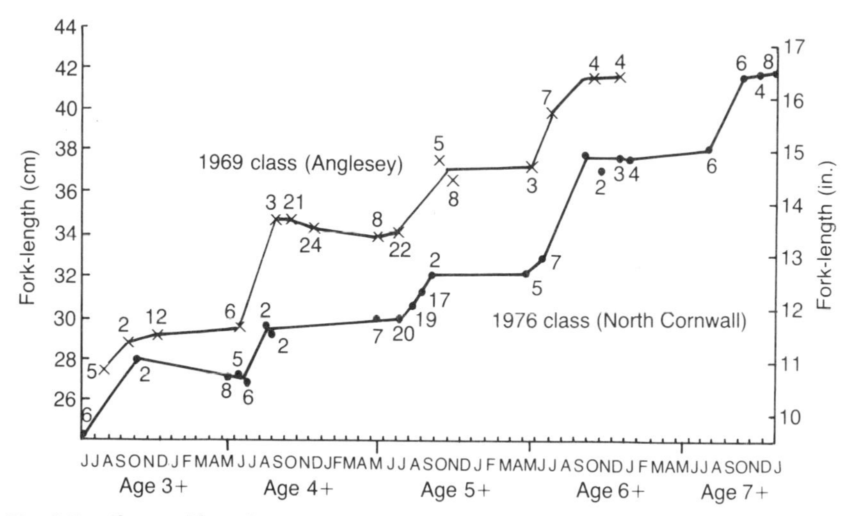

Fig. 6.7 Observed lengths-at-age, month by month over ages 3 + to 7, of the 1969 year-class in Anglesey and the 1976 year-class in N Cornwall. Figures: number in samples. Reproduced with permission from Kelley (1988b).

Irregularities in these growth curves beyond age 14 are probably due mainly to low sampling levels. The growth rate of females is significantly higher than that of males ($p < 0.05$) between the ages 7 and 15, as noted by Kelley (1988b). This type of presentation does not reveal any changes in growth that may have occurred over the sampling period, or of variation in growth between year classes. The back-calculated lengths of bass from DFR's samples, collected in 1983 to 1986, and shown in Fig. 6.9.

There is little difference between back-calculated lengths at age in samples from successive years, owing, in part, to the normalizing effect of combining regions, seasons and year classes in the sampling. The curves shown in Figs 6.8 and 6.9, therefore, probably represent the typical growth pattern of the UK bass population in the early 1980s.

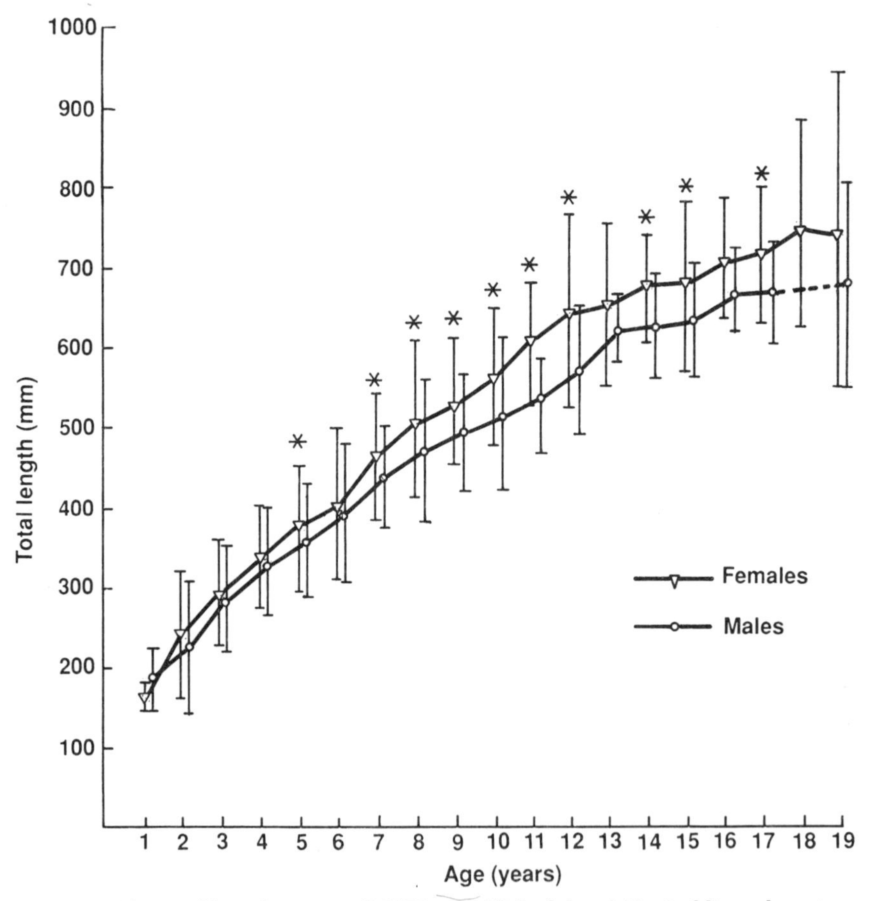

Fig. 6.8 Observed lengths at age. (MAFF unpublished data.) Vertical lines show two standard deviations from means. Stars indicate significantly different mean values for males and females ($p < 0.05$).

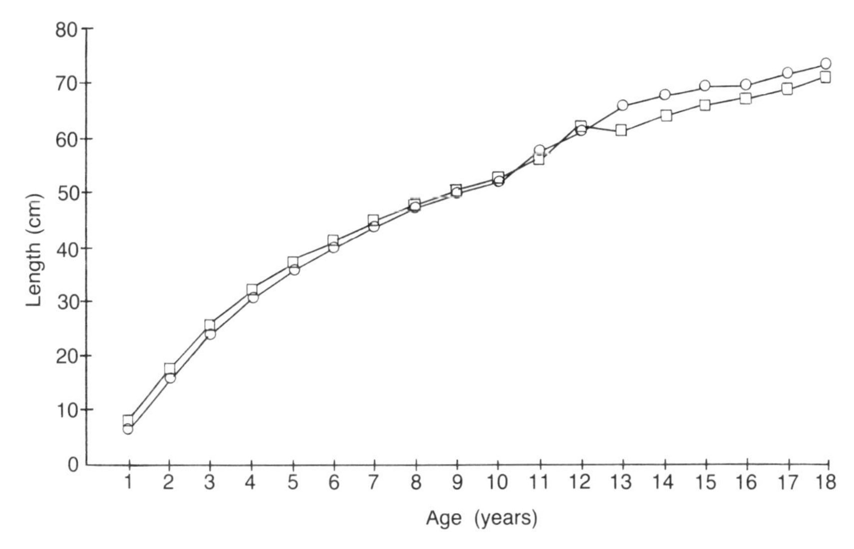

Fig. 6.9 Back-calculated lengths at age for bass of all ages collected in 1983, circles, and 1986, squares. (MAFF unpublished data.)

It is often desirable to express growth in a mathematical form, with which it should be possible to predict average size in length or weight at any given age. Generally, fisheries science is concerned with the rate of growth in the fishable population, as well as with the length or weight at age of individual fish. The growth equations preferred by most researchers are therefore those most easily fitted to growth rate data rather than simply to size at age (Gulland, 1969). The main uses of such models are to enable simple comparisons of growth of the same species to be made in different areas, and for use in calculating or modelling biomass production or yields from the fishery.

Most published work on the growth of fish has used the classic growth equation derived by von Bertlanffy (1938). This describes annual growth rate data for bass quite well and can be incorporated into stock assessment models. Several production models, e.g. Beverton and Holt (1957), use the von Bertlanffy growth equation, which is usually written as follows:

$$L_t = L_\infty(1 - \exp^{-K(t-t_0)} \qquad (6.2)$$

where: L_t is length at time t (in years); L_∞ is the length at which growth approaches zero, i.e. the theoretical maximal length attainable; K is a growth coefficient which essentially describes the rate at which L_t approaches L_∞; and t_0 positions the growth curve described by K and L_∞ on the time axis.

The values obtained for L_∞ and K give a simple means of comparing growth between populations. These growth parameters have been calculated for bass captured in the Irish Sea (Kennedy and Fitzmaurice, 1972), off Plymouth (Sabriye, 1983) and off South Wales (Parsons, 1982). The Plymouth samples (135 fish) were collected between May and November 1982, and Sabriye calculated growth parameters for males and females combined of $L_\infty = 72$ cm and K = 0.14, which were similar to those published by Kennedy and Fitzmaurice (1972) for Irish Sea bass ($L_\infty = 71$ cm, K = 0.14). Analysis of data from the Swansea area (Parsons, 1982) produced a range of growth parameters for bass, dependent on year class, sex and origin of the samples. Parsons suggested that the highest value of L_∞ (81 cm), which was produced when all year classes were considered together, was caused by the inclusion of a few old (13–21 years), but fast-growing, fish. This is contrary to expectations with the aforementioned Rosa Lee phenomenon, but Parsons found that, in general, the older the year class, the higher the L_∞ value. The average L_∞ for South Wales bass, both sexes and age classes combined, was 72 cm, similar to the values for Plymouth and Irish Sea bass. Values of K calculated by Parsons ranged from 0.14 to 0.22, with the average being 0.17.

6.9 GROWTH THROUGHOUT THE RANGE OF THE BASS

Growth in length

Although one should be critically cautious when comparing lengths at age and growth parameters reported in various studies, it is useful to show a range of results in order to illustrate the variability of growth in bass. Superficially, it appears that in European waters, slow growth occurs in the north and that there is faster growth to the south (Table 6.3).

A closer examination, however, shows that growth in bass from Egypt and Morocco is apparently slower than in those caught on the French Mediterranean coast, and that growth rates along the Atlantic seaboard; in Brittany, South-west France (Vendee) and Morocco, are similar. Barnabé (1976b) noted this, and suggested that both environmental factors, in particular temperature, and biological factors, such as feeding relationships and interspecific competition, controlled growth. It now appears that temperature is the main influence on growth in bass, although few published works present data on the temperature regimes experienced by the populations of bass being studied. It is likely that fish on the Atlantic coasts will generally experience lower temperatures than those in the shallow parts of the Mediterranean, even if they do live further to the south. We are therefore are at a loss to explain the low length at age given by Rafail (1971) for bass in Eygpt, unless temperatures there are too high, or the food supply really is limiting.

Kelley (1988b) reported the occasional bass that had grown faster than

Table 6.3 Mean observed total lengths of bass at ages 1–7 in different regions* (Adapted from Barnabé, 1976a)

Age (years)	Ireland[1]		S. Brittany[2]		Vendee[3]		Arachon[4]	Sète[5]		Morocco[6]	Egypt[7]	UK West Coast 1976 year class[8]		UK, all areas combined 1982–86[9]		UK East coast, 1991[10]
	Male	*Female*	*Male*	*Female*	*Lagoons*	*Sea*		*Male*	*Female*			*Male*	*Female*	*Male*	*Female*	
1	7.1	8.2	–		9.4	9.4	12.0	17.4		14.5	13.2	–	–	9.0		19.5
2	16.1	16.5	19.0		18.2	18.2	21.7	28.3		23.0	18.5	14.7	16.3	18.5		25.9
3	22.1	23.4	26.2		27.0	25.8	28.0	36.8	39.1	29.0	22.0	21.4	23.1	26.2	27.4	36.7
4	28.5	29.3	33.1	35.6	32.7	32.3	31.9	43.1	47.0	34.0	25.6	26.9	28.8	33.6	34.7	39.3
5	33.3	34.8	40.7	41.3	38.8	38.1	35.2	47.9	53.8	40.0	29.4	31.8	33.4	36.2	37.7	46.5
6	37.4	39.1	45.4	46.4	42.7	43.5	38.1	51.5	59.5	45.0	33.3	35.8	38.1	38.8	39.5	49.5
7	42.2	44.0	48.8	51.5	51.5	48.6	42.3	–	–	–	36.8	41.1	43.4	43.7	46.2	52.0

*Sources: 1, Kennedy and Fitzmaurice (1972); 2, Bertignac (1987); 3, Lam Hoai Thong (1970); 4, Stequert (19720 (converted from fork length); 5, Barnabé (1976a); 6, Gravier (1961); 7, Rafail (1971); 8, Kelley (1988b) (converted from fork length); 9, DFR unpublished; 10, DFR ALK (includes plus-growth).

normal for UK waters (e.g. with a length of 48 cm or more at age 6), and showed that back-calculated growth in the first 6 years of life in bass tends to be higher in populations sampled on the south-east coast than in other UK areas. Whilst it is not possible to discount Kelley's hypothesis that large year classes (such as that spawned in 1966) might have originated in Biscay and spread north at adolescence, extensive sampling by DFR in South-east England has revealed little difference between overall growth patterns of juvenile bass there and those reported for Biscay by Lam Hoai Thong (1970) and Stequert (1972) (Table 6.3). Even Kelley's own data (1988b, Table 6) suggest that there is no significant difference between the growth of bass in South-east England and in Biscay during the first 6 years of life. The UK growth pattern in the 1970s and 1980s (Fig. 6.8) is similar to that observed in South Brittany by Bertignac (1987).

There is, it appears, good evidence for a close correlation between the growth of juvenile bass and temperature (Pawson, 1992). The relationship between sea temperature during the main growth period (May–November) and the mean length of juvenile bass at the end of their third year in the Solent on the south coast of England during the 1980s is shown in Fig. 6.10.

The mild winters of 1988/89 and 1989/90 produced unseasonably warm sea temperatures and were followed by two warm summers. During the years 1988–90, growth of 0-, 1- and 2-group bass in particular was exceptionally high in England and Wales. This was most marked in the south-east, where length at age was similar to that reported by Barnabé (1976a) for the early 1970s in the South of France. It is likely, therefore, that growth in summer is faster in warm, shallow sea areas (e.g. the Thames Estuary, Solent Harbours, lagoons in southern France) than on oceanic coasts, where water temperatures fluctuate less between winter and summer. In autumn, temperatures in shallow water usually drop rapidly and, unless the bass migrate to deeper, warmer water, those that spend the winter in water with temperatures below about 7 or 8 °C, are unlikely to feed. Perhaps, in the warm winters at the end of the 1980s, temperatures were sufficiently high to allow normal feeding and maybe some growth to continue. It appears that, because the bass is towards the northern limit of its distribution in waters around England and Wales, its growth there is particularly affected by temperature, i.e. the relationship between growth and temperature is at the most responsive part of its range.

Longevity and maximum length

In parts of the Mediterranean Sea and the Atlantic on the North-west African coast, where the growth of bass is reported to be unusually fast,

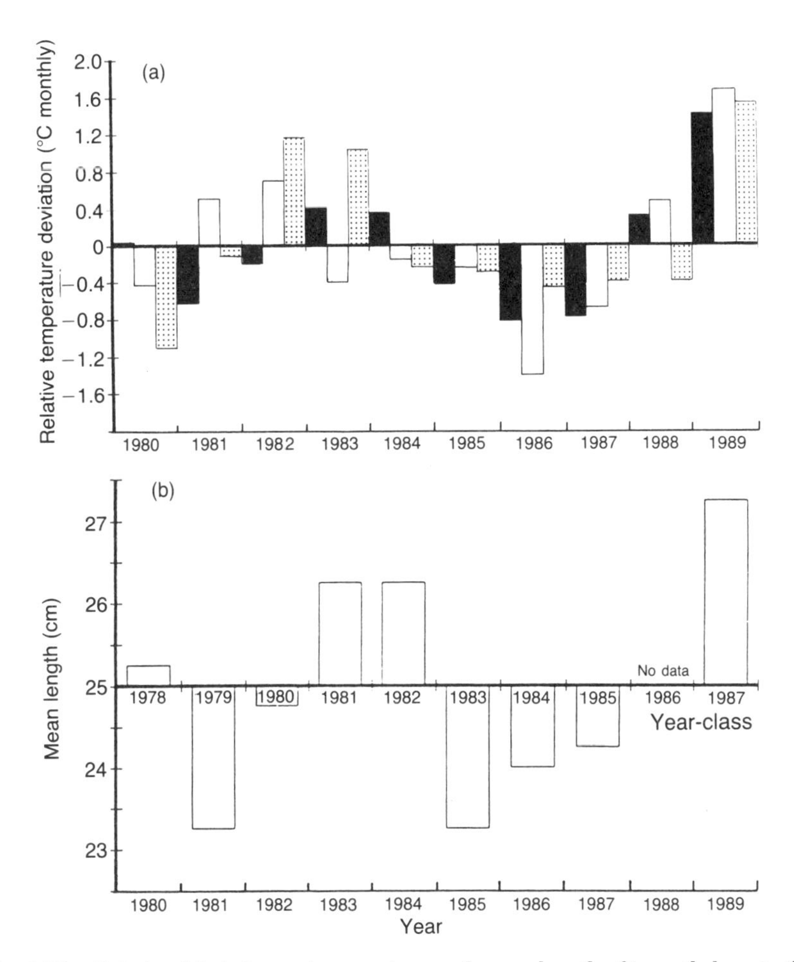

Fig. 6.10 Relationship between temperature and mean length of juvenile bass in the Solent (reproduced from Pawson, 1992). (a) Seasonal temperature deviations from the 1980–89 10 year mean in English coastal waters of the central English Channel: dark columns, November–March; light columns, March–May; stippled columns, May–November. (b) Deviations from the 1980–89 10 year mean of the length of bass in the Solent at the end of their third season of growth.

its natural life span is much shorter than it is around the UK. At Sète, for example, Barnabé (1976a) reports bass over 6 years of age as being rare, amounting to only about 1% of some 2000 specimens which he aged. He recorded only two bass at 10 years and one of 14–15 years old. Gravier (1961) states that the maximum age for the bass in Morocco is 13 years. Most authors agree that, around the UK and Ireland, the maximum age of bass is around 30 years. We have found several members of the 1959 year class in samples taken in 1988, with rather fewer (five) in 1989 and only one such fish in 1990. In all areas, male bass are thought not only to grow more slowly, but to have shorter lives than females (Kelley, 1988b). That most of the largest bass in any area are females is demonstrated by the UK data used in Fig. 6.8. Of those bass which were sexed and aged, 24 fish were 18 years or more, and only 5 were males, the oldest of which was 22 years. Despite having a much faster growth rate (100 cm in 8 years, according to Mansueti, 1961), the striped bass has a similar life span to that of *D. labrax*, up to 30 years or so. Again, most fish over 11 years old are females.

Although there are differences in maximum age from area to area, the maximum attainable length in bass does not appear to vary greatly. It seems that, in the south, the faster-growing bass reach their maximum size earlier, but do not become any larger than they do further north. The largest specimen that Barnabé encountered at Sète was 92.5 cm. Gravier suggested a maximum length of 85 cm for bass in Morocco. In the UK, specimens are regularly recorded from commercial catches at between 80 and 85 cm and, since 1983, several fish of this size have been caught that belonged to the 1966–69 year classes (MAFF and D.F. Kelley, unpublished data). The UK rod-and-line-caught bass records, set in 1988 and 1989 for boat and shore respectively, were 95 cm and 89 cm and aged 18 and 19 years respectively. Although there is anecdotal evidence of bass being landed in the UK commercial fishery at around 100 cm long, 95 cm might be taken as being the probable maximum length attainable by the species anywhere. The largest male bass we have encountered during the period 1982–1988 was 75 cm, and out of 43 sexed bass of over 70 cm, only three were males. In comparison, the spotted bass grows to about 60 cm, whereas the striped bass has been known to attain 155 cm in the United States (Holland and Yelverton, 1973).

Growth in weight

To fishermen and anglers, weight is probably a more important aspect of growth in sea fish than is length, as it is the unit with which landings are recorded, specimen and record weights are determined and prices are set. Because the bass is usually sold whole, ungutted, the price per unit weight, which is already high, becomes even higher relative to that of most other commercial marine species which are sold gutted or filleted.

Most fisheries are subject to technical measures, such as a minimum landing size (length), which are usually supported by a mesh-size regulation, and it is important to know how changes in the length at first capture will affect yields to the fishery. Most sampling data from catches in the UK bass fishery consist only of length measurements, and it is therefore necessary, when carrying out fisheries assessments, to have a knowledge of length–weight relationships. These factors have only recently been described in detail for bass in the UK, and are based on data collected since 1982 (MAFF unpublished data). The following is a brief review of some of this work.

Length and age–weight relationships

With increasing age, the variation in weight is likely to increase, and estimated mean weights at age may be of little use for calculating the actual weight of (large) fish. Weight at age is also likely to vary between populations in different regions, depending on the feeding conditions (food availability and temperature, mainly) experienced by the fish. Usually, however, the relationship between length and weight is more closely defined and does not vary so greatly between populations at a given age.

The overall distribution of weight as a function of length for males and female bass, based on a sample of 1153 fish, is shown in Fig. 6.11. When the sexes are combined, this pattern is very similar to Barnabé's (1976b) plot of weight against length for bass at Sète, although bass below 40 cm in the UK appear to be slightly heavier for a given length than those in the South of France (if Barnabé's data are typical). Female bass appear to be consistently and significantly heavier than males at any particular age between 5 and 13 years.

It will be seen in the example given in Fig. 6.11 that simple graphical presentations do not necessarily give a clear picture of differences in weight–length relationships. If they exist, these differences can often be demonstrated by mathematical expressions.

Many studies on fish growth have used the function:

$$W = \mathrm{a}\, L^{\mathrm{b}} \qquad (6.3)$$

where W is weight in grams, L is total length in cm, a is a constant which relates to the fish's 'condition', and b is an exponent which tends to have a value of around 3 – the cubic, linear-volumetric transformation.

Such equations are often used to predict weight at a given length in circumstances where accurate weight data are not available. For UK bass, the weight–length relationships for the separate sexes for the months of the year for which sufficient data were available are given in Table 6.4.

In all cases, the coefficient of determination of the correlation (r^2) is higher than 0.95, and the consistency of the result implies that neither sex nor size contribute much variation to the weight–length relationship. An overall weight–length relationship calculated for the combined data for both sexes for the months of August, September and October, when there is the least difference between the relationships for adult and juvenile fish and of the sexes, gave a value for b of 2.97. This suggests that bass might be regarded as having isometric growth, i.e. their weight increases in direct proportion to

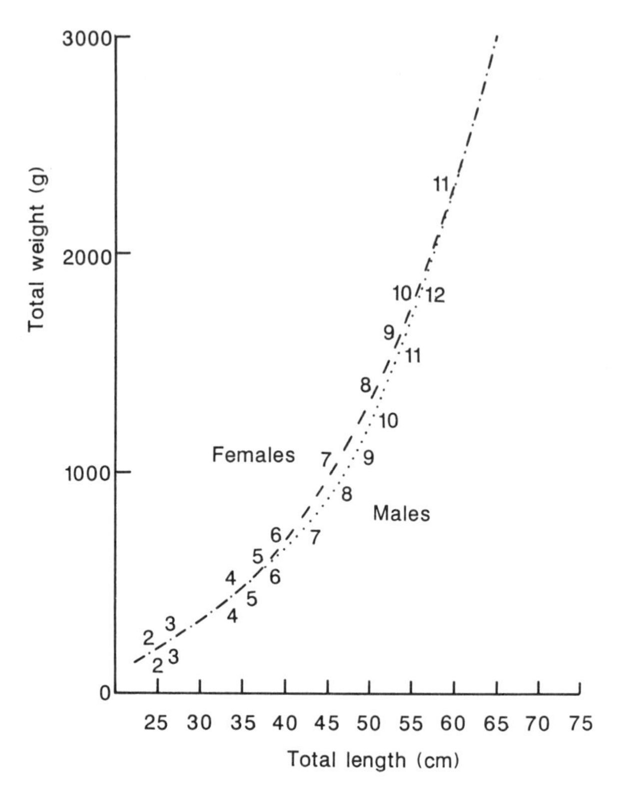

Fig. 6.11 Length–weight relationship for male and female bass. Numbers show ages of males and females. Reproduced from Pawson and Pickett, 1987.

the cube of their length and there are no changes in the species' body shape through life. Differences in weight–length relationships might be interpreted as being due to differences in growth and morphometry between regions (Barnabé, 1976a), but our recent work suggests a more general consistency.

Maximum weights

There is quite a large difference, as with length, between the maximum weights observed in bass and the theoretical maximum weight estimated for a population. This is because the latter is based on mean values, whereas observed maxima are obtained from records of particularly long-lived or fast-growing individuals.

Calculated maximum weights are of little interest to anglers, who are naturally optimistic and aim to catch fish which are as large as is practically possible. The chances of catching a very large bass (say over

Table 6.4 Seasonal length–weight regression parameters for bass using equation 6.3. MAFF unpublished data

Month	*Sex*	*Length range (mm)*	*Number in sample*	$a(\times 10^{-6})$	*b*	r^2
May	M	260–690	74	31.86	2.807	0.96
May	F	260–790	99	18.67	2.902	0.99
July	M	240–600	91	32.17	2.809	0.98
July	F	252–687	128	33.45	2.810	0.98
Aug.	M	193–660	25	6.15	3.082	0.99
Aug.	F	290–780	55	9.23	3.019	0.99
Sep.	M	159–700	108	18.72	2.907	0.99
Sep.	F	155–720	136	15.82	2.938	0.99
Oct.	M	273–725	72	11.95	2.978	0.98
Oct.	F	284–781	104	8.99	3.022	0.99
Nov.	M	213–696	24	6.00	3.091	0.99
Nov.	F	305–770	56	8.13	3.047	0.97
Sep.	M	159–420	59	20.22	2.892	0.99
Sep.	M	425–700	49	29.82	2.833	0.96
Sep.	F	155–420	75	15.50	2.941	0.99
Sep.	F	422–720	61	31.75	2.828	0.96
Aug., Sep., Oct.	M+F	94–781	592	12.96	2.969	0.99

15 pounds, 6.9 kg) are small, though the odds may vary with time owing to fluctuations in abundance and growth rates between year classes. Kelley (1988b) has compared the growth of individual fish from various year classes around the UK, and found that very few of the 1959 year class, which was probably the most numerically abundant and longest-lived cohort between 1950 and 1975, reached 7 kg. This is in sharp contrast to the much less abundant 1966 and 1969 year classes, which included proportionately much larger numbers of faster-growing specimens. Kelley has kept records of large bass caught on rod-and-line over the years 1943 to 1991, and these are summarized in Table 6.5 with age and year class.

Authenticated records of bass which were larger than the current UK rod-and-line-caught records (9 kg) are hard to come by, even in commercial catch records, although there have been reports (*Sea Angler*, March 1991) of one fish over 12 kg taken in a trawl catch in Cornwall in December 1990. Another fish, which weighed 10 kg, was taken near Chichester in 1985 in a gill net (Darracote, pers. comm.). Kelley has also produced a list of these very large bass caught commercially, with a note of their authenticity (Table 6.6).

It seems that 10 kg is the maximum attainable weight for bass around the UK. The European rod-caught record is 9.4 kg, though specialist spear fishermen in Languedoc consider that the maximum weight is 11.75 kg

Table 6.5 Bass of 17 lb (7.7 kg) and over, caught on rod-and-line in UK water (source, Don Kelley, unpublished data)

Weight		*Total length*	*Max. girth*	*Date*	*Place*	*Captor*	*Boat/ shore*	*Age*	*When spawned*	*Remarks*
(lbs–oz)	*(kg)*	*(cm)*	*(cm)*							
19–9	8.9	95.0	57.2	Aug. 1987	Off Herne Bay, Kent	P. McEwan	B	18+	1969	UK boat record
19–0½	8.6	89.0	55.2	Sept. 1988	Dover Breakwater, Kent	D. Bourne	S	19+	1969	UK shore record
18–6	8.4	93.7	50.2	Aug. 1975	Eddystone Reef, S. Devon	R. Salter	B	25+	1950	
18–2	8.3	92.4	49.5	Nov. 1943	Felixstowe, Suffolk	F.C. Borley	S		(No scales)	
17–15¾	8.2	–	–	Oct. 1987	Portland, Dorset	R. Milverton	S		(No scales)	
17–9¼	8.0	87.6	54.0	Oct. 1976	Off Bradwell, Essex	O. Woolley	B	13+	1963	
17–8	7.9	92.4	54.6	Aug. 1990	The Cobb, Lyme Regis, Dorset	P. Knight	S		(No scales)	
17–9	8.0	91.1	51.4	Sep. 1959	Minehead, Somerset	F.W. Lashbrook	S		(No scales)	
17–8	7.9	–	–	Aug. 1989	Off Dovercourt, Essex	P. Ellis	B		(No scales)	
17–8	7.9	91.0	53.3	Sep. 1980	Off Clacton, Essex	G. MacKintosh	B	20+	1960	
17–6	7.9	87.6	51.4	Sep. 1989	Off Southend, Essex	J. Arber	B		(No scales)	
17–4	7.8	91.0	48.3	Aug. 1983	Off Bradwell, Essex	P. Rowland	B	24+	1959	
17–4	7.8	–	–	July 1992	Telscombe Cliffs, E. Sussex	(Lewes angler)	S			
17–2½	7.8	86	53.3	Oct. 1975	Hinkley Point, Somerset	K. Flack	S	15+	1960	
17–1	7.7	–	–	June 1985	Blackwater Estuary, Essex	P. Prentice	B		(No scales)	
17–1	77	–	–	Aug. 1990	Off Canvey Island, Essex	N. Thomas	B		(No scales)	
17–0½	7.7	85.6	49.5	Aug. 1987	Lavernock Point, S. Wales	M. Katchi	S	21+	1966 (poor scales)	Welsh record
17–0	7.7	–	–	Sep. 1991	14 miles off Brighton, Sussex	J. Kearns	B			

(Barnabé, 1976a), and Aflalo (1910) reported fish of similar weight hooked and landed in the Bosphorus. These maximum weights are far lower than the sizes attainable by striped bass in the United States. Bigelow and Schroeder (1953) reported one of 57 kg and, before the recent stock decline, fish of 27–32 kg were not regarded as being exceptional in some areas.

Mean maximum weights, which are of interest to those modelling the dynamics and yields of bass populations, may be based on an estimate of L_∞ or on observed values of maximum weight. In most studies on bass, there are insufficient old fish caught (and weighed) to enable reliable calculations to be based solely on sampled weight. Most often, therefore, L_∞ is used as follows: $\log W_\infty = \log a + b \times \log L_\infty$.

With an L_∞ of 76 cm, calculated using data for large rod-caught bass, Parsons (1982) estimated a maximum weight of 4.2 kg for bass in South Wales, although the largest fish in his samples actually weighed 4.7 kg.

Length and weight increments with age

An examination of the annual additions to length and weight in some UK bass samples has revealed a consistent drop in the size of growth increments for both males and females in the fifth and sixth growing seasons. This corresponds to the onset of maturity, when energy and metabolites begin to be diverted to the growing gonads. Subsequently, however, the annual increments in weight can be quite large, and a bass can add 0.2–0.5 kg per year for a further 15 years or more. Kelley's (1988b) data do not reveal this phenomenon, but Barnabé (1976b) also records an inflection in the growth (length) curve at the onset of maturity, followed by a recovery in growth. It is possible that this pattern of growth could be an artefact of sampling, as an examination of the results of length back-calculations in the UK samples also shows no inflection at those ages in several separate year classes.

An interpretation of bass growth compared with some other marine fish species is shown in Fig. 6.12. It may be concluded that whereas bass are relatively slow starters, their growth performance (in weight) improves after age 6. Moreover, they have the advantage of continuing to add weight at ages which many other marine fish species do not reach or when they might be expected to be senescent. In this, bass show similarities with another long-lived percoid, the Nile perch (*Lates niloticus*, Centropomidae). Soriano *et al.* (1990) demonstrated a two-phase growth in this species, and suggested modifications to basic growth equations with which to describe this phenomenon. To date, no studies have applied similar methods to the growth of bass, although it is likely that there may be a common cause of the increased growth of Nile perch and bass in mid-life: a change to a mainly fish diet.

Table 6.6 Bass of 19 lb+ caught commercially around England and Wales (source, Don Kelley, unpublished data)

Weight		Date	Place	Method	Source*	Kelley's Remarks
(lb–oz)	(kg)					
29–0	13.2	1850s	Cornwall	Unknown	Couch I/192 (1862)	Vague verbal reports to Couch of Polperro
28–0	12.7	1850s	Unknown	Unknown	Yarrell II/118 (1859)	Vague verbal report to Yarrell (not mentioned in 1st edn 1836)
27–8	12.5	Unknown	Brixham	Unknown	Bickerdyke (1935 edn)	Hearsay. Pencil note on Kelley's copy (by a previous owner) states 'probably a stone bass'
26.–0	11.8	Nov. 1931	Wakering Stairs, Southend	Spiller (beach line)	FG 5/12/31, 26/12/31, 2/1/32	Reported to correspondent by another beach liner, auctioned at local club for 5/6 (27p)
25–0	11.3	Dec. 1990	Off Looe	Trawl		Seen by Fisheries Officer, four other 18–19 lb
22–8	10.2	1984	Hinkley Point, Power station	Trapped in intake	SA Jan. 1985	
22–8	10.2	*c.* 1982	Portland Race	Heavy commercial rod	K. Lyneham, & Authors 1/11/84	Verbal report
22–8	10.2	1926	Felixstowe	Fixed line	FG 9/7/27	Reported to correspondent by fishermen whose mate caught it and weighed it on fishmonger's scale
22–2	10.0	22 July 1902	Saltash	Stop-net	FG 2/8/02	FG repeats a *Western Morning News* report 25/7/02

22–0	10.0	Unknown	Herne Bay	Net	Day I 10 (1880–84)	'netted close to Herne Bay Pier'
22–0	10.0	1899	Maplin Sands, Essex	Net	SA Dec. 1977, citing SFC report	'10 from 16 to 22 lb' (Kent and Essex SFC 1st report 1903)
22–0	10.0	Nov. 1976	Off Falmouth	Longline	SA Feb. 1977	'4 totalling 80 lb'
20–8	9.3	1984	Eddystone	Net	Len Hawke,	Verbal report
20–8	9.3	1984	Off Chichester Harbour	Net	Authors	Verbal report + photograph, captor P. Darracote
20+	9.1+	*c.* 1972	Poole Harbour	Trammel	Unknown (possibly AT)	
19–12	9.0	Dec. 1981	Eddystone	Net	AM 16/12/81	
19–10	8.9	1979	Bristol Channel	Beach net	BASS No. 14/15 (1979)	Correspondent (Alan Kershaw)
19–8	8.8	1984	Off Rye, Kent	Cod-net 6" mesh	Authors 1/11/84	Verbal report
19–8	8.8	Aug. 1978	Conway Estuary	Net	AT 9/8.78	
19–7	8.8	1978	Eddystone	Gill net	Various sources	Possibly caught by Pio Rubassa Rivas
19–4	8.7	Aug. 1983	Off Bradwell	Net	AM 20/8/83	Reports by Owen Woolley
19–3	8.7	Nov. 1970	Portland Harbour	Spearfisher (Colin House, Weymouth)	AM 5/12/70	Was set up
19+	8.6	Nov. 1953	Mudeford Beach, Hants	Found dead on shore	AT 4/12/53	Washed ashore, found dead by Ashby
19–0	8.6	Oct. 1986	Off Beachy Head	Netted	AM 1/11/86	
19–0	8.6	1987	Bridgewater Bay	Stake net	Correspondence with captor (Brendan Sellick)	

FG, *Fishing Gazette*; SA, *Sea Angler*; AT, *Angling Times*; AM, *Anglers' Mail*.

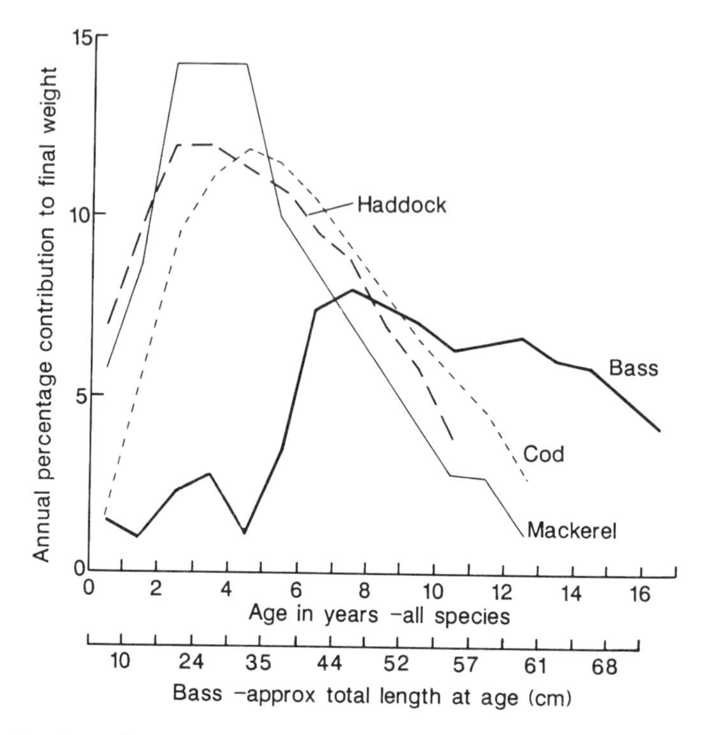

Fig. 6.12 Annual increments in weight of bass, cod, haddock and mackerel. From Pawson and Pickett, 1987.

6.10 SUMMARY

The ease with which bass scales can be aged, and the ability to 'back-calculate' the size of a fish at previous ages, have permitted us a confident insight to the variability in the species' growth patterns. The bass is often thought to be a slow growing fish, for a marine species, in relation to its potential maximum size. Around the UK, at least, the growth rate of juvenile bass is restricted by low water temperatures during the winter and spring, and their growth is also strongly influenced by temperatures in summer and autumn. The year-to-year variation in size increment around southern Britain can be as large as the range of mean growth rates observed at different latitudes along the Atlantic coastline. Once mature, however, bass appear to grow at a consistent rate everywhere, and tend to approach the same maximum size (9–10 kg) whether they are long-lived (30 years or more) fish from around the UK, or originated as much faster growing juveniles in the Mediterranean Sea.

Chapter seven

Condition and maturity patterns

7.1 INTRODUCTION

Despite rapid advances in breeding and rearing bass in captivity (Bromage *et al.*, 1988), comparatively little is known about the reproductive biology of bass in the wild. Published information on the 'natural' seasonal cycle of gonad maturity and its relationship with the fishes' size and age, and with feeding, somatic condition and environmental factors, has been scarce. This is chiefly because bass samples were only available from sport anglers' catches prior to 1982. Between 1982 and 1990, however, DFR has collected relevant data on bass sampled from commercial and research vessels' catches off the coasts of England and Wales, which have been analysed to describe the patterns of gonad growth, maturity and condition of the bass throughout its breeding range around Britain (MAFF, unpublished data). The majority of the 1657 fish which had gonads that could be sexed macroscopically (with the naked eye), also provided data on length, total weight and age, stomach fullness and weight (including contents), weight of mesenteric fat, gonad weight and maturity stage. Values for the maturity stage of gonads were assigned according to the characteristics given in Table 7.1. In general, it was found that bass under a total length of about 17 cm could not be sexed.

As with the tagging data (Chapter 5), the primary analysis was performed with fish in three length ranges–less than 32 cm, 32–42 cm and greater than 42 cm–chosen, in the first place, to correspond to the definitions of juveniles, adolescents and adults given by Pawson *et al.* (1987) for bass in UK waters. Variation attributable to differences in growth etc. between year classes and to the effect of annual weather patterns, was probably smoothed by using data covering a nine year period.

Information for some parts of the geographical distribution of bass around

Table 7.1 Macroscopic characteristics of the maturity stages of the ovary and testes of bass

Maturity stage		Ovary	Testis
I	Immature	Small thread-like ovary, reddish-pink	Small, colourless thread-like testis not practical to differentiate macroscopically below about 20 cm
II	Recovering spent	Ovaries one-third length of ventral cavity, opaque, pink, with thickened walls and may have atretic eggs	Testis one-third length of ventral cavity, often bloodshot with parts dark grey
III	Developing (early)	Ovaries one-half length of ventral cavity, orange-red; slight granular appearance, thin translucent walls	Testes thickness 10-20% of length, dirty white, tinged grey or pink
IV	Developing (late)	Ovaries one-half to one-third length of ventral cavity, orange-red; eggs clearly visible, but none hyaline	Testes flat–oval in cross section and thickness < 20% of length – half to two-thirds of ventral cavity. White colour and milt emitted from vent if pressure applied to abdomen
V	Gravid	Swollen ovaries two-thirds length of ventral cavity, pale yellow–orange; opaque eggs clearly visible, with some hyaline	Testes bright white and more rounded–oval in cross section. Only light pressure required to cause milt to flow from vent
VI	Running	Ovaries very swollen; both opaque and larger hyaline eggs clearly visible beneath thin, almost transparent ovary wall and expressed freely with light pressure	Testes becoming grey–white and less turgid. Milt extruded spontaneously
VII	Spent	Ovary flaccid but not fully empty, deep red; very thick ovary wall; dense yellow atretic eggs may be visible	Testes flattened and grey, flushed with red or pink, larger than those at stage II

England and Wales is still lacking. Sample sites were therefore nominated to three regions, based on a knowledge of the seasonal distribution and migrations of adult bass around the UK coast. The northern region is assumed to contain adult bass mainly during the warmer months (May–October), the central region to have adults migrating through in spring and autumn and a resident population in summer, and the southern region to have a mixture of winter immigrants from further north and adult bass that

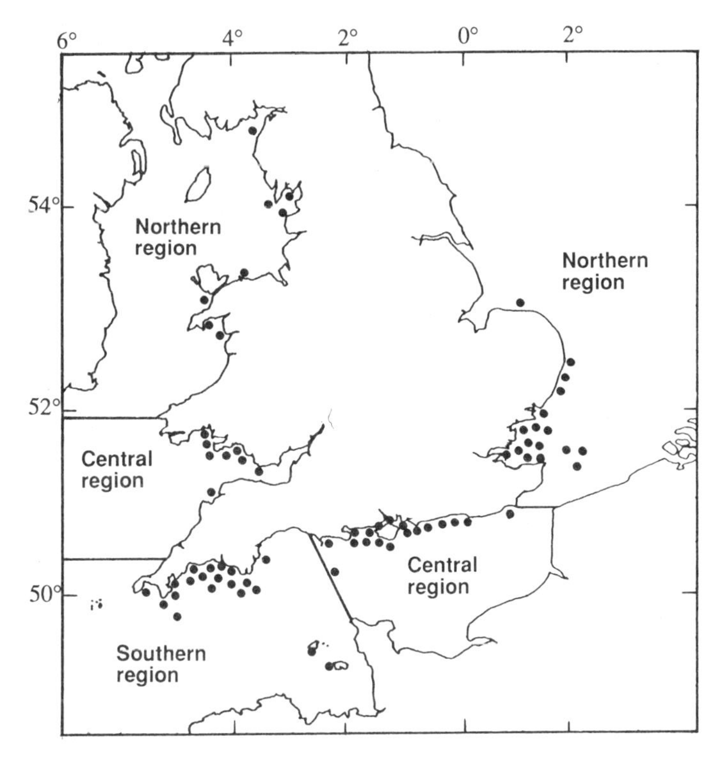

Fig. 7.1 Regions used for analysis of biological samples (sites indicated by filled circles). For definition of regions, see text.

are present all year round. Thus, the coasts of East Anglia, Kent and the Irish Sea formed the northern region; South Wales, Sussex, Hampshire and Dorset the central region and South Devon and Cornwall, and the Channel Islands the southern region (Fig. 7.1).

The monthly length-frequency distributions of bass sampled in each of these three regions indicate that there are characteristic seasonal patterns in the size distributions of bass. In the southern region, for example, large (> 45 cm) males and females predominated in samples taken from November through to April. From May until August the mean size of bass of both sexes decreased in the southern region, whilst elsewhere, the mean size in the samples increased through the summer and autumn. Although bass over 40 cm were relatively abundant in the autumn in the central region, fish of this size were generally scarce there from January until May, when most fish were in the size range 30–50 cm. In the northern region, very few male or female bass over 40 cm were recorded during the periods October–March and January–March respectively, the larger fish of both sexes were most prevalent in April–June and in September.

7.2 ABSOLUTE CONDITION FACTOR

The relationship between total weight and total length (Equation 6.3) was determined for each month in which adequate data were available for bass of each sex (Table 6.5). These showed that values of 1.3×10^{-5} for c (a constant chosen to give C values around unity in September) and 2.97 for b, could be used to calculate condition factors for the whole fish and the soma (body without gonads) using $C = c \times \text{Weight} \times (\text{Length}^{b})^{-1}$ (Le Cren, 1951) for all fish with relevant information in the database. The seasonal pattern of these values for the different sex and size groups of bass in each region are shown in Fig. 7.2.

Overall, the somatic condition of both female and male bass larger than 42 cm tended to reach a minimum between April and July, and improved throughout the summer and autumn to reach a maximum between September and January. The annual cycle of whole fish condition factor was also similar for both sexes, although from a maximum in October–January, it declined more rapidly in females and reached a minimum in May–June, whereas the minimum for males occurred in June–July. In the period July–October, there was least difference between the mean values of whole fish and somatic condition of either sex, reflecting the quiescence of gonad growth at this time. The contribution of the gonads to the whole body condition of bass of both sexes increased after October and reached a maximum in March and April.

On a regional basis, the best-conditioned females above 42 cm appeared in the northern region in November and December, whilst fish in the poorest condition in the central and north regions were found in May, after which they recovered condition through the summer. The condition of females in the southern region was lowest in June and July, and then improved more slowly than in the northern region through to January. The condition of males above 42 cm was higher in the southern region in December and January, declined everywhere until June or July, and improved earlier and most rapidly during the summer in the northern region. Males in the central region consistently showed the lowest condition factors throughout the year.

Bass of 32–42 cm showed considerable regional differences in their condition cycle, but the regional patterns were similar in both sexes. The small difference between whole fish and somatic condition throughout the year, suggests that there is little gonad growth in all female bass below 42 cm. With males of 32–42 cm, on the other hand, gonad growth appeared to

---

Fig. 7.2 Seasonal cycle of bass condition factors in three regions around England and Wales. (a) Females >42 cm TL, (b) males >42 cm TL, (c) females 32–42 cm TL, (d) males 32–42 cm TL and (e) males and females <32 cm TL (all regions). (MAFF unpublished data.)

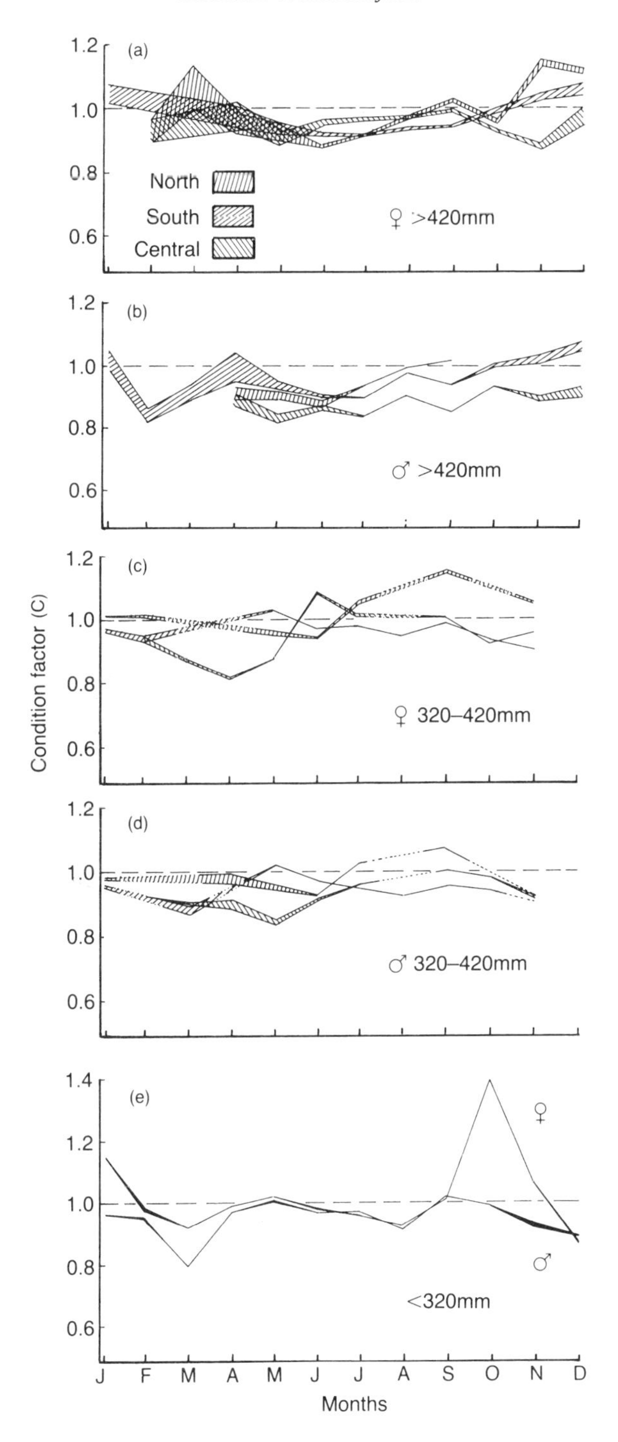
(a)
North
South
Central
♀ >420mm
(b)
♂ >420mm
(c)
♀ 320–420mm
(d)
♂ 320–420mm
(e)
♀
♂
<320mm
Condition factor (C)
1.2
1.0
0.8
0.6
1.4
J F M A M J J A S O N D
Months

start in January and reached a maximum in March and April. Evidently, their condition reached a minimum earlier (March) in the south, improving from May onwards in the central region and after June in the north. From July, the condition of both sexes in the 32–42 cm length range was higher further to the north, where it peaked in September, whereas it tended to decline from a peak in May or June in the southern region. The main difference between the sexes was seen in the central region, where the condition of females improved rapidly after April, whilst that of males increased more slowly from May onwards. The overall condition of bass of both sexes smaller than 32 cm reached a peak in May and again in September and October, and fell to an annual low between December and March. In neither sex was there any evidence of an annual cycle of gonadal growth.

7.3 FAT CONTENT

A reliable indication of the nutritional condition of fish can be obtained by examining their fat content, which represents an energy and metabolic substrate store. In oily-fleshed fish, such as mackerel and pilchard, the musculature can yield 20–25% fat at the end of the growing season, or less than 1% after spawning (Wallace and Hulme, 1977). With bass, fat is found mainly as solid, white deposits around the organs in the body cavity.

The seasonal cycle of fat content in bass of both sexes over 42 cm, which is shown in Fig. 7.3 as a percentage proportion of somatic weight, appeared to be at a minimum from March until May or June in all regions around England and Wales, then rose to a peak in September and remained high until January or February, after which it fell rapidly. In general, male bass above 42 cm had higher fat levels than did females in the same size range. The fat content of both sexes in the southern region appeared to be higher than elsewhere throughout the autumn, and, although fat levels increased most rapidly in the northern region through the summer, they fell from a peak in September towards zero from March until May (females) or June (males).

In female bass of between 32 and 42 cm, the fat content reached a minimum in April–June, with the highest levels from July until February. Males of this size also showed a regular seasonal cycle, with lowest fat content in April and May, and a maximum in July–January. As with larger fish, it is apparent that, from June onwards, fat accumulated most rapidly in fish in the northern region and reached much higher levels there in September than in fish further to the south. The main feature of fat levels in fish under 32 cm is that the highest mean values were recorded in

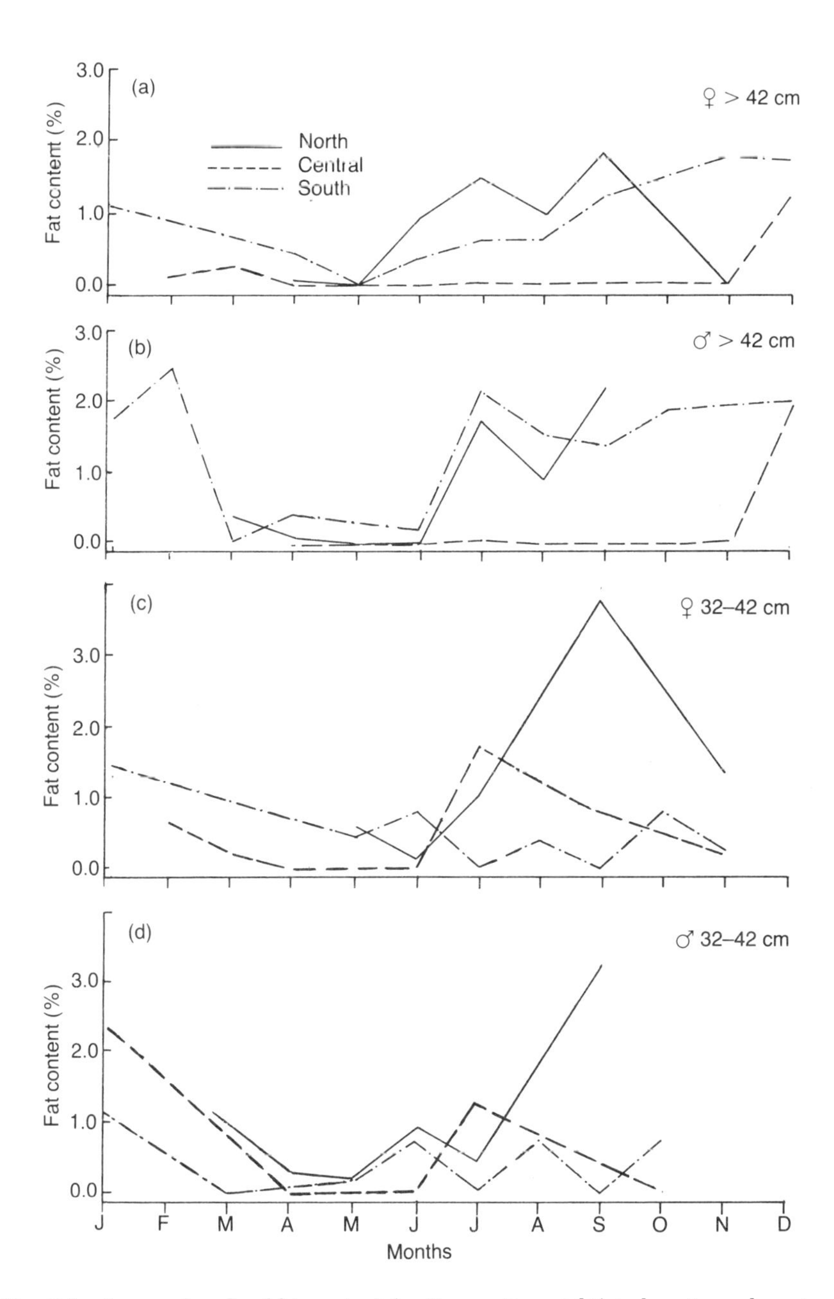

Fig. 7.3 Seasonal cycle of fat content (as % somatic weight) in bass in each region. (a) Females >42 cm TL, (b) males >42 cm TL, (c) females 32–42 cm TL and (d) males 32–42 cm TL. (MAFF unpublished data.)

January, May and July in the northern region. Fat levels in bass under 42 cm tended to be higher in females in than males.

7.4 GONADOSOMATIC INDEX (GSI) AND MATURITY

Although histological examination of the ovaries or testes of fish is necessary to ascertain with certainty the development and ripeness of gametes and the state of sexual maturity, the relative size and macroscopic appearance of the gonads provide a reliable indication of a fish's reproductive state (Table 7.1).

An easily determined measure of the state of maturity in fish is given by the relative size of the gonads. This can be expressed as the gonadosomatic index (GSI), which is the ratio of gonad weight to somatic weight given as a percentage; Fig. 7.4 shows how this varied for bass throughout the year. For most of the time, the mean GSI of females above 42 cm was higher than that for males above 42 cm. The lowest values for females (at around 0.5–1.5%) were recorded from July to September, and a peak at a mean of approximately 7% was reached in March and April. The highest GSI recorded was 28.3% for a female of 57.5 cm caught in April. In males, the lowest GSI values (all of less than 1%) also occurred from July to September, and the GSI then increased to a maximum mean level of around 6%, which was maintained from January through to April. The highest monthly mean GSI values were recorded in both sexes in April, May and June in the southern, central and northern regions respectively. Otherwise, seasonal trends in gonad growth for both sexes were similar everywhere.

For female bass of 32–42 cm, mean GSI values much above 1% were only observed in February in the southern and central regions and in May in the northern region. The two female fish with the highest GSI values (both at 6%) in this size range, were between 41 and 42 cm. Males of 32–42 cm showed a seasonal GSI pattern which was similar to, though slightly lower in amplitude than, that for larger males. Males under 42 cm were found with GSI values above 3% only in February and March in the southern region, and between March and May in the central northern regions, reflecting, perhaps, the late spawning as inshore waters warmed further to the north. All GSI values of female bass under 32 cm were less than 1.2%, with monthly means ranging between 0.4 and 0.6%. From the information available, there is evidence of a weak seasonal cycle only in males of this size

Fig. 7.4 Seasonal variation in gonadosomatic index (GSI) of bass by region. (a) Females >42 cm TL, (b) males >42 cm TL, (c) females 32–42 cm TL and (d) males 32–42 cm TL. (MAFF unpublished data.)

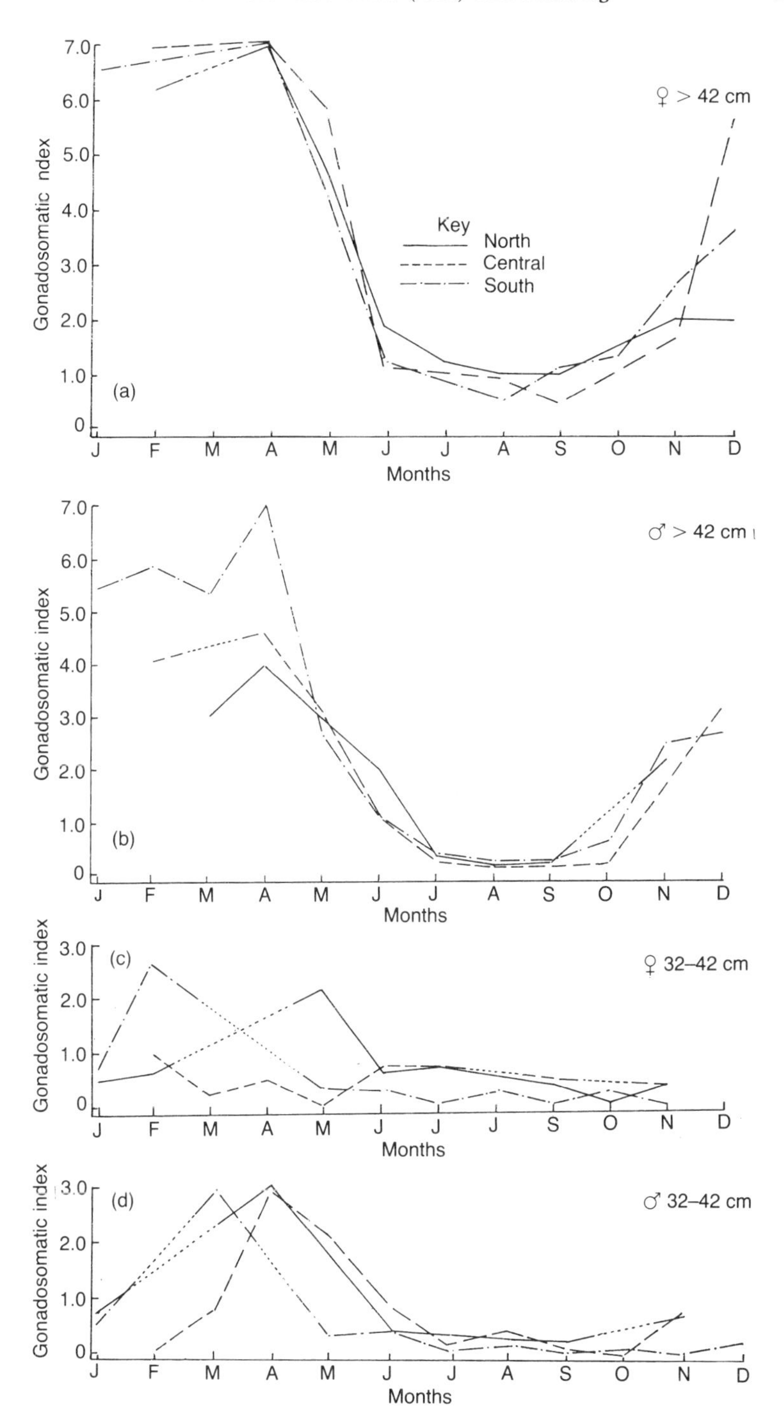
7.0
6.0
5.0
4.0
3.0
2.0
1.0
0
Gonadosomatic ndex
♀ > 42 cm
Key
North
Central
South
(a)
J F M A M J J A S O N D
Months
Gonadosomatic index
♂ > 42 cm
(b)
J F M A M J J A S O N D
Months
Gonadosomatic index
(c)
♀ 32–42 cm
J F M A M J J J S O N D
Months
Gonadosomatic index
(d)
♂ 32–42 cm
J F M A M J J A S O N D
Months

range: highest in January to June (0.4–0.6%), then declining slowly until December.

As might be expected, there is a good correlation between the GSI and the macroscopic maturity stage of bass. GSI tended to be greater in females than in males at all equivalent maturity stages, with a prominent peak in females at stage 6 (running ripe) and highest levels in males at stages 5 and 6. The mean GSI of both sexes decreased rapidly from stage 6 to 7 (spent) as the gametes were shed. Males under 42 cm had a pattern which was similar to that of larger fish, though with maximum values at stage 4. Few females under 42 cm, however, had gonads at maturity stages greater than 4, and their GSI values were generally much lower than those of larger fish. Females under 32 cm tended to have higher GSI values than did males of the same size, with means of around 0.4% and 0.3% respectively, at the maturity stages 1 and 2 recorded.

The progression of gonad development with fish size is illustrated in Fig. 7.5, which shows the relationship between maturity stage and total length for male and female bass separately. Whilst fish at stage 3 (developing virgin) began to be recorded at lower lengths (29–30 cm) for females than they were for males (31–33 cm), maturity stage 4 (and above) was recorded in only one female fish below 40 cm. In contrast, several males under 36 cm were classified as having maturity stages 4 through to 7. It would thus appear that, although females may show the outward signs of early gonad development whilst still less than 30 cm, they do not go on to spawn until they are at least 41 cm. From a knowledge of growth rates, it seems probable

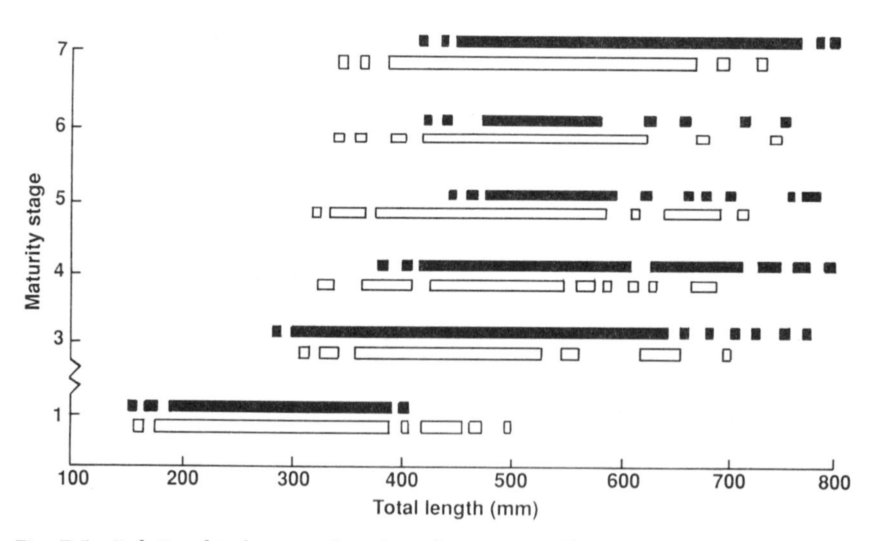

Fig. 7.5 Relationship between length and maturity of bass: open bars, males; solid bars, females (MAFF unpublished data.)

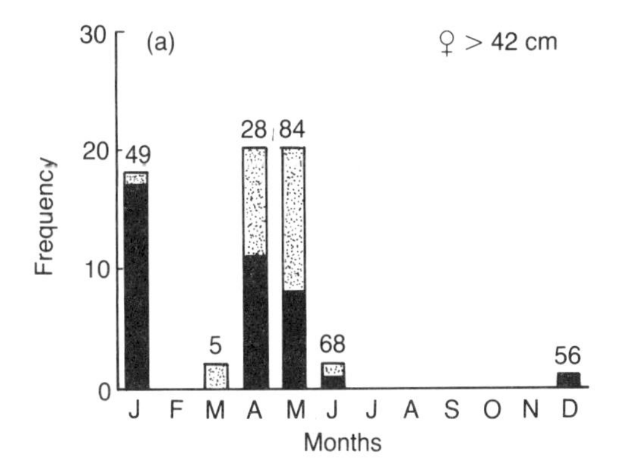

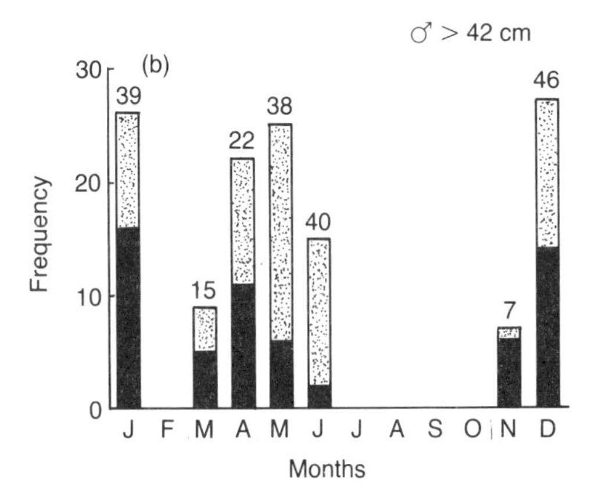

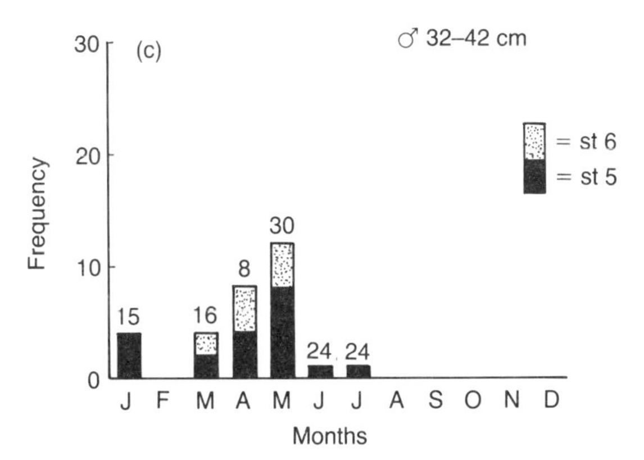

Fig. 7.6 Spawning period of bass, as shown by frequency of maturity stages 5 and 6. (a) Females >42 cm TL, (b) males >42 cm TL and (c) males 32–42 cm TL. Numbers are totals sampled in those months. (MAFF unpublished data.)

that many females remain at an early stage of maturity throughout two or more annual 'spawning' seasons, before their gonads develop to full maturity. The gonads of males, on the other hand, could probably become fully mature in one season.

To help define the spawning period of bass around England and Wales, the frequencies of fish with maturity stages 5 and 6 in samples taken at different times of the year are shown in Fig. 7.6 for all fish over 42 cm and for males between 32 and 42 cm. These stages were chosen on the assumption that once a fish is either gravid (stage 5) or running (stage 6), spawning can occur.

Spent fish were not included because they tend to remain in this condition for several weeks after spawning. Spent females were present in the UK bass population from May until November and spent males from May to December and also in March.

Ripe males above 42 cm began to occur in November and increased steadily in frequency until April, when all those sampled were ripe. Thereafter, the incidence of ripe males fell, and no males with maturity stages 5 or 6 were observed between July and October inclusive. No males between 32 and 42 cm appeared to be ripe before January, and these smaller fish showed a generally lower incidence of maturity stages 5 and 6 than larger bass. All ripe females were larger than 42 cm and did not appear in samples until December; as with males, there was an increase in frequency towards a peak in April, but a more rapid decline led to a virtual absence of fish with ovaries at stages 5 and 6 by June.

7.5 THE CONNECTION BETWEEN FEEDING, GROWTH, MATURITY AND TEMPERATURE

Along with several authors (Kennedy and Fitzmaurice, 1972; Barnabé, 1973; Kelley, 1988b) we have observed that female bass grow faster than males, and have shown that the size difference between the sexes at a particular age becomes statistically significant from age 6 onwards. Rather surprisingly, our data failed to show any significant difference in the weight of stomach contents throughout the maturity cycle. In fact, the proportional weight of stomach contents of fish above 42 cm in length was found to be generally higher in the months October to January, the main gonadal growth period, than at other times of the year (Chapter 3). Kelley (1987) recorded a fairly uniform pattern of feeding by adult bass in all seasons, but he suggested that his results may have been biased towards too high a value in winter, because the fish were caught on rod and line and his data were probably selective towards bass feeding at the time of capture. In cold water, however, the amount of food in fishes' stomachs might explain less about the frequency of feeding than the protracted retention of food in the stomach owing to the reduced digestion rates (Santulli *et al.*, 1993). Both Kelley and ourselves have found reduced feeding levels and low condition of juvenile bass everywhere in winter.

The results of DFR's study suggest that the adult bass population's movements around England and Wales may be attributed as much to feeding and growth benefits as to a reproductive necessity to migrate in autumn to areas where sea temperatures usually remain above 9 °C, as has previously been suggested (Pawson *et al.*, 1987). Feeding conditions, whether because of increased food availability or higher sea temperatures (Kelley, 1987), appear to be most conducive to an improvement in condition and accumulation of fat in summer in the North Sea and Irish Sea, but fish lingering there after September appear to experience a rapidly deteriorating feeding environment. In the winter, bass above 42 cm had more food in their stomachs, a higher somatic condition factor and more slowly dwindling fat reserves in the western English Channel than further to the east or north.

It appears, therefore, that because adult bass migrate between productive inshore feeding areas in summer and more southerly, relatively warm offshore pre-spawning areas in winter, their feeding and growth opportunities are considerably improved compared with the juveniles, which tend to remain in or close to inshore nursery areas throughout the year. As a consequence, the growth rate of bass is maintained or even enhanced upon achieving maturity, and the increased energy demands associated with migration and reproduction might be offset by feeding in relatively warm water in winter. This may explain the point of inflection in some growth curves (Chapter 6) at ages 5 and 6 for males and 6 for females which appears to correspond to the onset of maturity.

The analysis of regional data provides useful information on the seasonal patterns of condition and fat content in local populations of juvenile bass, but this is clearly not the case for adult bass which make seasonal movements between these regions. For the purpose of portraying these cycles within the main migratory population, data from the southern (Oct.–Mar.), central (Apr.) and northern (May–Sept.) regions have been combined in Fig. 7.7.

As might be expected, the seasonal patterns of condition, reproductive activity and fat content in bass appear to be strongly interrrelated. The indices of gonadal growth – GSI and condition factor (C) (comparing whole body with somatic values) – suggest that gonad development in adult males begins in November, and macroscopic examination of gonads shows that ripe fish predominate in the adult male bass population from January until June. Ripe females occur in a briefer, more definite season than males, and April seems to be the peak spawning month for bass around the UK. All of the relevant data suggest that the period of gonadal growth and

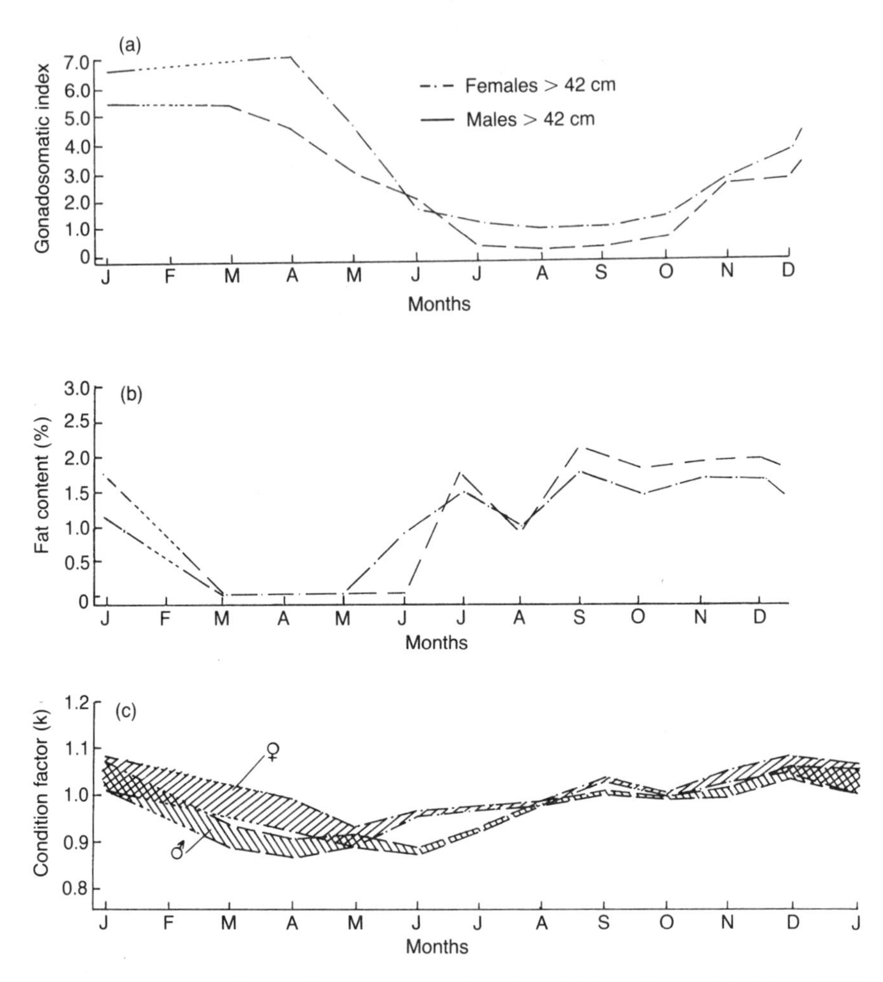

Fig. 7.7 Composite figure showing seasonal cycles of (c) condition, (a) GSI and (b) fat content (as % somatic weight) for adult bass within their normal range around the British Isles. (MAFF unpublished data.)

spawning in females ends by June, and by July in males, and that the decline of reproductive activity towards the end of the spawning season is most rapid in females.

Spawning in bass typically occurs earlier further to the south: in January off North Africa (Bou Ain, 1977) and in the Mediterranean Sea (Barnabé, 1990), in February in the Bay of Biscay (Stequert, 1972) and in March and April in the eastern Celtic Sea (Jennings and Pawson, 1992). Thompson and Harrop (1987) have shown that, in the period February to May, most bass spawning in the English Channel occurs offshore in deep water, with the activity spreading north-eastwards as the season advances. There is further evidence, from the gonad maturity data presented above, that spawning occurs later further to the north around Britain, and that the smaller mature males and females there do not ripen and spawn until April or May. These fish probably include first-time spawners which have not migrated as far south and west as large fish which, by that time, have already spawned offshore.

By the time that the gonads are fully mature in March, the mesenteric fat of adult bass has been almost completely utilized for energy and metabolites, though fat soon accumulates as a result of heavy feeding over the summer months and, by September, might constitute an average of 1.5–2% of the somatic weight of the fish, a level which is maintained until January. Fat levels appear to fall more rapidly and to rise again a month earlier in females than in males, reflecting, perhaps, a greater contribution of energy to gonadal growth and the more abrupt end to the spawning season in females. The mean values of fat content for both sexes of all sizes were in general similar, but the data for fish over 42 cm do not support Kennedy and Fitzmaurice's (1972) suggestion that female bass have proportionally more fat than males. The somatic condition of adult bass falls rapidly during February and remains low before beginning to rise again in May for females and June for males, coincident with the seasonal cycle of fat content.

We have paid particular attention to bass in the 32–42 cm range, the so-called 'adolescent' group. These fish have been found to be more likely to move away from nursery areas than bass less than 32 cm, though they still tend to remain close inshore, and not to adopt the regular migration pattern and offshore winter distribution of adults. Maturity stage, GSI, fat content and condition factors for males in this category all show definite seasonal cycles, which are similar in terms of both timing and amplitude to those of larger males. Many males of 32–42 cm had well-developed gonads and this category included ripe and spent fish. No fully mature females were found at lengths below 42 cm, though some showed signs of gonad development up to maturity stage 3 (and one at stage 4) and GSI values of 1% or more.

This observation, that the timing of sexual development in bass is different for males and females, has been noted by other authors (Kennedy and Fitzmaurice, 1972; Bruslé and Roblin, 1984). Kennedy and Fitzmaurice found mature males at total lengths of around 34 cm and females at about 38 cm, and upwards. Around the UK, a 36 cm bass of either sex could be from 3 to 6 years old, and it appears that there is little influence of age on the onset of maturity; the much-better-defined length effect suggests that size is a more important factor. The maturation of the ovaries in many teleosts is inhibited by low water temperatures (Bye, 1984), and because juvenile bass tend to remain in the same inshore

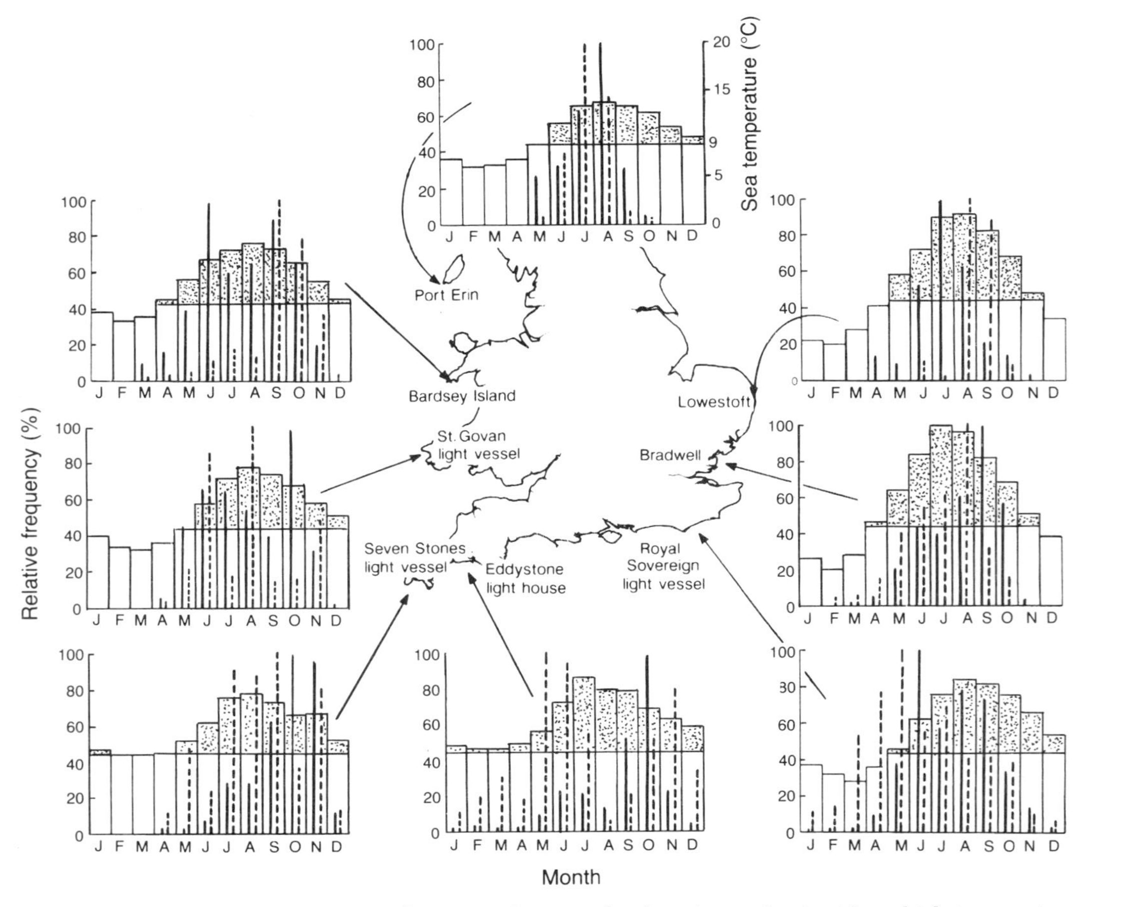

Fig. 7.8 Annual pattern of sea temperature around Britain, showing the duration and extent by which temperature exceeded 9 °C, and monthly commercial catches of bass above (solid lines) and below (dashed lines) 40 cm TL. (MAFF unpublished data.)

areas all year round, it is unlikely that full maturation and spawning in females will take place until they begin to migrate to warmer water in winter. This does not appear to apply so much to the maturation of the testes (MAFF, unpublished data), and might explain why there is a difference in the size (and age) of first maturity in male and female bass.

We have suggested that water temperature is the major influence on whether and when bass spawn. Figure 7.8 shows the annual pattern of sea temperatures recorded over the years 1982–86 at eight coastal sites within the normal range of adult bass around Britain, and indicates the duration and extent at which temperatures exceeded 9 °C. If temperatures below 9 °C inhibit maturation and spawning in female bass (Devauchelle, 1984), adult fish would be expected to leave the North Sea and northern Irish Sea by December and the West Wales coast and eastern English Channel by January, and congregate offshore around the Cornish Peninsula during January–March. This supposition is supported by tagging studies and by the daily catch records of commercial bass fishermen from which the seasonality of fisheries for bass above and below about 40 cm shown in Fig. 7.8 were derived. In the North Sea and northern Irish Sea, adult bass are likely to appear first in April, but are present in peak numbers only between June and September, and are largely gone by the end of October. The temporary influx of fish moving northwards (in June) and southwards (in October) is apparent in the eastern English Channel data, as is their arrival in southerly waters in October and November, before they move offshore in winter. Apart from in the English Channel, where bass under 40 cm appear to be accessible to the fishery throughout the year, weather permitting, the seasonal patterns of catches of fish above and below 40 cm were similar. Tagging, however, has failed to reveal any significant southwards or westwards movement of juvenile bass in the autumn, and those fish remaining to the north or east of the coasts of Devon or Cornwall, or living inshore, will normally experience extended periods of ambient temperature below 9 °C, which might be expected to inhibit maturation of ovaries.

It might be significant that during the unusually mild winters of 1988/89 and 1989/90, when ambient sea temperatures at Lowestoft never fell below 7 °C, catches of adult bass offshore in the central and western English Channel were reduced compared with previous years (MAFF, unpublished data). Subsequently, bass appeared much earlier than usual in fisheries to the east and north, and 0-group bass were observed well to the north of their usual distribution, from Suffolk to the River Tees on the east English coast, and in Morecambe Bay. The inference is that higher winter temperatures resulted in a northwards shift of the overwintering and spawning areas of bass. 1-Group and 2-group bass were still unusually abundant inshore at Lowestoft in Spring 1991, the result, seemingly, of local spawnings in 1989 and 1990.

Summary to Part One

The general biology of the bass is now well understood, but there is scope for much more study. Although we are beginning to reveal what happens during larval recruitment and the migration of older fish, the picture is incomplete as to how and why. The development of bass cultivation and the accompanying research on the biology of reproduction, nutrition and growth has provided some possible explanations for observations of natural phenomena. For example, the knowledge of the temperature range at which bass will mature and spawn has enabled us to postulate the underlying causes and effects of winter migrations of adult bass from the North Sea to the western English Channel. Little is known about the mechanism of this migration, though recent research on other marine fish, such as the plaice and cod (Arnold *et al.*, 1990), may provide some useful clues.

Many of the myths about the bass and its biology have been exploded or explained, and we are confident that there is a strong biological basis for the management of bass fisheries. Biological investigations on the bass should now be directed at aspects which may affect the species more strongly in the future. In particular, we know that the bass is able to adapt to a wide range of environments, and it has been shown to respond dramatically to hydrographic factors, especially changes in water temperature. The distribution and growth of bass appears to be strongly affected by even relatively short-term climatic changes and, at least at the fringes of its range, these effects should be monitored and studied.

Similarly, there is no denying the importance of the estuarine environment to the recruitment, survival and development of young bass and hence to the stock as a whole (this applies to several other species). Around much of North-west Europe, human activity in these areas is increasing considerably. It was estimated that, in 1992, some 40% of UK estuaries had the prospect of new developments such as marinas, barrages and land reclamation schemes (Jennings, 1992). It is likely that more power stations and gas and oil terminals will be sited on estuaries. These impacts are in addition to existing threats from pollution of various types, and it may be time to raise the profile of bass and other marine fish species which depend on the estuarine environment, in the minds of developers, government agencies and local authorities.

It is hoped that a sound knowledge of the species' biology, particularly of its young stages, and of the dynamics of local populations, will enable the bass (with the salmon) to be accorded the importance it deserves when such matters are debated. There appears to be little protection for young bass throughout the rest of Europe, although many nursery areas have already been described in reports (e.g. Desauney *et al.*, 1981; Le Masson, 1981) and their importance is beginning to be recognized.

Part Two

Exploitation

Catch of bass taken on traditional tackle. Photograph reproduced by kind permission of D. Kelley.

Introduction to Part Two

Records of the exploitation of the bass go back to ancient Greek and Roman times. Accounts by Aristotle and Oppian were probably first translated by Cuvier and Valenciennes (1828) into French and thence into English, but were most completely quoted by Thomas Couch (1862) in *A History of the Fishes of the British Islands*.

From these accounts we can see that hooks on lines and gill or seine nets were used for catching sea-bass around 2000 years ago. In that sense, little in the fishery has changed and, although it is doubtful whether a 'sport fishery' as such existed in Roman times, it is not beyond probability that this pleasure-seeking society did sometimes catch bass for fun. Equally remarkable is the observation that the Romans kept bass in tanks and managed to get them to spawn. It took a long time, subsequently, for the next serious attempts at intensive cultivation of this species to get under way around the Mediterranean (1970s!). So it seems that early attempts to exploit the bass were not too far removed from the three main ways pursued today: commercial fishing, recreational (sport) fisheries and cultivation (farming).

Although the third aspect may be considered to be separate from the others, aquaculture has many forms, and some practices, such as harvesting wild fry for ongrowing, may conflict with the interests of the commercial and sport fisheries. Around England and Wales, cultivation of bass has not yet been attempted commercially, and the main issues of recent times stem from the development of a modern commercial fishery alongside a well-established sport fishery. Until the 1970s, commercial fishing for bass was at a low level (in both the UK and France) and it was generally confined to isolated coastal communities, using traditional methods which were relatively inefficient and which had probably changed little since the early 19th century.

In this part of the book, each of these aspects of the UK bass fishery is described, with brief comments on how changes have taken place to bring about the present-day situation. Geographical details of the sport and commercial fisheries around England and Wales (Chapters 8 and 9) are based largely on DFR surveys, and a description is given of the methods used to catch bass, past and present. Other European readers will have to forgive our inability to give such comprehensive information for the bass fisheries in their particular countries. The cultivation methods used in each country are described in Chapter 10, which includes a resumé of information which is available in the literature. Much modern husbandry involving bass and other marine species is now in the hands of private companies, who wish to keep their breeding and rearing methods confidential, so some of this information will inevitably be out of date. The European market for bass is described in Chapter 11.

Evaluation of the bass fisheries, including landing statistics, values and catch and effort data, is described in Chapter 12 and 13, and assessment of the bass population, using DFR data as an example, is described in Chapter 14. By their nature, which is essentially local, involves small boats and has a large recreational component, bass fisheries are difficult to investigate. Traditional approaches are not necessarily appropriate, and novel ones have been developed. To our knowledge there have only been perhaps six or seven investigations of bass fisheries in Europe. The most detailed (apart from our own) is that of Bertignac (1987), who described the fishery in Morbras (South Brittany). Le Masson (1981) described the fishery at Etel (also in S. Brittany) and pointed to the need for management, but to date, only the recent DFR study, which was built upon that of Holden and Williams (1974), has led directly to management action intended to conserve the stocks of wild bass and sustain their fishery.

Chapter eight

The sport fisheries (a description)

8.1 THE ATTRACTION OF BASS FISHING

Rod-and-line fishing for bass takes many forms, and it is not difficult to find words to convey its attraction. Below we have reproduced an extract of an article penned by John Hudson, which appeared in 1963 in the first volume of a delightful, but long defunct, magazine called *Creel*, which was edited and illustrated by Bernard Venables.

> Bass, a magic name. Thousands of sea anglers know its fighting nature. A magnificent fish. But, strangely, mystery still shrouds certain aspects of its life cycle.
>
> . . . If, by some oddity of fate, I were permitted to fish only for one species, it would be bass that I should choose. To begin with, there are so many ways to fish for them – ledgering, spinning, driftlining, whiffing, fly-fishing. There are so many techniques to master – and not only ones directly related to angling, like distance casting. You have to be a long-distance walker, an expert at rock-scrambling, and skilled in the ways of smaller boats if you want to lay claim to the title of complete bass fisherman. You have to develop qualities which will enable you to get purposefully out of bed at 3 a.m., or to ignore sleep altogether. Your mind must be sharp enough to analyse and apply snippets of information gleaned from small boys and longshoremen, to sort out the grain from the chaff.
>
> Qualified in all these ways, it is possible that, if the fish are there, and if they are feeding, you will catch bass with some degree of consistency. And no fish is more worthy of the trouble.
>
> They are such a paradoxical fish that the angler never quite knows what to make of them. How can the tentative, plucking touches at the soft crab, which often herald the hooking of a seven-pounder, be

> reconciled with the smashing attack on the spoon-bait that the same fish might make? The bass is a fish that is often cautious, often the tearaway thug of inshore waters. And a big bass is a magnificent prize. Coarser, less delicate in coloration than the shining little school bass, there is a majesty about its thick-set body, and the gleam of a crown on its gold-bronze jowls. A kingly fish indeed.
>
> Such prizes are not commonplace, and a double-figure fish is not given to many. Some excellent bass fishermen go right through life without one. The record bass, the eighteen-pounder from Suffolk, often seems unassailable. And yet, and yet. . . what of that great fish from the Teifi estuary that was caught by a salmon angler, that weighed more than the present record but which was cut up and eaten before it could be fully authenticated? And what of the fore part of a huge bass that was washed up a year or two ago on the Cefn Sidan shore in Carmarthenshire, that itself weighed more than ten pounds? In all gravity, I do not consider a twenty-seven or twenty-eight pound fish from British waters an impossibility. If it comes, it may well be a 'fringe' area that produces it. The extreme limits of the bass's range seem to yield few, but very large, fish. The present record fish is a prime example. Many of the best Irish specimens have come from the coasts of Donegal and Antrim. North Wales produces big ones.
>
> Yet another paradox, then, and one for which there is no scientific explanation. Indeed, zoologists are singularly short of information on the species. Travis Jenkins, not giving anything away, says, that the bass spawns from May to August either in the sea, brackish or fresh water. The fact is that next to nothing is known of the bass's spawning habits. So far, no one has managed to obtain bass eggs, even though the eggs of almost every other species, including great rarities, have been taken in the zoologist's tow nets. And where most bass go in the winter is still a complete mystery.

Since that piece was written, at least two rod-and-line-caught bass larger than the Suffolk-caught record fish have been reported, and, as we have seen, many of the mysteries Hudson refers to have been solved.

8.2 HISTORY

Serious recreational fishing for bass appears to have started some time between 1820 and 1840, when a few game (i.e. salmon and trout, *Salmo trutta*) fishermen realized the sporting potential of this species. Much of the early sea angling literature seems keen to promote the sporting nature of the bass, presumably to attract people already skilled in the pursuit of game fish.

So, from the start, this was no 'poor man's fish' to be hunted with a heavy leaded handline and crude fishing methods.

Writers of the late 19th century advocated the use of flies and lures, similar to those employed in game fishing. The following passage is taken from John Bickerdyke's book *Angling in Salt Water*, published in 1887.

> . . . Fly fishing for bass, which has been practised for about half a century, is, when the fish are feeding close to the surface, by far the most sportsmanlike and pleasurable method of catching them. The sport afforded is, indeed, little inferior to salmon fishing, for the bass are almost as strong as salmon, and what little they lack in strength they fully make up for in numbers. The great difficulty is in finding the fish, for it is little use casting where they cannot be seen breaking the surface and playing, or rather feeding, in the surf. The time spent in searching need not be wasted if the angler is in a boat, for, while he is being pulled slowly along the shore, he can trail a dead sand-eel, a strip of mackerel skin, a spinning bait, or any of the thousand-and-one devices which pollack, bass, mackerel, and a few other sea fish seize when in motion. Immediately the bass are sighted the spinning rod is taken in, and the fly deftly cast into the middle of the shoal, the boat in the meantime having been sculled very quietly to windward of the fish. . . .The salmon fisher need not go far for flies, for any gaudy, small salmon or large sea trout fly will do admirably for bass.

Later, Holcombe (1921) lists several sporting methods of catching bass; top of these was fly fishing, for which he recommended using either a single-handed fly rod or a 4 m salmon rod.

We assume that most fishing with fly tackle for bass was practised in estuaries or lagoons and directed at shoals of near-surface 'school-bass' (fish up to 1.0 kg), rather than at the larger bass which are found off the open coast. Carter Platts (1940), however, laments the days gone by when bass were caught on fly from the rocks of Devon and Cornwall, and goes on to explain how a boat could be used to find a shoal of feeding bass, which could then be caught with light salmon gear.

At the turn of the century, writers such as Bickerdyke (1887), Wilcocks (1868) and Aflalo (1891, 1898) began to popularize sea fishing, and a range of methods for catching marine species was soon developed. Although general sea fishing tackle remained heavy and clumsy right through to the 1960s, bass tackle was always considered to be specialized equipment and, in addition to a heavy greenheart and hickory 'bottom rod', many sea anglers had a much lighter 'bass rod'. Throughout the late 19th century (from 1893) and up to the 1930s, the British Sea Anglers Society (BSAS) did much to encourage the development of better tackle and the use of sporting methods for all sea fish. Bass fishing has always stayed well ahead of trends in this respect, seeming to be more closely related to freshwater fishing even to this day. A testimony to the sporting quality of bass taken on salmon tackle was given by Gammon (1967).

> For the shore fisherman, there is no quarry like the bass. To begin with, it is such a sporting fish, perhaps the strongest fish, size for size, in British waters, fresh or salt. Not long ago, I spent a fishing holiday in Ireland with the spinning rod. The first day I spent salmon fishing on the Bandon, in Co. Cork. Then I travelled to Co. Kerry, to try for bass. Exactly the same tackle was used, down to the identical Toby spoon, for both species. The salmon in question were small springers, seven or eight pounds in weight, the bass running rather smaller in average size. But the bass fought longer and harder, even though they had no heavy river current to assist them.

More recently, the Bass Anglers Sportfishing Society (BASS), which was founded in 1973, has pulled together specialist bass anglers in the UK and encouraged the use of a variety of sporting methods. A few enthusiasts have rediscovered fly fishing for bass, and others have adopted American-style plug-fishing techniques, developed mainly in fresh water for largemouth bass, for use from UK shores. For many, it is not only the bass that they seek, but, as with game-fish anglers, it is the enjoyment gained from catching it with a particular technique that is important. Bass anglers can be divided into many groups. Within the ranks of shore anglers and boat anglers, there are advocates for live bait, trollers, spinner votaries, float fishermen, static-bait devotees, ardent plug and lure fishers, light-tackle enthusiasts and long-distance shore casters. Some emphatically refuse to use modern tackle, relying on split cane rods and braided natural fibre lines.

Others, such as charter boat skippers who, to make a living, must try to satisfy their customers with a good day's sport with any species, set out to catch bass with the most effective methods available. In recent years, a group of skippers from Essex has done much to conserve sea bass (and other prime species) by handling their quarry carefully and returning them to the sea after capture. Education of their paying guests is an important part of the exercise. This is in sharp contrast to the deckloads of fish shown in the angling press in the early 1970s, which at the time provoked considerable adverse comment (R.Y. Cox, pers. comm.).

There are also anglers who are less interested in opportunities for sport than the total quantity of bass caught. These people concentrate their efforts near power station cooling-water outfalls where large numbers of small bass are often found. At times the shoals are so dense that the more unscrupulous souls have stooped to 'snatching' or foul-hooking fish with the aid of treble hooks, to increase their catch rate. Fortunately, this illegal method seems to be in decline owing to the increased presence of fishery officers in the areas where it was most prevalent. Needless to say, few fish taken by rod-and-line fishermen of this ilk are eaten at home or given to friends. With increasing demand and correspondingly high prices in recent years, disposal of the catch for cash is neither difficult nor made with reluctance. This element of the fishery is the most difficult to control, as its activity takes place on an opportunistic, part-time basis at any time of the day or night. The practices of those who use rod and line as an overtly commercial method are described in Chapter 9.

8.3 TACKLE, METHODS AND BAIT

A vast amount has been written about angling for bass, probably more than for any other sea fish. Sea angling periodicals regularly feature articles on the species, including details of catching methods and items of tackle. If one refers to the early writers on bass fishing, it will be seen that although the basic methods have changed little, the design and materials used in tackle construction have advanced greatly. Reels, in particular, have improved since the First World War, when the universal wooden centre-pin was still the usual 'winch'. The modern bass angler can choose between centre-pin, multiplier or fixed-spool reels, with hundreds of different models available. However, the basic techniques of bottom fishing, spinning, float fishing and drift-lining with live or dead baits are still recognizable as those used 100 years ago. The baits used have changed little, except perhaps artificial lures, which are now made from better and more malleable materials, and the preservatives used to prolong the 'life' of natural baits. As so much detail is available elsewhere on modern angling methods we will confine ourselves to simple descriptions of the techniques in common use.

Bottom fishing

This method basically involves static fishing with a lead weight to aid casting and to hold the tackle and bait on or near the sea-bed. It can be employed from shore, pier or anchored boat and bites are registered on the rod-tip. Nowadays, shore fishing rods for bass are usually made from 3–3.5 m of

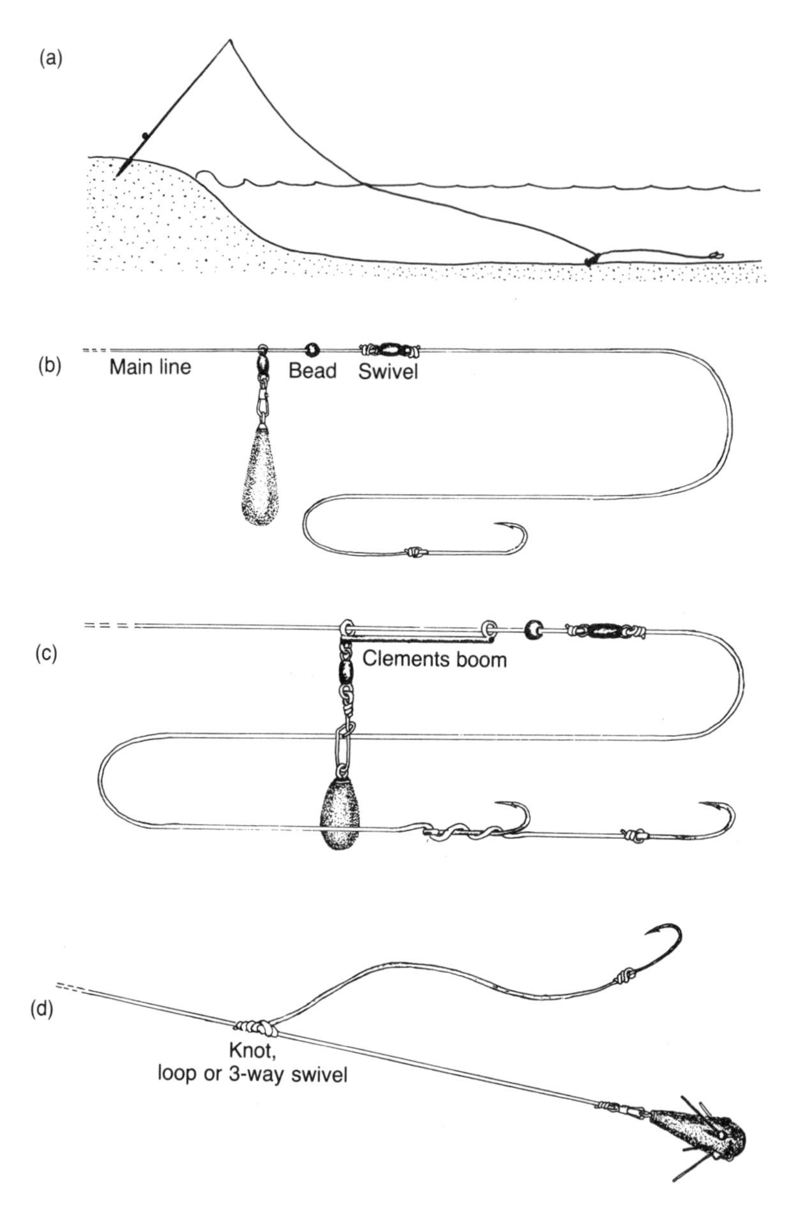

Fig. 8.1 Bottom-fishing rigs. (a) Bottom fishing from the shore. (b) Standard running ledger rig. (c) Running ledger with Clements boom and 2 hooks. (d) Fixed line paternoster with breakaway lead.

tapered hollow glass- or carbon-fibre tubing, designed to cast 50–200 g leads with 4–7 kg breaking strain (b.s.) line.

Bass anglers tend to use lighter line, and therefore need lighter leads to hold the bait on the sea-bed, than do anglers fishing for other marine species, such as cod. Normally, a single hook is used on a trace of up to 4.5 m in length, but when exceptionally large baits, such as whole squid, are used, two hooks may be employed. Hook sizes vary from size 2 to 4/0, depending on the size of bait used. The hook trace can either be fixed on the line, usually above the terminal lead, or left free to run on the line through the use of a small boom or swivel-eyed lead, some examples of which are shown in Fig. 8.1.

One very effective form of fixed-trace fishing is the simple nylon paternoster, comprising a 1–2 m length of nylon fixed to the main line at least 1 m above the weight. This method is useful for fishing near to piers or similar obstructions, where bass like to browse amongst the woodwork for food. Shore crabs or small, live fish are favourite baits to be used with this technique.

Float fishing

There are three basic styles of float fishing for bass. The first and most widely practised method is used off rocks or piers at fairly short range, in which the bait, traditionally a live prawn or a strip of dead fish, but today often a live mackerel or sandeel is suspended beneath a large cigar-shaped cork or plastic float (Fig. 8.2(a)). The bait may be fished at any depth, but around 2 m from the sea surface is most common. Casting is aided by adding either lead shot or a spiral lead to the line, which also serves to keep the bait sunk. This method has long been practised in Devon and Cornwall, and in the 19th century was known as the Nottingham style, after the float-fishing methods developed there for angling in the River Trent.

The second method, which John Bickerdyke recommended in the 1880s and seems to have been resurrected in recent years, is bottom float fishing or float ledgering (Fig. 8.2(b)). It is only really successful in relatively shallow water, say less than 6 m depth, where the tidal flow is not too strong. Nevertheless, the depth being fished is usually greater than the length of the rod, and a stop knot is placed at the required distance above the weight, between which the float is allowed to slide. A running trace is used which is usually 2 m long. Besides fixing the fishing depth, the float is employed to indicate bites, which are often quite delicate. This method is most effective when used in areas which have been ground-baited with fish offal or dead sandeels. The hook bait might be whole sandeel or strip of mackerel.

The third method of float fishing primarily uses a 'bubble' float as a means of casting and to keep the bait near the surface (Fig. 8.2(c)). In hot weather,

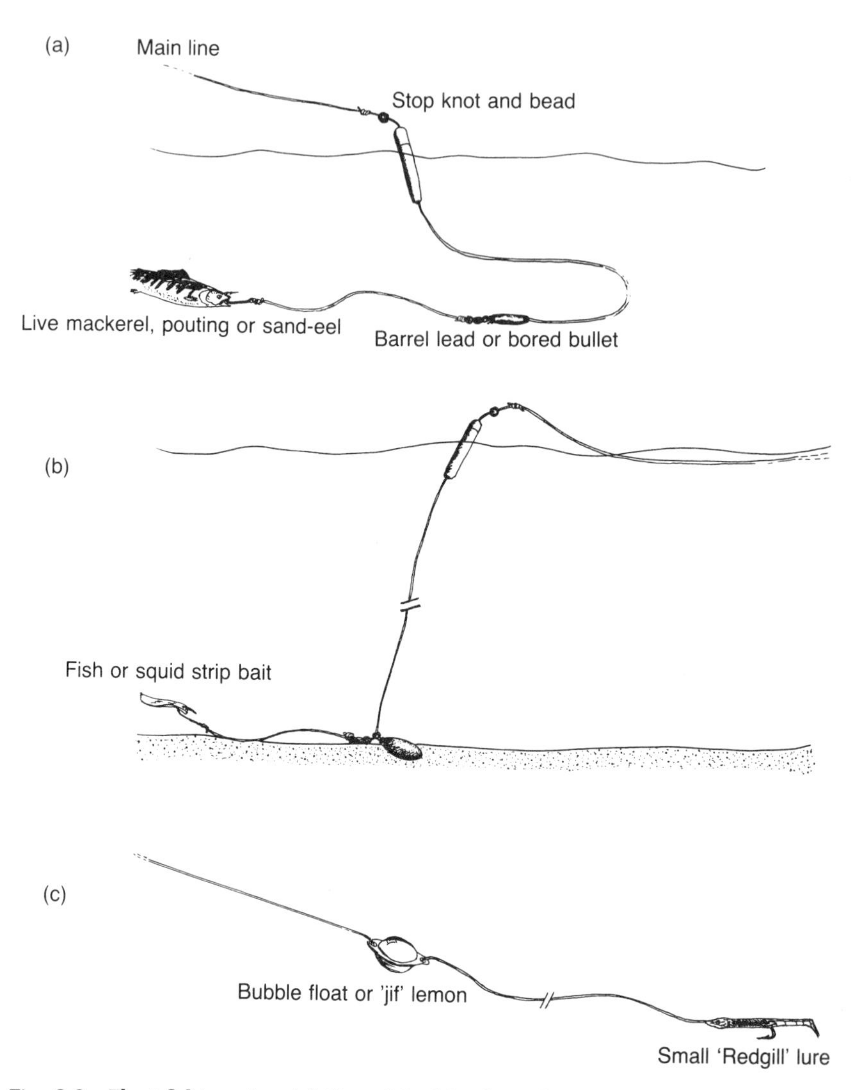

Fig. 8.2 Float-fishing rigs. (a) Float fished linebait. (b) Float-ledgered deadbait. (c) Bubble-float casting rig for surface feeding bass.

bass are often found near to the sea surface, feeding on small fish called 'brit'. These can be young sprat, herring, sand smelt (*Atherina presbyter*) or other fry. The most successful bait in these circumstances is a tiny, white version of the plastic or rubber imitation sandeels ('Redgills' or 'Deltas'). The bubble float is fixed on the line (of around 4 kg b.s.) some 2–2.5 m above the lure, and part filled with water. It is unusual for lead to be added to the line. The tackle is cast over or beyond the 'brit' shoals and wound back across the surface. Generally, light spinning rods and fixed-spool reels are used with this method as the bass taken tend to be small, though it has been particularly successful for larger bass in South Wales. Both this method and float ledgering are also used by commercial rod-and-line fishermen.

Driftlining (from a boat)

Nowadays all driftlining is carried out using live baits, usually sandeels or small mackerel. Generally the tackle is weighted with a bored lead ball threaded on the line. Below this is a trace of 4–7 kg b.s. nylon, which is generally as long as the rod being used. The lead is either allowed to drag along the sea-bed as the boat drifts with the tide or, if the bottom is rough, wound up so that the bait is fished just above it. It is normal for the rod to be held all the time in driftlining (to prevent its loss if the tackle becomes snagged on the sea-bed). Bites, which can be just a couple of 'knocks' on the rod tip, are struck hard, as there will often be a large amount of slack line between the rod and the hook.

Spinning and plug fishing

These techniques can be used either from the shore or from boats, and they rely on casting the lure and retrieving it at varying speeds and depths to simulate the action of a prey fish. Spinners are always fished sunk, but some plugs can be fished across the water surface and others made to dive when being wound in.

Spinners are usually made of metal and have silver, bronze or gold reflective surfaces. They can be fished unweighted or with lead on the line to aid sinking and casting. There are hundreds of different varieties. Over the years, and particularly in the 1960s and 70s, various sizes of Toby lure (Fig. 8.3(a)) made by ABU (Sweden) were popular and accounted for many good catches of bass.

A form of spinning, called pirking, involves fishing with heavy metallic lures (pirks) in deep water, mainly over wrecks. This method was initially used for cod and pollack, but was also found effective for bass in some places. As the tackle needs to be heavy, the method is now little used by bass sport anglers and it has become more of a commercial fishing method.

Plug fishing originated in the United States and was adopted in Europe in the 1920s, initially for freshwater fishing for pike, and by a few sea anglers for pollack, mackerel and bass. In the last ten years or so, a small band of enthusiasts has again shown that plugs can be fished successfully for bass

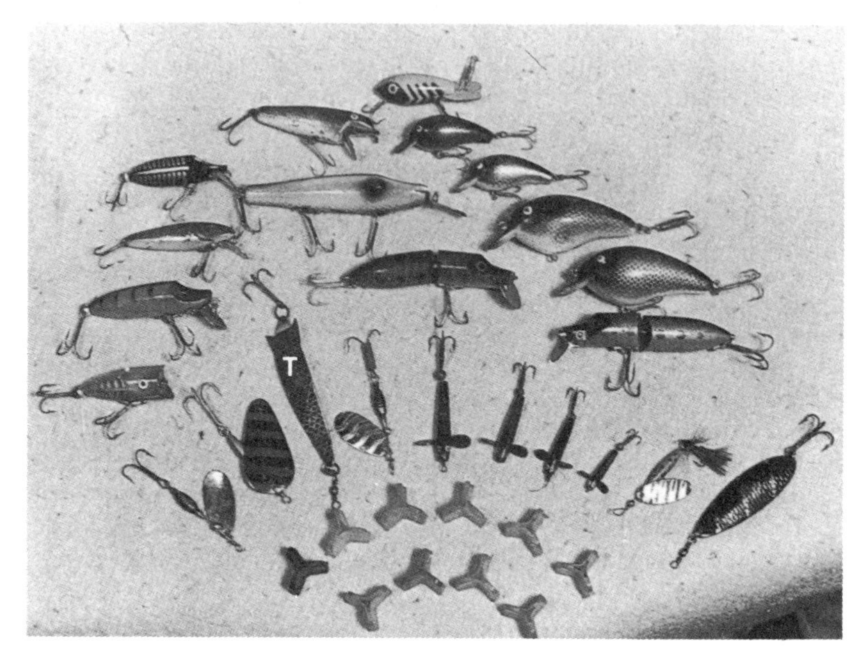

Fig. 8.3 Spinners, lures and plugs. Toby line marked w.-R'T'. Reproduced with permission from Pullen, 1990.

(Ladle and Vaughan, 1988). Most plugs used in Europe are of American or Scandinavian manufacture, and among the more successful patterns are the Finnish 'Rapalas', which are also regularly used by some commercial bass fishermen.

Another very successful lure, which had its origins in Cornwall in the late 19th century, is the artificial sandeel. Though this is neither a spinner nor a plug, it is fished in a similar fashion. In its simplest form, it consists of a piece of rubber tubing, usually black, red or green, attached around the shank of a hook with 5–7 cm left trailing and mitred to represent a fish's tail. The first imitation sandeel manufactured commercially was the Mevagissey sandeel, available in blue and white or green and white. These have largely been superseded by the now-famous Redgill lure (Fig. 8.4), which is available in a wider range of sizes and colour patterns. Its attraction to bass is derived from the movement of the flexible tail, which wiggles from side to side as the sandeel is pulled through the water. The balance obtained by the hook protruding from the lure's body is important, as it must fish upright to appear natural.

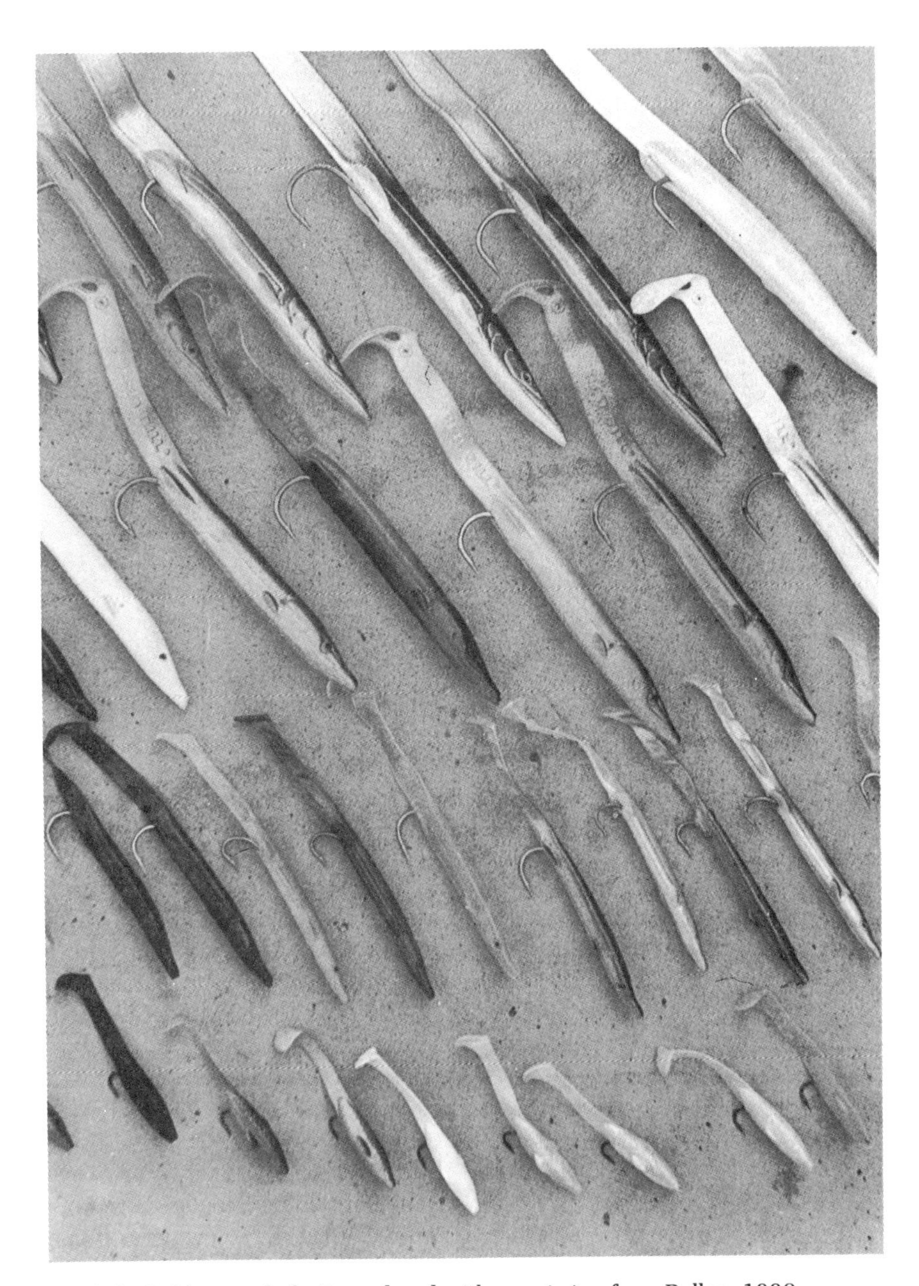

Fig. 8.4 Rubber sandeels. Reproduced with permission from Pullen, 1990.

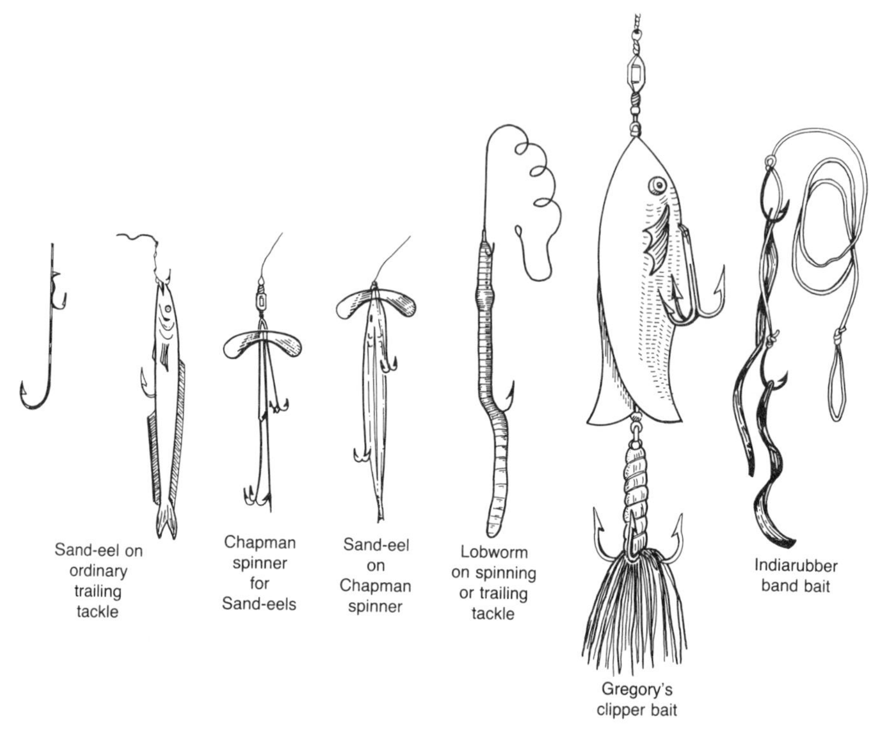

Fig. 8.5 Traditional bait flights and spinning baits, from Bickerdyke (1887).

Trailing and trolling

Trailing is an old term used to describe a method which involves towing spinners or dead fish mounted on bait flights (Fig. 8.5) behind a vessel under way (rowing, sail or motor), which is usually propelled across the tide rather than downtide as in drifting.

No casting is necessary; the lure is just lowered over the side of the boat and 30–40 m of line is paid out. The amount of lead added depends on the depth at which bass are feeding. A favourite trick is to bait the hook of a spinner with two or three worms or a strip of fish, when it is then known as a baited spoon.

Trolling for bass involves the use of heavy leads to fish lures at a fixed depth. These lures are usually Mevagissey sandeels or Redgills. Although regarded as a traditional method, trolling was actually developed much later than trailing, being similar to the boat fishing method used for salmon or trout in lakes. As this practice generally employs handlines for bass, rather than a rod and reel, it can be regarded as being a fully commercial (non-sport) method and is used in both England and France. A more complete description is given in Chapter 10.

Fly fishing

This method is now used for bass only by specialists, and we have no reason to add to Bickerdyke's description given on page 155.

8.4 BASS ANGLING AREAS

Angling for bass is popular in England, Wales, the Channel Isles, Ireland, France (especially Brittany), Spain, Portugal and Italy. Since the warmer spell at the end of the 1980s, anglers have started to catch them intentionally in Scotland and Holland (D. Lewis, pers. comm.).

England, Wales and Scotland

Figure 8.6 shows the summer and winter range of angling activity directed at bass around England and Wales.

Bass are also caught incidentally whilst fishing for other species, both inside and outside the limits shown. From time to time bass are deliberately sought on the Yorkshire coast in the Holderness area (BASS, 1990) and for many years have been taken in Luce Bay, in South-west

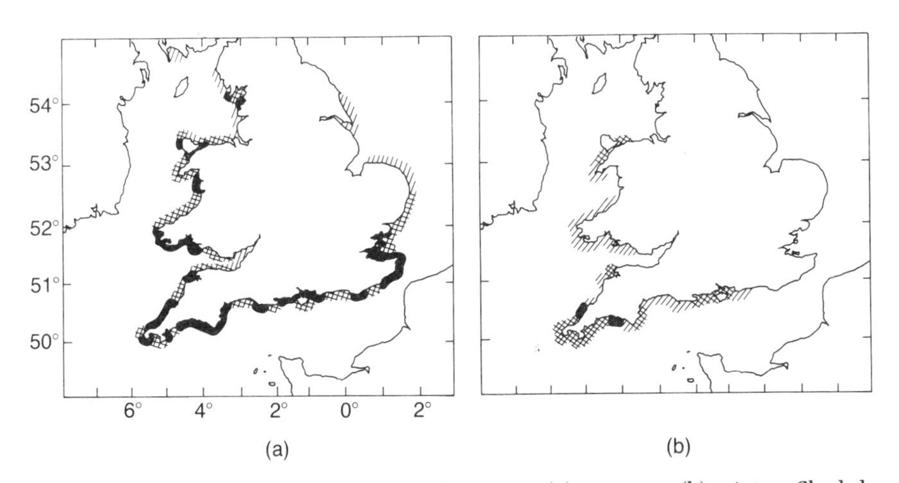

Fig. 8.6 Bass angling activity around the UK in (a) summer, (b) winter. Shaded areas show activity levels: hatched = low; cross hatched = moderate and blocked in = concentrated bass angling. Reproduced from Pawson *et al.*, 1987

Scotland (Niall, 1964). In warm summers, bass have been taken further to the north, even near Cape Wrath and in Dunnet Bay, though these fish must be regarded as being on the periphery of the species' normal range.

Bass can be caught, in season, from almost any part of the English and Welsh coasts, from Norfolk clockwise to Cumbria, but certain beaches, piers, estuaries and rock outcrops seem to produce most of the fish. Local conditions may change over the years, however, and a venue that was productive a few years ago may no longer provide good bass fishing. In the Introduction we mentioned some of the famous surf beaches of western England and Wales, and steeper, gravelly beaches, such as those on the North Suffolk coast, have produced several good bass in the 1980s. The gravel beaches of Kent and Sussex have a similar reputation and shore angling for bass in Dorset and Devon is still worthwhile. A bass angling festival is held each year on the shores of Colwyn Bay, North Wales.

Bass have a liking for the cover provided by pier pilings and jetties and, on otherwise featureless stretches of coast, these places have long been known as hotspots for bass. The most famous of these is the pier at Felixstowe, Suffolk, from which the one-time rod-caught record bass (18 lb 2 oz; 8.3 kg) was taken. Along with piers must be included stone breakwaters, and it was from Dover Harbour breakwater in September 1988 that the largest authenticated shore-caught UK bass was taken by Mr D. Bourne; it weighed 19 lb 0½ oz (8.6 kg).

Fishing off rocks or into deep gullies is practised mainly in South-west England and in Wales, although the chalky cliff bases and rock gullies near Beachy Head in Sussex have been fished successfully for bass for many years. Other well-known marks are the rocky ledges around the Isle of Purbeck and Portland Bill, in Dorset, Berry Head in Devon, around Land's End, Cornwall and on the Gower Penisula in South Wales.

Boat angling for bass tends to be much more localized than shore fishing, mainly owing to the need for good access from ports and safe sea conditions, but also because of the behaviour of the bass themselves. Generally, boat angling falls into two main types, (1) fishing known hotspots where bass congregate, such as around power station cooling-water outflows, reefs, wrecks, gullies and sand bars, or (2) searching for feeding shoals of bass, which are often betrayed by sea birds diving on their mutual prey, and then fishing with lures or live fish bait.

Channel Islands

There are many shore angling sites for bass around the rocky coasts of both Jersey and Guernsey. The l'Ancresse Bay area of the latter has recently received considerable attention from local and visiting anglers. Much of the boat fishing is carried out off the rocky reefs of Grouville Bay in the south-east corner of Jersey. The many tide races and overfalls around Alderney attract both charter and private angling boats, in addition to supporting a commercial line fishery for bass. A bass festival is held in Guernsey almost every year.

France

Although French sea anglers are not as numerous, nor as likely to be bass specialists as are their UK counterparts, bass angling from the shore is popular in many locations from the Cherbourg Peninsula to South Brittany. Anglers often congregate where good catches of small bass can be made on piers and jetties at the mouth of estuaries. Tackle and techniques are generally less refined than those used in the UK. Long rods, of 4–5 m, are standard and fixed-spool reels have replaced centre-pin reels. The tackle often consists of a basic ledger rig, baited with lugworm, though sandeels are also a popular bait in some places. Most of the boat angling for bass in France takes place around Brittany and trolling with artificial sandeel baits is widely practised. Where bass are feeding near the surface, some form of bubble float is often employed to assist casting. There is specialized hunting for large bass on the Mediterranean coast, particularly in the Languedoc region, and the use of fish live baits has produced some very big bass (> 9 kg) in this area in recent years.

Ireland

The best of the bass sport fishery in Ireland extends from the west of County Wexford around the southern coast to County Clare. Few bass are caught to the north of the River Boyne in the east, or north-east of the Moy Estuary in the north-west. Favourite angling spots are where bass shoal off rocky headlands and reefs and along open storm beaches, particularly in the west and south-west. The Splaugh Rock, near Rosslare, County Wexford is a well known bass hot-spot, and the beaches of Youghal Bay, County Cork, Dungarvan Bay, County Waterford and Rosscarberry, County Cork are all fished for bass. As in England and Wales, estuary mouths are popular bass-angling sites.

Other European countries

Portugal has a long-held reputation for fine sea angling and was given considerable publicity for angling holidays when Bernard Venables and others wrote articles following visits in the early 1960s (e.g. *Creel* Vol. 1, no. 7, 1964). It seems that most local fishing is for commercial purposes, but visiting anglers are well catered for in many areas. It is surprising that

Portugal is not more popular with bass anglers, given the reputed size and quantity of the bass to be caught there. According to a booklet produced for Portuguese Airways (TAP) in the 1960s, bass of 4–6 kg were commonly caught from the shore, and 8–11 kg fish were taken from boats! The surf beaches of northern Portugal at Viara do Costelo, Póvoa de Varzim, Matozinhos, Miramar and Figueiri da Foz were particularly recommended. In recent years directed angling for bass has become worthwhile in the Netherlands, particularly in and around the Sheldt Estuary.

8.5 CONCLUSION

Less than twenty years ago, a description of the fishery for bass around the British Isles would have dwelt almost entirely on that pursued by sport anglers. This is still the main picture for the majority of those with an interest in bass, and the species continues to be highly regarded as a sea-angler's quarry. If anything, the development of specialized tackle and tactics, by people who regard themselves as 'bass anglers', has advanced more in the last ten years than in any similar period previously. Aside from the technological progress in the materials used in tackle manufacture; carbon fibre, plastics, etc., there are two main reasons for this.

Continued exploitation has reduced the abundance of big bass available to inshore anglers, and this has encouraged experiment and innovation by those who wish to catch bass successfully. In addition, the high market price for bass has led some anglers to become commercial fishermen, or the latter to adopt rod-and-line tactics. In either case, there is a strong incentive to increase the effectiveness of their fishing effort, both for pleasure and for profit.

Angling is the only catching method which is considered to legitimately fulfil the requirements of fishing as a recreational pursuit in Britain and, even then, some rod-and-line techniques are held to be more 'sporting' than others. The commercial fisherman, however, inevitably looks to the methods which produce the greatest income for a given operational cost, and the next chapter describes the wide variety of gear which is used by the commercial fishery in the UK to exploit bass.

Chapter nine

The commercial fisheries (a description)

9.1 INTRODUCTION

The commercial fishery for bass has been most thoroughly investigated around the UK in recent years. There have been other intensive studies of small-boat fisheries, e.g. in southern France by Farrugio and Le Corre (1985a), but these tended to be multispecies in scope and were not aimed specifically at bass.

Prior to 1982, documented information on the inshore fisheries in England and Wales was rather patchy, compared with that for the middle- and distant-water fleets. This came about partly because the inshore fleets were made up of large numbers of small, widely distributed, and often unregistered vessels, which landed mainly to minor ports and harbours and onto beaches. The offshore fisheries, on the other hand, had attracted most attention because they traditionally accounted for the biggest part of the national landings, and were the subject of catch quota regulations under the EC's Common Fisheries Policy. For little-studied species, such as the bass and grey mullets, there was little available in the way of reliable catch statistics. By 1981/82, however, the increase in the relative importance of small-boat fisheries of England and Wales had been recognized, and a description of the coastal fisheries was compiled (Pawson and Benford, 1983), based on visits and interviews with fishery officers covering every port around the coast. This formed the basis for follow-up investigations aimed specifically at bass and mullet fisheries (Pickett, 1990), which provided much of the information available today. The following is largely descriptive, and details of fleet size, structure and an evaluation of effort are contained in Chapter 12.

9.2 FISHERY COMPONENTS

It is possible to identify two principal components in the commercial bass fishery: a directed fishery, and one where bass are taken as a bycatch. The distinction between these categories is not always clear cut, as many fishermen aim their efforts at an array of four to six species on any fishing trip, and it is difficult to quantify a bycatch as such. Bass are rarely exploited as the main target species throughout the whole year in any fishing area and, in some places, the bass season lasts for only 2–3 months before the boats turn to other species, often by using different fishing methods. Our definition of a bass fishery is therefore fairly broad, encompassing all who deliberately fish for bass or catch them incidentally.

The bass fishery also contains two more, usually distinguishable, elements. These are the inshore fishery and the offshore fishery. The inshore fishery comprises small boats which go out on daily trips and use a wide variety of fishing methods. This fishery persists throughout the year in some parts of the UK, but it is mainly a summer activity, owing to the inshore movement of bass in warm weather and to increased weather restrictions in winter. The offshore fishery is based mainly on midwater trawling, often by two boats fishing as a pair, and takes place over the cooler half of the year (November–May). Much of this fishing is now directed solely at bass, but in the past, black bream (*Spondiliosoma cantharis*) and the small pelagic species, such as mackerel, pilchard and scad, were the main target species, with bass a valuable bycatch. This same distinction can be made within the fleets of England, France, Spain and Portugal; all have inshore and offshore elements which are characterized by the size of vessels involved and the catching gear used. General descriptions of the various fishing gears used in the UK fishery are given below. Where there are differences in those used by boats of other nationalities, this will be indicated under the respective regional fisheries.

9.3 THE FISHERY IN ENGLAND AND WALES

Regions

For assessment purposes, the UK coastline has been divided into five regions, based on the appropriate sea areas delineated by ICES* divisions. Conveniently, these five divisions have been found to encompass the major differences in seasonality, exploitation patterns and stock characteristics of the bass fishery around the UK. Over 140 ports or other places at which bass are landed around England and Wales have been identified (see fleet census,

*International Council for the Exploration of the Sea.

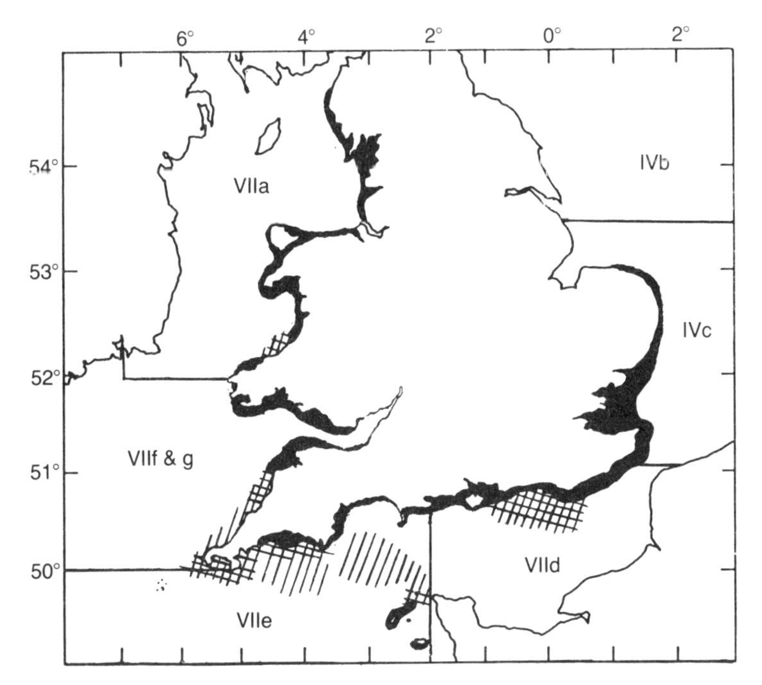

Fig. 9.1 Main commercial bass fisheries around England and Wales, summer and winter. Solid shading represents areas fished in summer, hatching represents areas fished in winter and cross-hatching represents areas fished in summer and winter.

Chapter 12) and the main fisheries in summer and winter are shown in Fig. 9.1.

Most of the fishing methods now used to catch bass are practised in a similar manner in all areas in which they are used, and it is appropriate to describe the main techniques before going on to detail the character of the fishery in each region.

Fishing methods

Bottom trawling – single boat

Directed trawling for bass from inshore boats has become profitable, following developments made in the early 1970s by some fishermen on the south coast of England, most notably by Mr Jack Pallot. The regional variations in trawl design are influenced largely by the nature of the sea-bed over which the gear has to be towed. The basic requirements are that the trawl should be light, its footrope should fish on the sea-bed, but not dig in, and its headline should be as high off the bottom as possible. Large trawls cannot be towed quickly enough to catch bass, and most measure 12–20 m along the footrope, compared with 30 m for those which are used to catch cod or plaice. It is helpful to have some form of mesh funnel or 'flopper' in front of the cod end, to prevent the bass swimming out after the tow is complete. Bass are also caught incidentally in other types of trawl, including heavier otter trawls being used for gadoids or flatfish, beam trawls and eel (*Anguilla anguilla*) pair trawls.

Pair trawling

A single bottom or midwater trawl towed between two boats can be used for a wide range of species, although bass have only been targeted in this way by UK vessels since about 1988. Pair trawling has several advantages over single-boat trawling, the most important being that much larger nets can be towed at a greater speed with two vessels, and there is no need for otter boards to keep the trawl spread. The net is kept open by the tension on two sets of long bridles, which transmit vibrations from the boats and are thought to herd the fish into the path of the oncoming trawl. This is a particular advantage in shallow water.

Pair trawling developed in Portugal, where large nets were used to catch pelagic species such as mackerel and scad, and bass were taken as a bycatch. The method was adapted by French skippers, working out of ports on the Biscay coast, in the autumn and winter for black bream and bass. In the early 1980s, the Lorient fleet found black bream scarce in the English Channel and began to concentrate on bass. Since that time, the French fleet has established a seasonal pattern of bass fishing, which begins in Biscay in the autumn and moves north, finishing at the end of March near the Channel Islands.

Gill nets

The term 'gill net' is often applied only to nets set on the sea-bed which catch the fish by trapping them in a single mesh around the body or behind the gills. Compared with other fishing gears, gill nets are highly selective of the size of fish which are caught. A fish with a maximum girth which is less than the mesh circumference is able to swim straight through the net, while a fish that is so large that it can barely get its head into a mesh, is unlikely to become firmly wedged and will escape (Fig. 9.2) (Potter and Pawson, 1992). Although fish may also become entangled by spines, teeth or other body projections, the selection range for bass is remarkably narrow (Reis and Pawson, 1992).

Up to the mid 1970s, opaque braided or plaited yarns were used in the construction of gill nets, originally of treated cotton and more latterly of

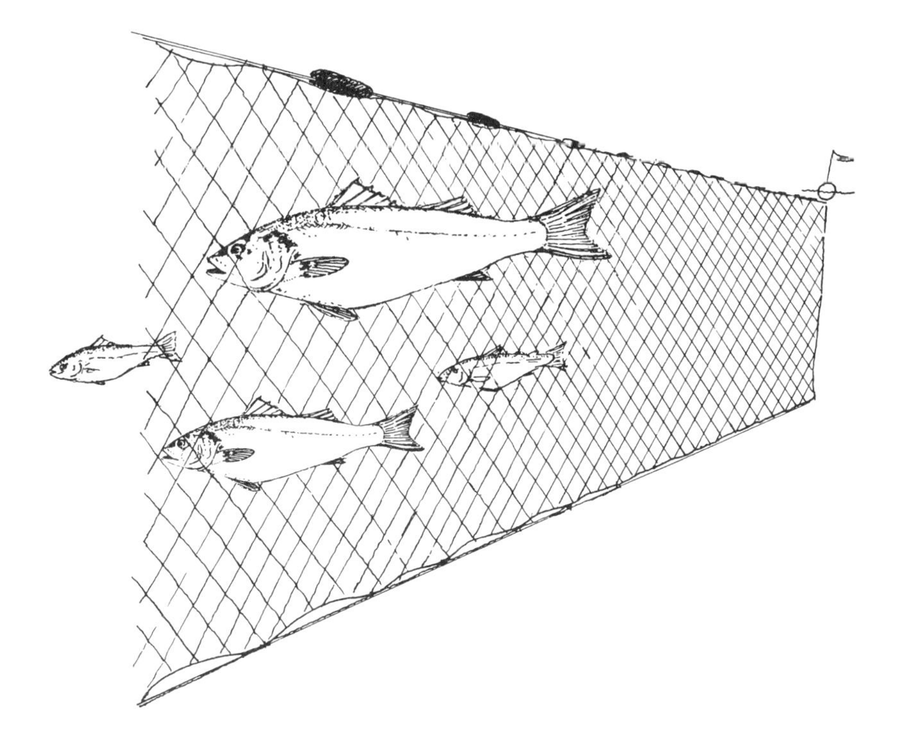

Fig. 9.2 Enmeshing action of a gill net. Reproduced from Potter and Pawson, 1992.

terylene. These had the disadvantage of being heavy, voluminous and visible to fish, and were therefore relatively ineffective in clear water during daylight. Most gill nets today are constructed of monofilament nylon, which has a low visibility in water. Its disadvantages are that it can be hard and stiff and can become brittle following exposure to sunlight. Recently, softer multifilament yarns have become available, which are more easily handled and less susceptible to fouling and being clogged by seaweed than monofilament, though they are more opaque. For greater strength, multi-mono, which consists of three to five braided strands of fine monofilament yarn, is often used. Compared with natural-fibre nets, modern synthetic nets are light and can be compacted into a small space for storage. They are also relatively cheap, an attribute which has attracted many people to this method of fishing.

Fixed gill nets are rigged so that, when the tide is not running, the net stands as a wall in the water, and they are usually set along the line of, or diagonally across, the tide to minimize their tendency to be flattened towards the sea-bed in strong tides. Individual lead weights or weighted-core ground-lines are used to prevent the fish from swimming under the net, and to

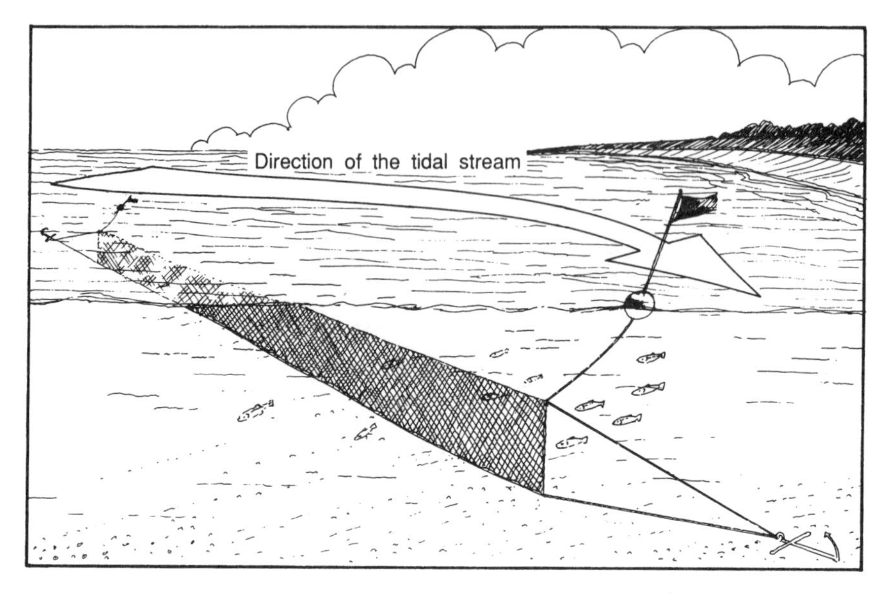

Fig. 9.3 Operation of a fixed net. Reproduced from Potter and Pawson, 1992.

counteract the buoyancy of the floats on the headline. An anchor or other weight is attached to each end of the net, and at intervals along the footrope if long lengths of net are being set. One or two marker buoys complete the rig (Fig. 9.3).

One of the main factors in determining the species and size of fish which will be caught in a gill net is its mesh size. In 1990 the first national legislation was introduced in the UK which restricts the size of mesh that can be used in gill nets and other enmeshing nets (Great Britain–Parliament, 1990), though the South Wales and the North Wales and North Western District Sea Fisheries Committees previously had bye-laws imposing minimum stretched mesh sizes of 100 mm and 89 mm, respectively. (Stretched mesh is the distance from knot to diagonally opposite knot when a single diamond of mesh is pulled tight, Fig. 9.4.) Further restrictions on the setting and use of fixed nets in tidal waters have been imposed in various areas, under the Salmon Act (1986) (Great Britain–Parliament, 1986).

Drift nets

Traditionally, drift net fisheries involved the use of single-walled gill nets made from natural fibres such as cotton or hemp to catch pelagic species, such as herring, mackerel, salmon and sea trout (*Salmo trutta*). In the late 1970s, some enterprising fishermen on the South-east coast of England

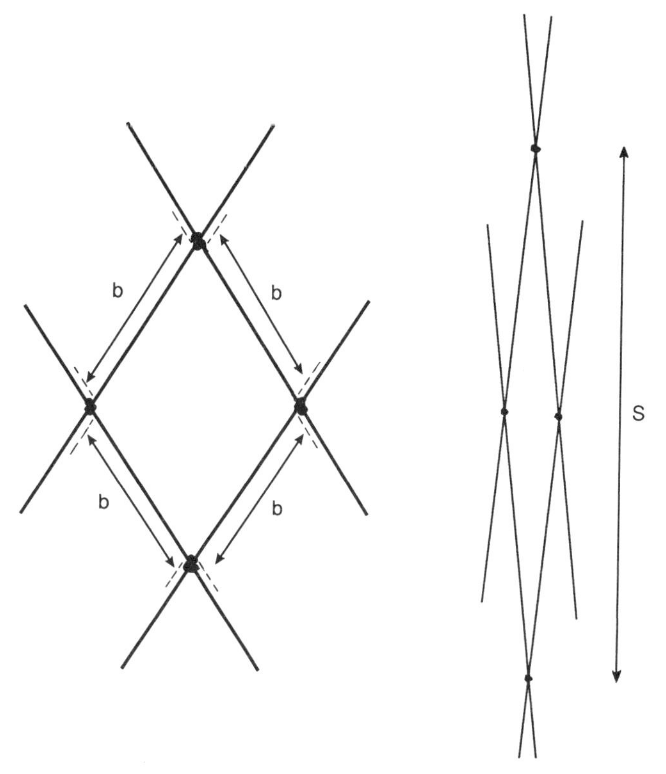

Fig. 9.4 Mesh size measurements. The length of material between knots is the bar length (b), but the stretched-mesh gap (S) between knot(s) is the dimension normally used in regulations. Reproduced from Potter and Pawson, 1992.

experimented with monofilament nylon netting with meshes of 80–100 mm (stretched mesh) which were drifted with the tide in shallow water. This was found to be an effective method for catching bass and grey mullet. Drift nets are marked with tall dahn buoys at each end, fitted with a light when fishing at night.

Drift netting for bass is normally practised over clean ground with an even depth (Fig. 9.5), as the aim is to have the footrope tripping along the bottom, and snags will cause the ends of the net to pull together. If possible, the net is set so as to cover the full depth of the water being fished, with the buoyant headline remaining on the surface to prevent fish swimming over the net, and its floats acting as indicators if the net becomes snagged. Shallow sandy areas, such as are found in the outer Thames Estuary, have produced the best results for drift netting, and it is perhaps no coincidence that fixed gear

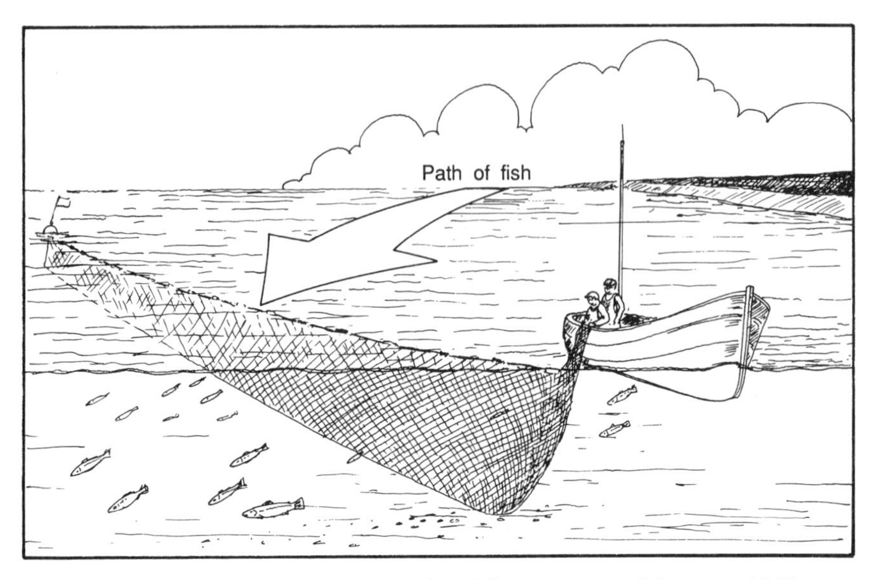

Fig. 9.5 Operation of a drift net. Reproduced from Potter and Pawson, 1992.

such as crab and lobster pots is unlikely to be encountered there. In the less cluttered areas off the Kent coast, some of the larger boats have shot fleets of up to 10 nets at a time, each of which may be 500 m long.

Seine and ring nets

Seining and ring netting involve encircling the fish with a curtain of net so that they cannot escape. In the past, fine meshes were used in these gears and the fish were simply retained within the wall of net. Today, the netting used for seining and ring netting is similar to that employed in fixed and drifted nets. It is important that the whole depth of the water column is fished, with the headline floating on the surface. In areas where it is allowed, seining with fine meshes, say 20–40 mm, is still practised for pelagic species such as mackerel and also for sandeels, bass and mullet. The advantage of the fine mesh is that if the fish are too large to become enmeshed, they lay on the meshes and are less damaged and easier to retrieve than gilled fish. The disadvantage, of course, is that the fine mesh may catch undersized bass or other species. Seining always takes place in shallow water, often in estuaries and creeks, and the net can be pulled out by hand or shot in a semicircle from a small boat, rowed or propelled with an outboard motor. Ropes from one end of the net are secured or held on the shore and the net is paid out around the fish, making sure that there is no room for them to escape at each end of the net. The ends of the headline and leadline are drawn onto

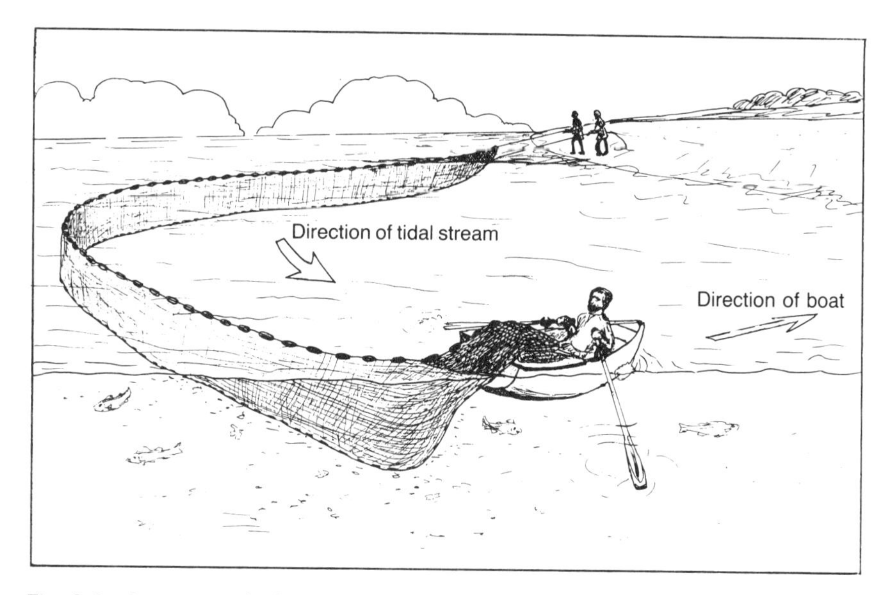

Fig. 9.6 Operation of a beach seine.

the shore, and by working from each end towards the middle, the bag of net and the fish contained within it are beached. Some seines are fitted with a cod end in the middle as this makes handling the catch easier. The method of working a beach seine is illustrated in Fig. 9.6.

Ring netting for bass differs from seining in that a shoal of fish is completely encircled within a wall of net, sometimes using two boats, and the net is not drawn ashore. The intention is to gill the fish, so they are usually scared into the net by rowing one boat into the middle of the net circle and banging the oars on the boat bottom or splashing the water. The net is then hauled into the boat from one end, and the gilled fish are extracted.

Trammel nets

Trammel nets are constructed of three walls of netting, and were originally designed to be fixed to the sea-bed. The two outer walls are called the armouring and are formed of large mesh, usually 25 cm or more from knot to knot. The inner wall, the 'lint', is of much smaller mesh and is set very slack. The method of capture by the inner wall is similar to that of a gill net, but fish too large to become enmeshed will push the loose netting through the opposite wall of the net and will become pocketed (Fig. 9.7).

Trammels are rarely set deliberately for bass. They are fished for sole (*Solea solea*) and cod, but a large number of bass do get taken by this method.

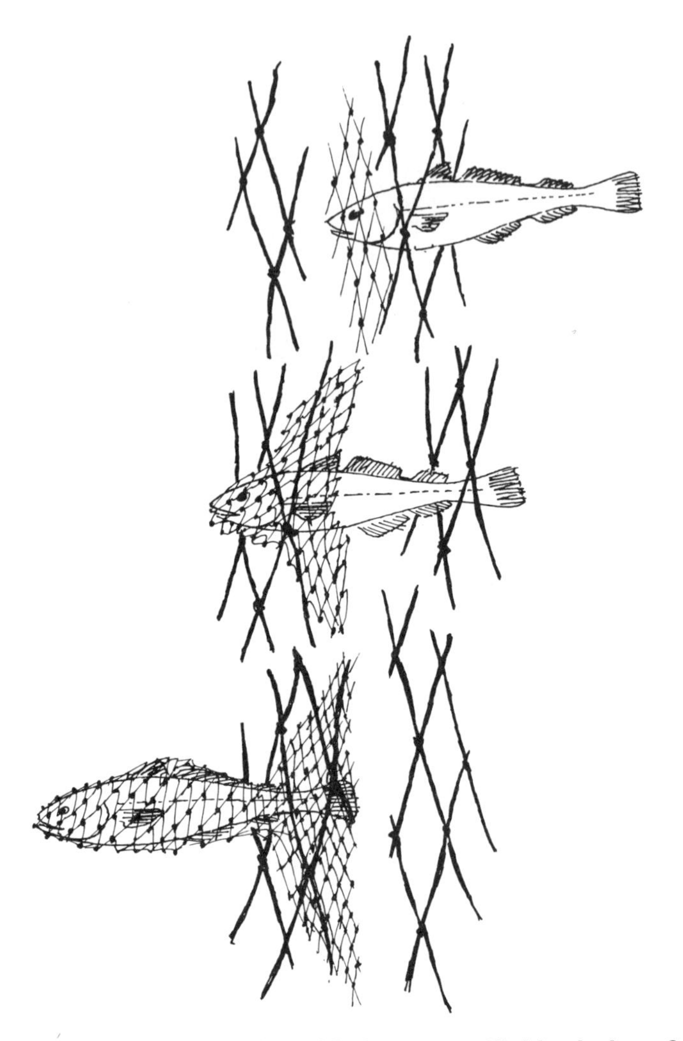

Fig. 9.7 Trammel net, showing how fish that are not filled by the loose fine mesh ('lint') may became pocketed by pushing it through the coarse mesh ('armouring'). Reproduced with permission from Millner, 1985.

Trammels are not as selective for fish size as are single-walled gill nets and produce a wider size range of fish in the catch.

Tangle nets

Tangle nets are large-meshed gill nets that are usually set in deep water for a variety of fish, particularly rays (*Raja* spp.), anglerfish (*Lophius* spp.), turbot

(*Scophthalmus maximus*) and crawfish (*Palinurus vulgaris*). They do not have floats on the headrope and are set very loosely, so that the net folds onto the sea-bed in layers, in which the fish becomes entangled. Occasionally, good catches of bass are taken in them when they are set inshore.

Longlines

Longlining is an age-old method of catching fish which, in recent years, has become widely adopted for catching bass, particularly in southern England and in Brittany. The principle is to use baited hooks on short 'snoods' attached at intervals to a stronger main line. Each line may have 10–100 or more snoods and each boat may fish 5–20 lines, which are hauled and shot up to three times a day. Longlines may be set on the bottom or drifted near the sea surface (Fig. 9.8).

Baits vary from region to region and with the season, but fishermen believe that the most effective are sandeel (live or dead), cuttlefish (*Sepia officinalis*), squid, whelk (*Buccinum undatum*), shore crab, worm (*Arenicola and Nereis* spp.) and fish pieces.

Because of the tension on the line whilst shooting and hauling the gear, and the presence of hundreds of sharp hooks, longlining can be a more than usually dangerous fishing occupation. Wooden boards with holes, or a modified plastic bucket with a perforated rim, may be used to secure baited and non-baited hooks on board the boat and to prevent tangles and snags. The lines are usually only fished for a few hours at a time, as crabs will soon devour any hook bait that is not taken by fish. On hauling, which can be done mechanically or by hand, a landing net is used to take the bass from the water, as they are usually still very lively and many are only lightly hooked in the mouth. Bass catches taken on longlines tend to contain a wider range of fish sizes (and often larger bass) than gill net catches taken on the same ground.

Handlines

Although the manner of hooking fish is similar with handlines and rod and line, the two are distinguished for assessment purposes. This is primarily because handlining is an entirely commercial method for bass, whereas rod and line is used both by recreational and by commercial fishermen. Handline methods have changed very little over the years and therefore provide the most useful basis upon which to evaluate trends in catch rates. Most other methods of catching bass have been developed or continue to be improved in efficiency with the recent expansion of the bass fishery.

Handlining methods include trolling (Chapter 8), when it is practised using hand winders and 'outriggers' or poles, jigging, in which feathered

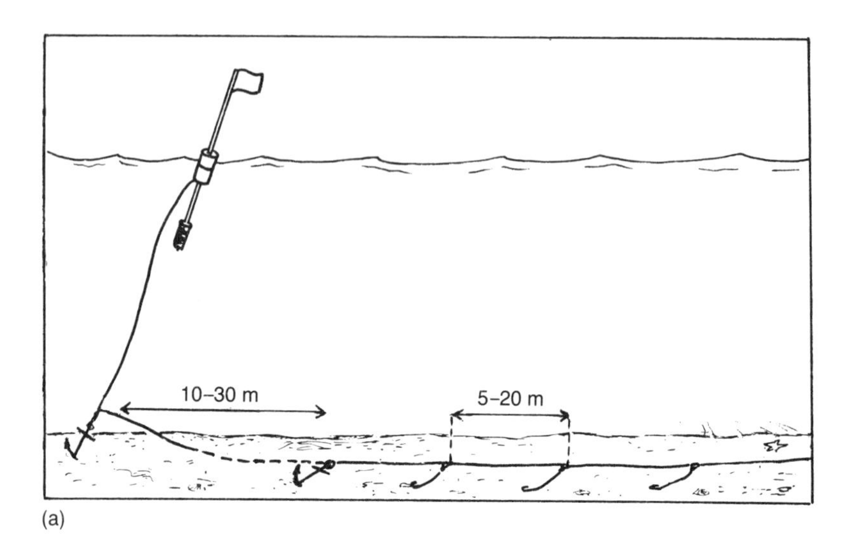

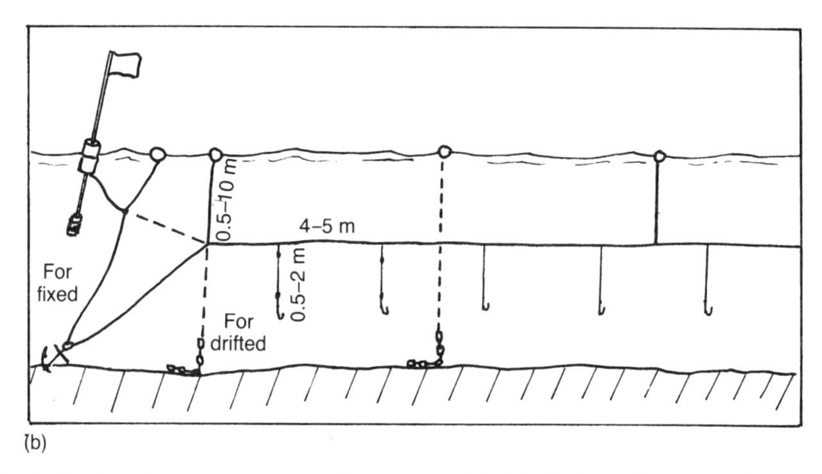

Fig. 9.8 Longlines, redrawn from Muyard (1978). (a) Method of setting on bottom. (b) Method of setting in mid-water or near the surface.

hooks or other lures on weighted lines are worked up and down in the water by hand, and mechanically operated lines which are wound onto large spools (or gurdy reels).

Trolling fisheries are in decline in the UK, and only in Cornwall is this traditional method still practised. In general, the method is less effective than gill netting, for example, and has consequently become less popular in

Brittany (Bertignac, 1986). The main difference between English and French trolling fisheries is that the English boats most commonly use three lines and the French boats four or five. Each boat is equipped with small manual winders to hold the lines, which may each be 100 m or more long and of about 10–15 kg b.s. monofilament nylon. The lures, which are usually artificial rubber sandeels, are attached singly to swivelled traces, which are 5–15 m long, according to tide and sea conditions, and of 7–9 kg b.s. nylon. The depth at which the lure fishes is controlled by the size of the lead weight clipped to a link swivel on the line above the trace swivel, and by varying the speed of the boat and the amount of line paid out. When a line is being fished over the stern of the boat it is made secure round a thwart, and the fisherman will often hold the line to feel for 'takes'. Other lines pass through a metal ring attached to the end of the outrigger pole by a piece of elastic or rubber shock cord, which stretches when a fish is hooked. The main line passes through another ring which is attached to a separate line. This is used to pull the main line within reach of the fisherman so that he may retrieve each lure or hooked fish. The outriggers are usually made of 4–5 m bamboo poles, which project at right angles to the vessel, lie parallel to

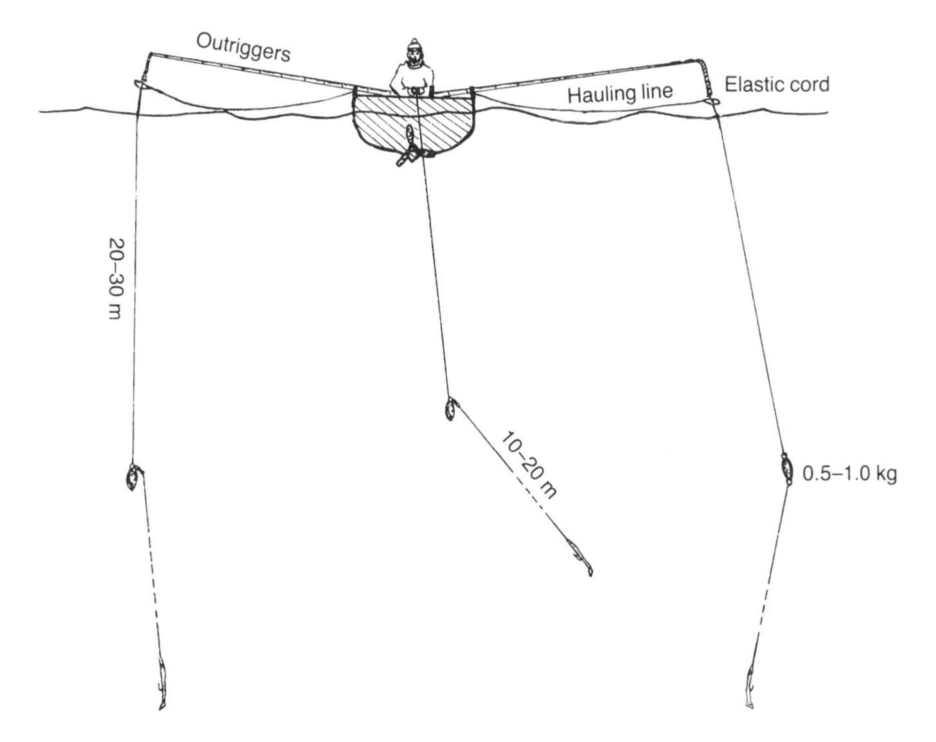

Fig. 9.9 Trolling (SW England).

Fig. 9.10 Commercial rod-and-line fishing for bass. Note the hand-line winders (gurdy reels) used in strong tides.

the water surface and keep the lines apart. A simple trolling set-up is shown in Fig. 9.9.

Rod and line

The techniques employed in commercial angling are often little different from those used in sport fishing. The principal exception is where deep or turbulent water is being fished, when the tackle used by commercial fishermen may be considered by anglers to be unsporting, i.e. a heavy glass fibre rod, large multiplying reel and 20–30 kg b.s. line. Such tackle is used in deep water drift-line fisheries, such as that at the Portland Race in Dorset, where strong tides necessitate large leads of 0.5–1.0 kg to get the bait down to the fish. The bait is usually a live sandeel or small mackerel impaled on a large treble hook mounted on a trace of 6–9 m of 15–20 kg b.s. nylon. In the early 1980s, bass at Portland Bill could be very large (up to 9 kg), and even on such heavy tackle were difficult to bring in. Sometimes it was necessary to resort to the use of hand line winders to bring the fish up to the surface (Fig. 9.10).

Other methods

Bass are often taken as a bycatch in eel fyke nets set in estuaries or the sea itself, and not infrequently turn up among lobsters or brown crabs (*Cancer pagurus*) in pots. Recently, they have been pursued for sport by divers using spear guns in marinas and around wrecks.

Description of the fishery in England and Wales by region

Southern North Sea – ICES divisions IV *b & c*

There is little directed commercial bass fishing north of Great Yarmouth on the east coast of England. Trawl-caught bass are occasionally landed into the Yorkshire ports of Scarborough and Whitby, and there is a seasonal gill net fishery in northern Norfolk, which targets sea trout and grey mullet, but also regularly takes bass. Southwards to Essex, bass are often caught as part of a mixed fishery in drift nets, fixed nets, trawls and longlines. However, bass may be specifically targeted when they are known to occur in worthwhile quantities. Commercial rod and lining takes place in the vicinity of Sizewell power station and around wrecks off Harwich. From Harwich south to Deal in Kent, bass fishing is conducted primarily in the Thames Estuary's complex of sandbanks and channels. The offshore banks, the Kentish Knock, the Gabbards and Galloper, have produced catches of large bass in drift nets and on rod and line. A few fishermen have made a speciality of fishing for bass over and around offshore wrecks, often necessitating journeys of 60 km or more. A plethora of areas closer inshore are fished commercially, including all the local estuaries. Much of the bass catch in the Thames Estuary is now the result of directed fishing during a season which normally lasts from May until October or November.

In the early 1980s, several boats from the East Kent area, particularly Ramsgate, started drift netting for bass in the outer Thames Estuary, off the Goodwin Sands and even offshore into the eastern English Channel. Catches were initially very large, but declined considerably by 1990. A number of other gill netting techniques are used by boats from the Mersea Island area of Essex and the East Kent ports. These include fixed netting, ring netting and trammel netting, which may be alternated seasonally according to which method is found to be most effective at the time. The Thames area is well known for blooms of algae which either clog the netting or, by adhering to the yarn, make it visible to fish, thus frustrating gill-net fishing. Blooms often occur early in the bass fishing season, and the sudden water clarity when they die back is also not conducive to good fishing. The Thames Estuary is also an important charter boat angling area for many species, and specialist bass trips are conducted from Bradwell, Southend, Canvey Island in Essex, the Medway Ports and Ramsgate and Deal in Kent.

Eastern English Channel – ICES division VII d

This region extends from Dungeness in Kent to Swanage in Dorset, and includes the Isle of Wight. The inshore bass fishery is largely a part-time activity to the east of Selsey Bill in Sussex, prosecuted from a large fleet of beach-launched day-boats, which fish trammel and gill nets for a mixture of species including sole, plaice, cod and grey mullet.

A commercial rod-and-line fishery has existed alongside sport anglers for many years off Beachy Head in Sussex, using the ports of Newhaven and Eastbourne. More recently, commercial rod and lining has begun at Brighton and off Selsey, concentrating on the Owers Bank and around the Nab Tower. Boats from as far afield as Littlehampton and Portsmouth fish these grounds in good weather in summer, using live sandeel and sometimes pouting or mackerel as bait for large bass.

In 1989 and 1990, a midwater trawl fishery for bass, black bream and other species developed in Sussex, based on three or four pairs of boats working mainly out of Shoreham and Newhaven. A few large catches of bass have been made, in addition to those taken by single boats trawling out of the same ports.

Longlining specifically for bass has been carried out regularly since about 1983. This followed the success of some Vietnamese refugees who, using monofilament lines and small hooks baited with pieces of squid or cuttlefish, landed many tonnes of bass at Littlehampton and Worthing. Several local boats from ports between Brighton and Lymington (Hampshire) have tried similar methods and, in summer 1987, up to 40 boats were longlining for bass in the Solent alone.

Much of the bass fishing between Selsey and Swanage used to be carried out from part-time vessels in the many natural harbours found in this area, although bass fishing is now restricted in these areas (Chapter 17). In the Solent, fixed gill netting and drift netting were favoured for catching bass in the early 1980s, but these methods had largely given way to longlining and commercial angling by 1990. In addition, several small trawlers have used locally designed, high-headline bottom trawls for bass fishing.

Commercial rod-and-line fishing takes place in the 'runs' or entrances to all the local harbours, and Poole Harbour has, in addition, a fleet of charter angling boats which takes out groups of anglers specifically to catch bass. Other charter vessels go further offshore to particular bass 'marks', including Christchurch Ledge and the Needles area to the west of the Isle of Wight. A small group of commercial rod and liners visit marks throughout Poole Bay, as far west as St Albans' Ledge, in a season lasting from May to November.

Western English Channel – ICES division VII e

This region encompasses most of the Dorset coast, South Devon and South Cornwall, from just west of Swanage to the Lizard. A diverse range of bass fisheries operate throughout the area. In the east, between Swanage and Portland Bill, offshore rod and lining predominates, the favoured marks being the Portland Race, Shambles Bank, Lulworth Banks and St Albans' Ledge. Most of the boats work from Weymouth or Portland Harbour, where some fixed gill netting takes place for both bass and grey mullet. Several angling charter boats operate from Weymouth, but most do not specialize in bass fishing.

There is little commercial fishing for bass between Portland Bill and the Exe Estuary in Devon, though there is some gill netting in the Beer area. The estuary mouths of the rivers Exe, Dart and Teign, in South Devon, all once supported prolific bass fisheries, with much of the effort being by part-timers, but these have declined in recent years. Gill netting there has tended to be a mixture of both legal and illegal (according to salmonid fishing legislation) fishing for bass, grey mullet, salmon and sea trout, though a recent tightening of regulations – both by the Salmon Act of 1986 and the 1990 bass conservation measures (Chapter 17) – has led to a reduction in netting in estuaries. Commercial rod and lining and angling charter trips for bass are popular, both often using live sandeels as bait from a boat drifting off the estuary mouths and headlands such as Berry Head and Start Point. Substantial landings of trawl-caught bass are sometimes made into Brixham and Plymouth harbours by both local and visiting (in winter) boats. Danish pair trawling teams and Scottish midwater trawlers, which are drawn to the area by the winter mackerel and pilchard fishing, have made large landings of bass at Plymouth from time to time, which, prior to the mid 1980s, were not recognized by these fishermen for the value they represented on Continental markets.

Some of the most famous bass fishing in Europe is near the Eddystone Reef and has involved boats working out of Plymouth. The original commercial bass fishery there used longlines, and this was followed in the 1960s by traditional trolling with 'Eddystone' rubber sandeels. For many years it was thought by the vast majority of anglers that these catches were recreational, but most of the bass were sold and the fishery was, in fact, a commercial activity. In the late 1970s and early 1980s, one particular fisherman developed gill-netting methods to catch bass at the Eddystone Reef. After much fishing with lures and detailed study of the area from the vantage point of the Eddystone lighthouse, he determined the timing and movements of the large bass shoals which visited the reef. He found that only a few, critical, days in the month were suitable for fishing and, weather permitting, he made some very large catches (Rivas, pers. comm.).

Anglers fishing the area became alarmed at the stories, many of them wildly exaggerated, which began to circulate about the size and frequency of these catches, and this rather too effective fisherman was forced to switch his operations elsewhere. There is still some commercial fishing at the Eddystone Reef, but little gill netting takes place there now.

In the early 1990s, South Cornwall was still home to a few trolling fleets, many of which used handlines for mackerel, but switched to bass when these were likely to be a paying proposition. Trollers operated from Looe, Fowey, Mevagissey, Porthoustock and Coverack, and offshore rocks and reefs, such as the Gwineas Rock and the Manacles, were among favourite areas. At that time, the use of live sandeels or mackerel had begun to displace trolled lures as bait for bass. Despite these attempts to increase efficiency, catches steadily declined, particularly in the Mevagissey area. In contrast – or maybe as a causal factor – it is off this stretch of coast that some very large catches have been made in winter by local and visiting trawlers that have been adapted for pair trawling for bass. Some of these boats are based at Looe and 1990 was their most successful season to date, with at least 50 tonnes of bass being taken from fish shoaling offshore prior to spawning.

There is relatively little gill netting for bass in this region, although some does takes place in the estuary of the River Fal, an area which includes the Carrick Roads, once well known for big bass. Some semi-commercial rod and lining takes place alongside the usual recreational fishing. With the exception of February and March, bass are caught here all year round, but today, the average size of bass in the Fal Estuary is small, and a large part of the upper reaches has been closed to bass fishing.

Included in this region are the Channel Isles, where there is a small commercial bass fishery. The largest catches have been taken in the Alderney 'race' area by local and Guernsey-based boats, using drifting longlines baited with live sandeels. A winter rod fishery also operates from Alderney, and commercial trolling and some gill netting takes place on the south coast of Jersey.

Celtic Sea – ICES divisions VII f & g

Vessels in the Newlyn/Penzance/Land's End area work in a similar fishery to the rest of the fleet in the western English Channel, but some larger gill netting vessels out of Newlyn travel round to the North Cornwall coast into St Ives Bay to catch bass in winter. The period that the fish run may be brief and, as the bass nets are set close inshore and the journey round Land's End to the fishing grounds is through an exposed and strongly tidal sea area, this is a hazardous business. For much of the year, these vessels are involved with fisheries in deeper water using gill nets and tangle nets for other species such as hake (*Merluccius merluccius*), spurdog (*Squalus acanthias*), turbot, ling (*Molva molva*) and anglerfish.

Newlyn bottom trawlers land bass as a bycatch and, when the mackerel fishery is operating, visiting purse seiners and midwater trawlers have also landed bass as bycatch, or as a targeted species if the opportunity arose. Bass landings of several tonnes at a time by midwater trawlers are a feature in most winters.

The small-boat fisheries for bass in this area generally employ rod and line, trolling and fixed gill netting in the bays. A trolling fleet fishes near Land's End and involved 12–16 boats in the 1980s, the largest remaining fishery of this type in the UK. Most of the boats are launched from isolated coves, particularly Penberth, with a few boats from Lamorna, Mousehole and Newlyn sometimes joining the fleet. The season usually lasts from October to January and relies, it is believed, on bass migrating round Land's End to overwintering/pre-spawning areas to the south of Cornwall. The fishery once centred on the Runnel-stone, one of the huge offshore rocks around which the bass congregate.

Most of these boats still fish handlines for mackerel during the summer months, and some may also catch a few large bass at this time. One or two vessels also use similar methods on the North Cornwall coast in St Ives Bay and around the Stones of Godrevy – an area of dangerous tides and eddies.

Occasional bass catches were still made near Land's End in the 1980s, using large seine nets fished as described earlier in this chapter, but these traditional fisheries have declined.

The bass fishery along the north coasts of Cornwall and Devon is sporadic nowadays, having once relied on a considerable run of large fish close to the coast during the autumn. Summer fishing in the estuaries (Hayle, Camel, Taw and Torridge) has been the fishery's mainstay in recent years. Netting here tends to be a part-time activity, concentrating on small bass and grey mullet, or sea trout and salmon when fishing with the appropriate licence. Most of the commercial rod-and-line fishing in North Devon takes place from ports in the Taw/Torridge joint estuary. Fishing marks may be within the lower estuary, just outside the bar, or further afield in the tide rips off Hartland Point. The latter venue is also well known to bass anglers and a few charter boats from Bideford, Appledore, Instow and Westward Ho! fish there along with the commercial and recreational boats.

In this area the main trawling activity is by boats from Padstow and Bideford, and a bycatch of bass is landed throughout the year, with occasional directed catches. In winter, vessels working tangle nets for turbot and rays in the Bristol Channel have occasionally landed good catches of bass into Bideford or Clovelly for transport overland to Brixham market.

Further east in the Bristol Channel, from Ilfracombe to Avonmouth, the high tidal range makes fishing difficult. Most fishing for bass is by small-scale gill netting, often using stake nets, or angling from charter boats which work from Ilfracombe, Minehead and Watchet. Some of these boats fish near Hinkley Point nuclear power station, where large bass are sometimes caught.

On the northern, Welsh, side of the Bristol Channel, few bass are caught by commercial fishermen east of Cardiff, though there are increasing numbers of part-time fishermen in the Barry, Porthcawl and Port Talbot areas and bass is one of their target species. Various netting methods are used, including drift netting, and there are several active angling charter boats. Some of the sand banks off the South Wales' coast, such as the Nash Bank and Helwick Bank, are also fished by the more committed drift netters from the Swansea area.

The coast from Swansea to Tenby, particularly along the Gower Peninsula and in Carmarthen Bay, is the main bass fishing area in South Wales. Until the mid 1980s, commercial fishing for bass in the area was underdeveloped compared with that in other regions, and most of the catch there was taken by rod and line. Since 1986, however, there has been large-scale investment in fast, dory-type gill netting boats, and for a while drift netting was very successful in the Gower–Burry Port area. Along with the introduction of modern netting methods came more specialized rod-and-line techniques, including that using bubble floats with small Redgill-type lures.

A small group of fast trawlers work specially designed trawl nets in Carmarthen Bay, sailing from Saundersfoot and Burry Port. These seem to have displaced the slower trawlers which used to work from Swansea and occasionally had good catches of bass whilst fishing for cod, flatfish and rays. Stake nets are still set for bass and grey mullet on the beaches of Carmarthen Bay, particularly within the Taf, Towy, Gwendraeth Estuary (the Three Rivers area) and within the Ministry of Defence firing range at Pendine Sands. This method was originally used for flatfish and employed cotton netting yarns. Modern monofilament nylon has helped to extend its use, which does not require a boat, to species such as bass. A fleet of small trolling and rod-and-line boats regularly works from Saundersfoot and Tenby, fishing rock and reef marks near to Caldey Island. A few charter vessels also run bass angling trips in this locality.

Further west, the only regular bass fishery is within the confines of Milford Haven, where gill netting and rod and lining have tended to concentrate around Pembroke power station's warm-water outfall, which attracts bass. Some of the larger boats from the Haven have recently taken to fishing for bass near the St Govans shoals, an area of tidal overfalls some 7 km offshore, where catches of large bass have been taken using rod-and-line-fished lures. Although traditionally a good shore-angling area, the coastal waters from Milford to Fishguard are quite dangerous for small boats and there is little commercial bass fishing activity. The main bass fishing season throughout this region is usually from May to November, though some fixed gill netting takes place in winter at New Quay.

Irish Sea – ICES division VII a

This region covers the West and North Wales coasts, Anglesey and the coasts of Cheshire, Lancashire, Cumbria and South-west Scotland.

The bass fishery in Cardigan Bay, from Fishguard northwards, is characterized by small boats and shore-set gill, trammel and stake nets, all operated mainly on a part-time basis. Catches taken by trawlers in this area have increasingly included bass, but owing to the lack of good ports and poor market facilities, they are landed mainly into Milford Haven. The area has several shallow, sandy estuaries in which the smaller bass predominate in summer, but recent National Rivers Authority (NRA) and national legislation, introduced to protect bass and migratory salmonids, has reduced fishing opportunities in these places.

The net fishery extends along the Lleyn Peninsula and around the Anglesey coast, but there are many more commercial rod and liners working this area than further south, as tidal conditions make netting difficult in many places. The Menai Strait, for example, has been a famous angling area for large bass for many years, and it is almost impossible to use any other catching method because of strong, turbulent currents, rocks and abundant seaweed. The broader areas of sand bank at each end of the Strait also attract bass shoals, and these are sought by netters, commercial rod and liners, boat anglers and several charter boats, which work from Caernarfon, Beaumaris (Anglesey) and Bangor. A particularly favoured angling station at the eastern end of the Menai Strait is Puffin Island, where large shoals of bass may sometimes be found.

From Bangor eastwards, along the North Wales coast and into the Dee Estuary, fixed netting has for many years been the main method of catching bass (along with grey mullet, flounders and the occasional sea trout), but effort has decreased under strict NRA and SFC bye-law controls. Some commercial rod-and-line fishing for bass takes place in the lower Conwy Estuary and off Great Orme's Head. In the Dee, some fishermen from the area between Connah's Quay and Deeside have used trammel nets for many years and, so as to avoid the restrictions of bye-laws prohibiting 'fixed engines', they have successfully adapted these nets for drift-net fishing for flounder, grey mullet and bass.

There is little bass fishing on the Cheshire and Lancashire coasts as far north as Fleetwood, just a few boats at Southport, Lytham and in the Ribble Estuary using fixed and drifting nets. Very few bass are landed into the major port of Fleetwood, probably because many of the vessels there are large trawlers, which fish for demersal species mainly outside the shallow inshore waters.

Northwards, from Morecambe Bay around the Cumbrian coast, bass are caught in fixed and drifted gill nets, stake nets and on fixed or 'band' lines (set by using bicycles or tractors to get out onto the sand banks and retreat before the rapidly moving tides), and rod and line. One important venue for small bass has been the warm-water outfall of Heysham power station, just to the south of Morecambe town, but this is now closed to boat fishing for bass between 1 June and 30 September each year.

There is often good bass fishing to be had in summer near Walney Island, to the west of Barrow-in-Furness, and although this originally developed as a sport fishery, local fish merchants now buy much of the catch. Bass are caught in set nets in the Duddon Estuary and occasionally at Whitehaven. To the north of here, however, most bass are taken as a bycatch in nets or traps set for flatfish or salmon throughout the Solway Firth.

Scotland has no commercial bass fishing, although salmon netsmen and sport anglers catch them from time to time, especially in the Luce Bay area of West Galloway.

9.4 OTHER EUROPEAN FISHERIES FOR BASS

The bass fishery in Ireland

In 1990, commercial fishing for bass in the Republic of Ireland was banned and even recreational bass fishing became severely restricted by the Department of the Marine (Chapter 17). A commercial bass fishery had developed in the 1960s (Fitzmaurice, 1978; Fahy, 1981) but, according to Fahy, it had declined in Wexford (South-east Ireland) by the mid 1970s; the amount of bass exported each year from this fishery had dropped from near 10 t in 1964 to 2 t in 1976. It is probable that the Irish commercial fishery for bass was so much reduced by 1990 that it was quite easy to close it down.

The commercial bass fishery, although widespread, occurred mainly in the south-eastern and, to a lesser extent, south-western parts of Ireland. Various rod-and-line and handline techniques accounted for most of the fish. Bass were also taken as bycatch in the salmon draft (beach seine) and drift net fisheries and in the grey mullet gill net fishery (Fahy, 1981). Stake nets and ring nets were also used to catch bass, which were often immature fish in the 30–40 cm range.

Bass were taken in estuaries such as Wexford Harbour (the estuary of the River Slaney) and also on the open coast. Occasional trawl catches of bass (always as a bycatch) have been landed into various ports, including a large catch taken between Rosslare and Wexford harbours in January 1953 (Fitzmaurice, 1978). Catches were made in all months of the year, but the peak season for bass was between May and October.

The bass fishery in France

France has the largest bass fishery in Europe and this operates in three distinct sea areas: English Channel, Atlantic and Mediterranean. In many respects, the character of these fisheries is similar to that found in Britain; it has inshore and offshore components which, respectively, employ many of the methods used to catch bass in the UK. We are unaware of any collective description of these fisheries, although studies of specific areas have been made on the French west and south coasts. An additional feature in France is the lagoon fisheries, which may be operated under natural or semi-cultivated conditions.

North Sea and English Channel along the North French coast – ICES divisions IV c and VII d & e

Bass are caught throughout the English Channel and in the southern North Sea by French boats operating from Dunkirk and Boulogne and using bottom trawls. These bass catches are sometimes made just outside the Kentish Knock and Goodwin Sand Banks. According to the Institut Francais de Recherche pour L'Exploitation de la Mer (IFREMER), little effort is directed at bass in the eastern sector of the Channel. There are, however, records of several big catches of bass made by vessels from Dieppe and Boulogne, trawling 20 km or more south-east of the Isle of Wight during the late 1980s.

There is much more effort by French boats on bass to the west of the Cherbourg Peninsula. This fishery concerns both inshore, artisanal, boats and offshore trawling. There are often three or more target species, of which the bass has become the most valuable. The main method in this area is handlining, which encompasses all the variants described for the English fishery on the opposite side of the Channel, including trolling. Most of the boats are small, between 4 and 10 m long, and carry one or two men, of which a high proportion work part-time only. Pollack, mackerel and ling are caught, as well as bass, in a fishery which takes place mainly in summer. At other times, these boats may either use fixed nets or set pots for crab and lobster. The main ports are Brest, Morlaix and Paimpol.

As in southern England, longlines are used to catch bass, mainly by boats fishing out of Brest and Paimpol. This tends to be the full-time activity of this fleet of small boats, which work mainly in spring and summer. The longlines used in this area each have 80–140 hooks and are set for 2–4 hours a day with each boat using up to four lines.

Fixed gill nets are becoming more widely used in France, particularly in the western Channel. In some areas, bass and sole are caught together in nets with meshes of 110–140 mm, and in others, bass and pollack. This is usually a summer fishery, exploited by boats working out of Paimpol, Granville, Brest, Morlaix, St Brieuc and St Malo. The average boat size is 10 m and this fleet has similar characteristics to the English inshore gill net fisheries in the Channel.

The largest catches of bass in the western English Channel, however, are taken with pair trawls. Few boats from local ports (Cherbourg and Audierne) participate in this fishery, but up to 30 vessels from ports in the Bay of Biscay visit the area. They fish mainly in ICES division VII e from February until April, and return to fish in Biscay during the rest of the year. Very large catches (for a bass fishery) of 20–30 tonnes are occasionally reported. This fishery is thought to have started towards the end of the 1970s, when vessels previously targeting black bream encountered more profitable quantities of bass. The local boats prosecute the fishery for around five months of the year, landing mainly to the market at Cherbourg. Boats from Scotland, England and Denmark have also used midwater trawling techniques for bass in this area.

Atlantic coast (Biscay) – ICES divisions VIII a & b

Although the bass fisheries in this region do not appear to have been described collectively, studies of individual fisheries have been reported (Audousset, 1978; Muyard, 1978; Le Masson, 1981; Bertignac, 1986). The lagoon fisheries of the Arcachon region have also been described by Barnabé (1990). Le Masson (1981) gave a brief description of the fleets operating from ports in southern Brittany in the late 1970s. Perhaps the most detailed, recent description is given by Bertignac (1986), but this covers only the Morbras region near Lorient.

As on the north coast of Brittany, handlining, particularly trolling, is still carried out, although, since the development of longlining and trawling in the early 1970s, catches by this method have declined and fishermen have converted to the more profitable methods. Records for the fishery at Etel go back to 1963 and show clearly that the trolling fishery continued to expand until 1973. The handline fishery also operates out of Audierne, le Guilvinec, St Guénolé, Loctudy, Lesconil, Concarneau, Etel and Auray. In the south of this region, around Pertuis Charentais, the spotted bass (*D. punctatus*) is taken alongside *D. labrax* in commercial catches (Muyard, 1978). Pair trawlers also operate from Loriént, La Turballe and Le Croisic, mainly in late summer and autumn.

Mediterranean coast

Bass fishing on the Mediterranean coast of France employs a large number of small boats, which operate on a wide range of species and can be found working on the open coast or in the saline lagoons around the Golfe du Lion. These fisheries have been described by Farrugio and Le Corre (1985a, b, 1986), and Barnabé (1976a) gave an account of bass fishing in the vicinity of Sète. Bass have been caught for food in the south of France since Roman times and, since the 19th Century, various methods including trolling, trawling and fixed netting have been used to capture them in the lagoons. In the winter, most bass are taken by handlines (and rod and line) over the rock outcrops or reefs on the open coasts, mainly at night by full-time fishermen. In summer, small bass are caught in the lagoons by fixed nets and specially adapted small trawls. The main landing places and markets are at Perpignan, Sète, Marseilles and Nice. Commercial landings of the spotted bass are also made here irregularly.

Spain

The Spanish fishery for bass is concentrated mainly on the Biscay coast of northern Spain (ICES division VIII c). As in France and the UK, the fleet seems to be divided into small boats which work inshore and in estuaries, and larger boats which work offshore. Many small boats (less than 10 m in length) work in the Santander and La Coruna areas, prosecuting a mixed fishery for bream, grey mullet, squid and shellfish. Longlines and fine-meshed gill nets, seines and ring nets are used, mainly in summer. The winter fishery accounts for catches of large bass taken on longlines and in midwater trawls.

Portugal

Portugal has a large fleet of midwater pair trawlers which often target bass. In some areas, bass fry are trapped in inlets or creeks, where they are netted and exported to Italy for ongrowing.

Italy

Although much of the bass produced in Italy is cultivated, normal net-fishing methods are often needed to recapture the fish. This is because much of the farming of bass is by extensive, *valli*-culture (Chapter 10), where large estuaries or bays are impounded to prevent the fish escaping, though the fish are not as readily accessible as they are in intensive cultivation.

9.5 SUMMARY

Although there are commercial landings of bass in Greece, Turkey, Israel, Egypt, Morocco and Tunisia, and of spotted bass in Mauritania, Senegal and Romania, there are no descriptions of these fisheries in the scientific literature. It is quite likely, however, that lines and gillnets are also used to take bass in these countries' waters, probably as a bycatch rather than in directed fisheries. The latter are predominantly a feature of small-boat, inshore fisheries in North-west Europe, in which bass are targeted at the appropriate season. To our knowledge, the French fleet of mid-water pair trawlers, which works in northern Biscay and the western English Channel, is the only example of large (>10 m) vessels which intentionally pursue bass every year.

This may, in part, be due to the market requirement for fresh, whole bass, which encourages daily landings and rapid marketing, and discounts at-sea processing, preservation and bulk sales. This quality requirement also puts a premium on bass caught in gears which result in less damage to the fish, i.e. lines rather than gill nets and both before trawls. The habits of bass also favour flexible, small-boat fleets, which can fish close inshore, in estuaries, around reefs and in tidal races, and where the ability to work with the tide or weather conditions, might be more important than sea worthiness or towing power. In general, bass will seldom be taken by any method in quantities which are sufficiently large or consistent to repay the sort of investment in gear and vessel capacity which beam trawling or purse seining require. Such fleets can earn a living on sole and plaice, or mackerel and herring, respectively, which could not be contemplated with landings of bass.

Chapter ten

Aquaculture

10.1 INTRODUCTION

The cultivation of bass was probably first practised by the Romans, who caught wild fry and reared them in ponds and tanks. Over the centuries, extensive lagoon systems have been used for ongrowing with a reasonable degree of success, particularly around the Mediterranean, but also on the Atlantic coast of France, between South Brittany and the Arcachon basin (Clement, 1990). In the last 20 years or so, the intensive cultivation of bass has developed rapidly, based mainly upon government-backed experimental systems. Today, economically viable commercial production has become a reality, and the early problems encountered with producing juvenile bass from eggs spawned in captivity have been largely overcome. As a consequence, farms can now guarantee regular supplies of high-quality size-selected fish at a stable price, a service which cannot be rivalled by wholesalers of wild bass, who have to rely on a seasonal and rather unpredictable fishery. Much of the knowledge gained on the aquaculture of bass (and other species) in recent years has been reviewed and summarized by Barnabé (1990), and we have limited the following account to a basic description of the ways in which bass are farmed.

Bass cultivation falls into four broad categories:

1. extensive – using the natural environment with an unenhanced food supply;
2. semi-extensive – at ambient temperatures, but in ponds or cages with a supplemented food supply;
3. intensive – in artificially warmed water, e.g. from power station cooling-water discharges, with a wholly manufactured food supply;
4. production of fry from hatcheries – for ongrowing in (2) and (3) above.

Wild-caught fry are also used in categories (1) to (3).

10.2 EXTENSIVE CULTIVATION

Local communities in France, Spain, Italy and Israel have practised extensive cultivation of bass throughout the 20th century. The origins of this form of farming go back many centuries, and often involved species other than bass, particularly grey mullet and sea-bream. The technique is similar in each country, although the scale may vary according to local geographical features. In essence, the *valli* (tidal lagoons or inlets in Italy) are naturally stocked with wild bass larvae and juveniles, which move in with the tide. Screens and traps at sluice gates prevent their escape, and the bass feed solely on wild food during the 2–3 year period before harvesting begins (Barnabé, 1990). The fish are then caught with normal fishing methods, such as seine nets, or are hand-netted following artificial lowering of the water level. *Valli*-culture in Italy is carried out in river basins such as the Po Delta and is on a large scale compared with the lagoons and flooded marsh pools and dykes used near Arcachon in France, or the *esteros* and *salinas* (fish ponds) near Cadiz in Spain.

In an attempt to increase production, fish farmers in Italy have often imported young bass from other areas. A fishery for bass fry and fingerlings developed in Portugal to supply this requirement and, in 1993, it was still exempt from the EC minimum landing size regulations. The success of these efforts has demonstrated the suitability of bass for more intensive culture.

10.3 SEMI-INTENSIVE CULTIVATION

Interest in intensive bass farming increased during the early 1970s following the successful spawning of captive fish (Barnabé and Rene, 1972; Barahona-Fernandes *et al.*, 1977). In addition, the increasing price of wild fish has encouraged considerable investment in bass farming.

Adult bass adjust readily to captivity and may spawn naturally. Alternatively, they may be induced to spawn by the injection of human gonadotrophin (Lumare and Villani, 1973), carp (*Cyprinus carpio*) pituitary extract (Barnabé, 1974) or luteinizing hormone-releasing hormone (Barnabé and Barnabé-Quet, 1985). Because ripe eggs are released and fertilized naturally by bass following hormone injection, artificial stripping (as with salmonids or turbot (Howell, 1979)) is unnecessary and egg viability is usually very good (Barnabé, 1990). The time of spawning may be shifted by photoperiod and temperature control (Bromage *et al.*, 1988; Carrillo *et al.*, 1989), but this method is not used commercially because bass farms may be able to obtain eggs spawned naturally during six months of the year.

Each female bass spawns approximately 0.3 million eggs per kg of body weight. The eggs are incubated in submerged net cones, and air or water is arranged to flow upwards from the base of the cone to ensure that eggs remain in suspension (Barnabé, 1990). The bass larvae hatch after 5–9 days, depending on water temperature, and are transferred to rearing tanks which have a continuous flow of fresh seawater. Light intensity is kept low at first, as strong light has been found to be lethal to young larvae in these circumstances (Barahona-Fernandes, 1979). As the larvae develop, light intensity is increased. Continuous lighting enhances survival and a 14–16 h light period each day promotes the fastest growth. Initially, the larvae are nourished by their yolksacs, but as the yolk becomes exhausted they are fed on microscopic live foods.

Cylindrical tanks with conical bases are favoured because they may be cleaned easily, and water quality is maintained by continuous flow-through which passes to waste through a large surface area of fine netting (Barnabé, 1990). It has been demonstrated that suspended unicellular algae can reduce the ammonia concentration in static systems when turbot are being reared (Alderson and Howell, 1973), but although 'green water' systems, which contain high densities of algae, have been used experimentally for rearing bass (Barnabé, 1974), they have not yet been used commercially.

The rotifer, *Branchionus plicatilis*, has been widely used for the initial feeding of bass larvae (Barahona-Fernandes, 1978; Jennings, 1990), although it is also possible to feed them with 50–160 μm commercial pellets (Barnabé, 1976b) or homogenized natural food (Lumare and Villani, 1973). Subsequently, larger larvae are fed with brine shrimp nauplii, (Bedier, 1981), pelleted food of 160–250 μm particle size (Barnabé, 1976b) or plankton collected from the wild (Kentouri, 1980).

Cultured live foods commonly contain relatively low levels of those highly unsaturated fatty acids (HUFAs) (Watanabe, 1982) which have been shown to be essential to the health of many fish species (Bell *et al.*, 1986; Bolla, 1989), and feeding solely with rotifers and brine shrimp has created nutritional problems in many fish species. Despite their potentially poor nutritive value, rotifers (Howell, 1973; Persoone *et al.*, 1980) are easily cultured at high densities, whereas many problems still remain with culturing the nutritionally preferable copepods (Ohno and Okamura, 1988). Katavic (1986) noted the accumulation of uneaten, starved brine shrimp during bass culture, and observed that the nutritional value of these brine shrimps was low. The HUFA content of brine shrimp is related to their own diet (Katavic *et al.*, 1985) and can be improved by feeding them with marine algae (which synthesize HUFAs) and microencapsulated diets containing the essential HUFAs (Gatesoupe and Robin, 1982). There is little specific information on the HUFA requirements of bass, and rearing techniques have evolved in practice and without a sound knowledge of the biology of bass nutrition.

Bass larvae are weaned from live foods to more cost-effective pelleted foods from 30 days onwards (Barnabé, 1990). A delay in growth may be observed at this time, which does not occur if feeding continues with live food. Alkane yeast and other plant proteins may be substituted for animal protein without affecting growth, conversion efficiency or protein digestibility in bass (Alliot *et al.*, 1979), and growth may be improved by addition of L-Cartinine (Santulli, 1985) or certain hormones to the diet. There is an endogenous annual cycle of changes in growth, feeding rate and conversion efficiency (Zanuy and Carrillo, 1985), and growth of bass has been found to cease entirely below 10 °C (Tesseyre, 1979). This is why bass farming is primarily confined to warmer waters in the Mediterranean region and there has been no commercial development of bass farming in Britain.

Ongrowing

The young bass are usually transferred to sea cages at a length of 3–4 cm, though they grow successfully in a wide range of salinity and temperature regimes (Dendrinos and Thorpe, 1985) and ongrowing may even be conducted in fresh water with natural zooplankton as food (Chervinski and Lahav, 1979). Satisfactory growth is achieved at high salinities (30‰ to 37‰) and declines at lower salinity, though at the extremes of a temperature range of 15–22 °C, Alliot *et al.* (1983) also recorded rapid growth at 6‰ salinity, whilst Barnabé (1990) reports optimum salinities of 10‰ to 30‰. Factors such as the stocking density of fish and their diet may interact to influence the apparent optimal temperature and salinity for growth, although these have not been studied in bass. When the water temperature is 22–25 °C, well-fed bass take less than a year to reach the minimum acceptable marketing size of 250 g.

10.4 AQUACULTURE IN HEATED WATER (INTENSIVE CULTIVATION)

Since the early 1970s, research and development of marine fish culture has been carried out using the heated water from power station cooling-plant effluents. A pilot farm at the nuclear power station at Braud St Louis, on the Gironde Estuary, has been used to assess bass cultivation (along with sturgeon, *Acipenser baeri*), but the only significant quantities of warmed-water bass to appear on the market have come from an experimental farming system at Gravelines in Normandy. Market-size bass and fry have also been produced experimentally in small quantities at Hunterston power station in Scotland, though not, apparently, with sufficient commercial potential.

At Gravelines in 1982, a syndicate of aquaculturists began to develop a pilot farm for intensive culture of bass and sea bream. Six 900 MW reactors produced cooling water heated to 15–32 °C, which was used in tanks of 30 m^3 and 65 m^3 (for different sizes of fish). This system has proved suitable for bass and bream, which were purchased as 2 g fry from hatcheries and grown to a marketable size of 300 g. This was achieved in 14–16 months, on a diet of dry pellets and a mixture of wet fish and meal. Fifty tonnes of bass were produced in 1987, and it has been estimated that an annual output of 80 tonnes for the unit would be profitable. The Gravelines site could produce not only fish for direct sale to markets, but also juveniles for ongrowing (Brunel and Fuzeau, 1990).

Chapter eleven

Marketing, value and production

11.1 INTRODUCTION

Much of our knowledge of the economics of the bass market in Europe is derived from two sources: the proceedings of a workshop on the markets for the main cultivated species in the Mediterranean (bass, grey mullet, bream and eels), held in Italy in 1986 (New *et al.*, 1987), and a study commissioned (by MAFF) on the economics of the UK bass fishery (Dunn *et al.*, 1989). In addition, some qualitative information has been acquired during the course of DFR's studies, and monthly prices are recorded by MAFF's Sea Fisheries Inspectorate (SFI). The fishing trade press also publishes the weekly bass prices quoted at the main markets. Although MAFF produces annual statistical tables (published through HMSO) of the quantity and values of sea fish landed in the UK, these do not include bass.

There are several reasons why a knowledge of marketing and economic evaluations are important in the management of a fishery:

1. When introducing measures to protect a fish stock, some catch or effort limitation may be required which will affect people's livelihoods. It is useful to know how important the exploitation of that particular stock is to (a) the participants, and (b) the local economy, and what will be the financial impact or burden of changes in the fishery owing to the introduction of management measures.
2. With a high-value species such as the bass, it is important to know whether a reduction or increase in the supply of bass (in the short term and the long term) will produce a rise or fall in market price, how this will affect demand, and what any consequent economic impact is likely to be. Will cultivation ultimately reduce pressure on the wild stock as it appears to have done with European salmon fisheries?

3. Historical price trends and their relationship to various indices, such as the supply of and demand for fish, can be used to understand how the market operates, and hence to predict whether it can be sustained or be developed further. Such analyses must include other influences on the demand for bass, such as the general level of income, the supply and acceptability of alternative fish species, eating habits, etc.
4. In terms of the national economy, and therefore the resource's relative importance, we need to have some idea of its earnings potential as an export commodity.

The relative importance of various species or particular fisheries was at one time based solely on 'first sale values', i.e. the total annual payment to the fishermen for their catch. Not only did this ignore the costs of making these landings, but first sale values alone give little indication of the true economic importance of a fishery. This is particularly true with fish such as bass and salmon, where lucrative recreational fisheries also depend on the welfare of the resource. Similarly, reliance on the weight of national landings as an indicator of relative importance will place the bass low on the ranking list compared with lower-priced but much more abundant pelagic species; mackerel, for example. Total value (weight × price) gives a more meaningful economic ranking, and bass emerges as being much more significant, owing to its high unit price.

Though the commercial bass fishery in the UK was insignificant in all respects in the early 1970s, compared with those for many other marine species, recreational anglers were already lobbying the Government for more research on the species. The expansion of research on bass took place in the early 1980s in parallel with an expansion of the commercial fishery, and has probably only been sustained because of the latter's economic value and consequent political need for scientific information and advice for management.

11.2 MARKETING IN EUROPE

France, Italy, Spain, UK, Portugal, Greece and Turkey are all producers of bass, both from wild fisheries and through various forms of cultivation. Much of the production in excess of home demand in the UK and France now goes to Italy, which imports around half of its domestic bass consumption all the year round (New *et al.*, 1987). The UK also exports to France, Belgium and Spain, mainly during the summer, whilst France imports only a small proportion of its total consumption of bass. In the winter, the UK now imports bass from France and other countries such as Egypt and Senegal, to satisfy an increased home demand. Holland imports bass and then exports some of this fish to other markets, including the USA.

Greece and Turkey have also been unable to satisfy their home markets with bass, and have begun to develop intensive cultivation of the species, as well as expanding their wild fisheries. Turkey, in particular, had little internal demand for bass until the late 1980s, and consequently there had been little fishing for this species. Throughout Europe, the demand for bass is expected to continue to increase (New *et al.*, 1987). All this has come about because of the wide geography of demand in Europe, and complex marketing networks have been developed which can use many alternative sources of supply. Bass are transported by road, sea and air from coastal auctions or specialist buyers to the large inland markets in London, Paris, Brussels, Madrid and Rome. Some buyers sell direct to restaurants, hotels or large food consortia. Local markets are sustained from the same sources, or through a vigorous 'back door' trade, which may involve part-time and unregistered fishermen or recreational anglers who sell their catch directly or through local dealers, who often operate from mobile premises. Bass landed under the legal minimum size might be disposed of in this way to satisfy the demand for small bass by the restaurants that serve whole, plate-sized fish (G.D. Pickett, pers. obs.).

In the UK, bass are offered on the menu in all manner of restaurants, hotels and bars, especially at large resorts or sea-side towns. The Spanish tapas bars, which in the Bilbao–Santander area often serve bass (along with traditional dishes of hake (*Merlucius merlucius*), sea bream, squid, cod and shellfish) are now being copied elsewhere in Europe. The bass is in demand everywhere.

11.3 PRICES

The price paid to the fisherman for fresh bass varies according to where it was caught and to the strength of demand at any particular time, although it is always near the top of the unit price range for fish. Bass is one of the few fish where the price paid is for the whole fish. We have not heard of dressed or filleted bass being offered for sale, other than in restaurants. The main requirement of the high-value markets is freshness. Fish with a high fat content rapidly deteriorate once the body is cut open, and it is easier to tell if a bass is really fresh if it is ungutted. There have been several instances where not-so-fresh bass of UK origin were rejected by French and Italian markets. Additionally, many Mediterranean recipes require the presentation of the whole fish and, to this end, gutting is achieved by cutting through one side of the fish. High wholesale prices are only available if the fish are packed in ice, lying on their backs to prevent squashing internal organs, usually in insulated polystyrene boxes, and dispatched quickly.

In France and Italy, line-caught bass are generally found to be in better condition than trawled or gill netted fish, and as a consequence the former attract a higher price. Cultured bass are often sold as being line-caught and have a similar value. The following list gives a comparison of the international variation of mean unit prices of bass in 1985/86 (from New *et al.*, 1987).

France: Ff 65–80 per kg (US$9.90–12.20)
Greece: Gr.Dr. 1600 per kg (US$11.40) (as sold to supermarkets and restaurants)
Italy (Venice fish market) Lit. 25000 per kg (US$17.90)
Spain Ptd. 1176 per kg (US$8.70)
Tunisia (local) US$6.30–10.00 per kg, (export) US$8.00–9.50 per kg

In France, the first sale price of bass rose steadily between 1972 and 1984, from 30 Ff per kg to 50 Ff per kg (Bertignac, 1987). By August 1988 bass had reached 180 Ff per kg in Bordeaux (Jennings, 1990).

UK fish prices are usually quoted as first-sale values. Not only have those for bass risen over the last 15 years, but the fisherman's share of the wholesale, selling-on price has also increased (Dunn *et al.*, 1989). This means that the profit margin for the first buyer (the marketing margin) has been reduced. The unit price trends of bass at first sale in the UK between 1976 and 1987, relative to the basic rate of inflation, are shown in Fig. 11.1. It is apparent that the value of bass as a commodity rose steadily during this period.

A study of the economics of the UK bass fishery concluded that, in the UK at least, prices tend to be a function more of the standard of living, and thereby consumer incomes and appreciation, than of the supply of bass available to the market (Dunn *et al.*, 1989). The market can therefore be said to be demand-led. A 10% increase in landings, for example, may only depress the unit price by 1% or less, whereas a 10% increase in consumer incomes can be expected to induce a rise of over 22% in bass prices.

The local variability of price depends greatly on the level of competition between buyers. In the past, and in areas away from the main distribution networks, a buyer who had little competition may have paid fishermen only 50–70% of the national average price for bass. As more specialist buyers came into the market in the late 1980s, fishermen have shopped around for better deals. Some will now transport their fish up to 300 km overland, if there is sufficient financial incentive, rather than sell to their local markets. The market at Newlyn, in Cornwall, is generally associated with high prices for bass, and during winter scarcities in 1989 and 1990

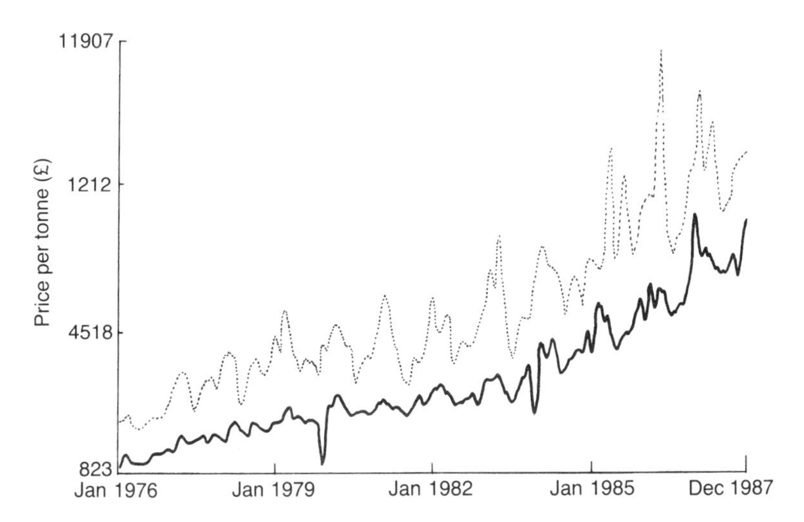

Fig. 11.1 Price trends of bass, showing marketing margin, in the UK 1976–1987. Dotted line indicates relative wholesale prices and solid line indicates quayside prices. Redrawn from Dunn *et al.*, 1989.

some catches made £22 per kg (Jennings, 1990). During 1990, prices averaged £9–12 per kg, well above the £5 per kg reported in 1987 (Pawson and Pickett, 1987).

Nevertheless, sudden local gluts of bass landings may cause markets or merchants to lower their prices. In December 1990, for example, several landings of trawl-caught bass, including one single catch of 10 tonnes, were made at Looe, in Cornwall. Prices fell from around £13 per kg to £7 per kg (Lockley, 1991). Such differences in price may be due to the fishes' condition or the size of fish currently in demand, but they often reflect local demand, cash flow or transportation limitations which arise when unexpectedly large quantities of bass arrive on the market. However, once on the European continent, there are usually no problems in finding buyers for such quantities, and bass prices there are remarkably stable.

The 1987/88 study by CEMARE correctly predicted that prices would continue to rise along with improving economic conditions in the UK. It will be interesting to see whether the economic recession in Britain (at the time of writing, 1992–93) will reduce home demand, and whether bass disappears, albeit briefly, from some menus because the European market is still sufficiently strong to make it too expensive in the UK.

11.4 RELATIVE IMPORTANCE

Bass was considered to be a minor commercial species in Europe in the early 1970s and, even in France, it ranked seventeenth in terms of first sale value in 1974 (Le Masson, 1981). It had entered the top ten most valuable species in France by 1980, behind cod, hake, tuna (*Thunnus alalunga*), saithe (*Pollachius virens*), anglerfish, whiting, sole and gilthead bream (*Sparus auratus*). Bertignac (1987) records bass as being the eighth most valuable species in 1985. In the UK, bass landings statistics are not published, but catch estimates (Pickett, 1990) show that it first entered the top ten there in 1986.

A comparison of the unit prices of the main commercial species landed in the UK (Table 11.1) shows that those for bass come within the range of, or even exceed, those of other prime fish. This list is based on monthly figures supplied for the main UK markets during 1990 (source, MAFF SFI data).

Although there are no official UK statistics for total catch and value of bass landings, some data are collected by the MAFF's SFI. To obtain total values, we have multiplied the national average unit values for each year by DFR's estimates of total catch, obtained chiefly through a fishermen's logbook scheme (Chapter 12). At the major ports, where large, middle-distance trawlers land, the bass catch is an insignificant part of the earnings by these fleets, which depend more on large tonnages of the main commercial species such as cod, sole, plaice and mackerel. Because of the high price paid for bass and the displacement of fishing effort from other overexploited or quota-restricted stocks, there is a trend towards more directed exploitation of bass by some of these boats, in areas where bass are available.

Table 11.1 Unit price ranges of bass and other prime species in 1990 (source, MAFF, Sea Fisheries Inspectorate)

Species	*Price per kg (mean monthly price range)*
Bass	£6.81–8.70
Black bream	£0.94–4.25
Turbot	£5.40–7.46
Salmon (wild Scottish)	£3.00–9.00*
Red mullet	£1.17–8.30
Lobsters	£6.70–11.86

*I. Russell, pers. comm.

11.5 PRODUCTION

Accurate figures of the production of bass by any country are impossible to obtain, owing to difficulties in compiling statistics for a fish which is sold through a wide variety of outlets. Some countries, such as France, do provide statistics on bass which are useful indicators of trends, although they almost certainly under-estimate total production. The annual national production figures for those countries for which data are available are summarized in Table 11.2. These figures also include landings of spotted bass, which is generally not distinguished from sea bass in statistical tables, and indicate the currently low production of striped bass.

It is interesting to note that the total recorded Mediterranean and Atlantic/North Sea productions of bass, as recorded by the Food and

Table 11.2 Reported world production (in tonnes) of sea basses, 1984–90 (source, FAO yearbooks, 1988–90)

Species and country	1984	1985	1986	1987	1988	1989	1990
Sea bass *Dicentrarchus labrax*							
France	3667	3116	3481	3867	3066	3181	–
Ireland	–	–	–	4	4	1	0
Netherlands	–	–	–	–	–	2	2
Portugal	427	310	462	325	167	6	93
Spain	430	364	388	402	402	451	451
Channel Islands	7	18	15	14	29	48	50
England and Wales	145	106	129	128	171	189	1
Species total	4676	3914	4475	4740	3999	3878	3778
Sea basses *Dicentrarchus* spp.							
France	551	765	635	751	755	752	752
Italy	1545	2815	2427	2570	1997	1614	1540
Spain	–	–	8	–	–	–	–
Tunisia	398	329	344	341	336	459	442
Turkey	317	653	814	699	1339	1727	1246
Yugoslavia	95	22	187	51	76	239	208
Greece	101	162	254	177	176	458	440
Mauritania	–	347	457	582	427	430	420
Species total	3007	5093	5226	5166	4457	5679	5048
Striped bass *Morone saxatilis*							
Canada	20	21	4	–	–	–	–
USA	1224	545	152	196	184	129	478
Portugal	–	–	–	–	–	5	–
Species total	1244	566	156	196	184	134	478

Table 11.3 Total catch and aquaculture production (estimated) of bass and sea breams in the Mediterranean (t $year^{-1}$) (source, New *et al.*, 1987)

Country	*Total catch*	*Aquaculture*	*Cultured (%)*	*Year of estimate**
Cyprus	0.64	–	0	1984
France	5000	175	4	1986/85
Greece	515	280	54	1984
Italy	4100	2700	66	1985
Portugal	750	252	34	1985/86
Spain	1400	45	3	1985
Tunisia	740	140	19	1986
Turkey	2263	835	37	1985/82

*Where two dates are given, the first refers to total catch data and the second to aquaculture data. Countries where only incomplete estimates are available have been omitted.

Agriculture Organisation (FAO, 1992), are similar at around 4000–5000 t in most years.

An impression of the relative production of wild-caught and cultivated bass in the Mediterranean in the mid 1980s may be gained from Table 11.3, which shows figures for bass and sea bream combined (after New *et al.*, 1987).

Sea bream generally make up less than 30% of these values, and a rough estimate for the total European/Mediterranean production of bass in 1985 is around 13 000 t, of which cultivation (intensive and extensive) accounted for 18%. New *et al.* (1987) forecast that the production of cultivated bass and bream in all Mediterranean countries would increase from a total of around 4000 t in 1985/86 to over 27 000 t by 1992.

Table 11.4 Annual 'official' bass landings in England and Wales 1972–91 (source, MAFF Sea Fisheries Inspectorate)

Year	*Weight (t)*	*Year*	*Weight (t)*
1972	20.7	1982	134.0
1973	28.2	1883	234.1
1974	52.4	1984	138.1
1975	79.3	1985	105.8
1976	101.7	1986	103.0
1977	104.9	1987	125.1
1978	115.3	1988	176.9
1979	103.7	1989	173.6
1980	118.0	1990	188.7
1981	130.9	1991	236.5

The UK contributes only landings from the wild fishery, which have risen substantially since the 1970s. Nevertheless, according to 'official' statistics (Table 11.4), the quantity involved is small compared with the European and Mediterranean totals.

With the retail price of bass frequently exceeding £10 per kg in 1989 and 1990, and landings of wild fish unlikely to increase much further, farming is expected to have an important role in the market in future years. This is clearly recognized by the larger farms such as Pisbarca SA in Spain and Cephalonia Farms in Greece, which planned production of several hundred tonnes per year from 1992. The increased production of farmed bass will be handled through an advanced marketing network. Esperanza Siglo XIX SA of Cadiz, Spain guarantees delivery of freshly harvested bass to any site in Europe or America within 24 h.

11.6 IMPLICATIONS FOR MANAGEMENT OF THE BASS FISHERY

An econometric analysis of quayside prices by CEMARE suggested that the level of domestic bass landings has only a minor influence on prices in the UK, but that exports and imports may have some indirect effect. One implication for fisheries management is that if, for example, there was an enforced cutback in landings of bass, there would be a nearly proportional fall in gross earnings by the fishery, which would not be compensated by an equivalent price rise. Alternatively, an increase in yield from the fishery would bring about real financial gain, because the unit price would fall only a little. Expanding international markets and increases in supplies of cultivated bass or cheaper supplies from outside the EC, e.g. Egypt or Tunisia, may eventually depress domestic prices for wild fish irrespective of the level of local landings. There is, however, no evidence for this happening in the UK, although some merchants are now receiving imports of bass from North Africa in winter. Much of these are small fish, which are probably destined for different sections of the market from those supplied by UK vessels, which land mainly large fish at that time of year. It is unusual for bass to be frozen and stored for later sale.

If the demand for bass continues to grow, it is likely that aquaculture will provide for much of the market's new requirements. The total European production of cultured bass was equal in quantity to the entire wild catch in England and Wales in 1990, and by 1992 was predicted to exceed it. Until there is oversupply of bass and prices begin to fall, however, it seems unlikely that the expansion of aquaculture will discourage fishermen from pursuing the wild fish, as appears to be happening with Atlantic salmon.

Chapter twelve

Commercial fishery assessment and evaluation

12.1 METHODS AND RESULTS

Assessment strategy

Before starting to collect catch and effort data in a fishery, it is important to have a clear view of how the results will be used. Historically, it seems that catch statistics for many English and Welsh inshore fisheries have been collected in a rather *ad hoc* manner and, as a consequence, they have seldom provided a reliable foundation upon which to base management decisions. In DFR's bass study, however, the objectives were clear. Data were collected to help evaluate the state of the bass stock and its response to exploitation, and to help predict the effects that the introduction of particular management measures might have on the fishery.

The method of data collection is also determined by the time scale of the study and, for the UK bass fishery, a work programme of around 5 years was originally envisaged. Comprehensive assessments of large or widely distributed fisheries can be labour intensive, and the expense of setting them up, which in itself might take one or two years, tends to overshadow annual running costs. Given the objectives of the bass investigation, it was necessary to establish a time series of data on the number of bass caught at each age or size group, region by region, throughout the UK fishery. This was required to determine the variations in total annual catch in the fishery, to estimate mortality rates in the stock, elucidate exploitation patterns and yield dynamics, and to provide stock abundance indices. In the event, useful catch and fishing effort statistics have been collected from the boat fishery for bass from 1985 onwards.

To carry out an effective evaluation of catch and effort in a fishery, it is necessary to develop a sampling strategy which takes the characteristics of the fishery's distribution and its participants into consideration. Consequently, a knowledge of the UK bass fishery, its operating methods, seasonality and geographical extent was needed. Semi-quantitative data were obtained from a review of the coastal fisheries (for all commercially exploited species) in England and Wales (Pawson and Benford, 1983); the UK bass fishery itself is described in more detail in Chapters 8 and 9 of this book. To summarize:

The main commercial fishery for bass extends from the Thames Estuary along the coast of southern England to South Wales, with rather less activity around the rest of Wales northwards to Cumbria and off the coast of East Anglia. The effort of individual boats is often directed solely at bass in season: June–August in the north, all year round in South-west England, and generally from May to October in between, although much of the commercial catch of bass is taken as a bycatch in trawls and gill nets. Over 20 varieties of fishing gear are recorded as being used for taking bass, and these can be grouped into seven main types, which catch fish in different ways and with varying levels of efficiency. Some 2000 boats, most being under 10 m in length, are involved in the commercial inshore fishery, the majority being used on a part-time basis. There are also around 50 specialized angling charter boats, and a few inshore trawlers, which are regularly used to catch bass.

The chosen option

MAFF's SFI conducts a national census of landings of marine commercial species, which covers around 70 of the major ports in England and Wales. Because bass are usually landed by small boats at other harbours and on beaches, this system is inappropriate for such a widespread, fragmented and diverse fishery. It proves useful, however, to continue to use SFI data for the few ports where coverage of the local bass fishery (usually the larger boats fishing offshore) is considered adequate. Additionally, an independent economic evaluation of the commercial sector of the bass fishery has provided some catch and effort data (Dunn *et al.*, 1989).

Whilst abandoning the concept of a total catch census in favour of a sampling approach, a census of fleet size and structure is both possible and necessary. Pawson and Benford's (1983) survey included information on the English and Welsh fishery at a much larger number of landing places (300) than those covered by SFI statistics. A knowledge of the number of vessels that might be used to catch bass provides a measure of potential effort, which is stratified into catching gear and boat size or fishing power categories on the assumption that these have characteristic levels and patterns of exploitation, region by region. Additionally, effort is designated as being part-time or full-time, the latter classification applying only where bass are the main target species in the appropriate season.

To quantify the level of fishing effort aimed at bass, and to enable an estimate to be made of total actual effort and catch, individual boats are sampled within the various effort strata for daily catch and effort data. Samples are obtained by giving logbooks to a representative selection of fishermen, and asking them to record catches for each day fished during the local bass fishing season. The total annual catch for each effort stratum is derived by raising the catch recorded for each sample (catch per fishing boat each year) by the total estimated effort in that stratum. Because of the diversity of gears and variations in their catching efficiency, and to permit direct comparisons of catch rates, fishing activity is expressed simply as boat days. Boat days thus become the standard unit of fishing effort and, by stratifying effort by fishing power, it is possible to avoid the complications of attempting to standardize units of effort between various gears.

A more detailed description of the system was given in Pickett (1990), but the main elements are as follows:

Effort census

The number of boats fishing for bass in the various effort categories (strata) is recorded annually at each port, against a prescribed list covering five main geographical regions based on ICES divisions abutting the English and Welsh coasts (Fig. 12.1).

Data for boats working in the current year are obtained during site visits by DFR staff and through follow-up correspondence with local fisheries officers. A summary of the data obtained for full-time and part-time boats in the years 1985–88 is given in Table 12.1.

Catch sampling

The system recognizes 210 effort strata (5 regions × 7 gears × 6 fishing powers, based on number of crew, vessel size and part/full-time designation), and therefore a sampling level of one logbook per effort stratum would meet the statistical requirement of covering about 10% of the fleet. Unfortunately, this sampling level is impracticable, and the strategy adopted is to sample randomly across strata, with no attempt to satisfy statistical criteria, other than to achieve a similar minimum level of cover in each of the five regions. It is not always necessary to select vessels by main gear type, because many skippers use several methods to catch bass within the course of one season, and therefore data on catches by various gears are often available from a single boat's logbook. By

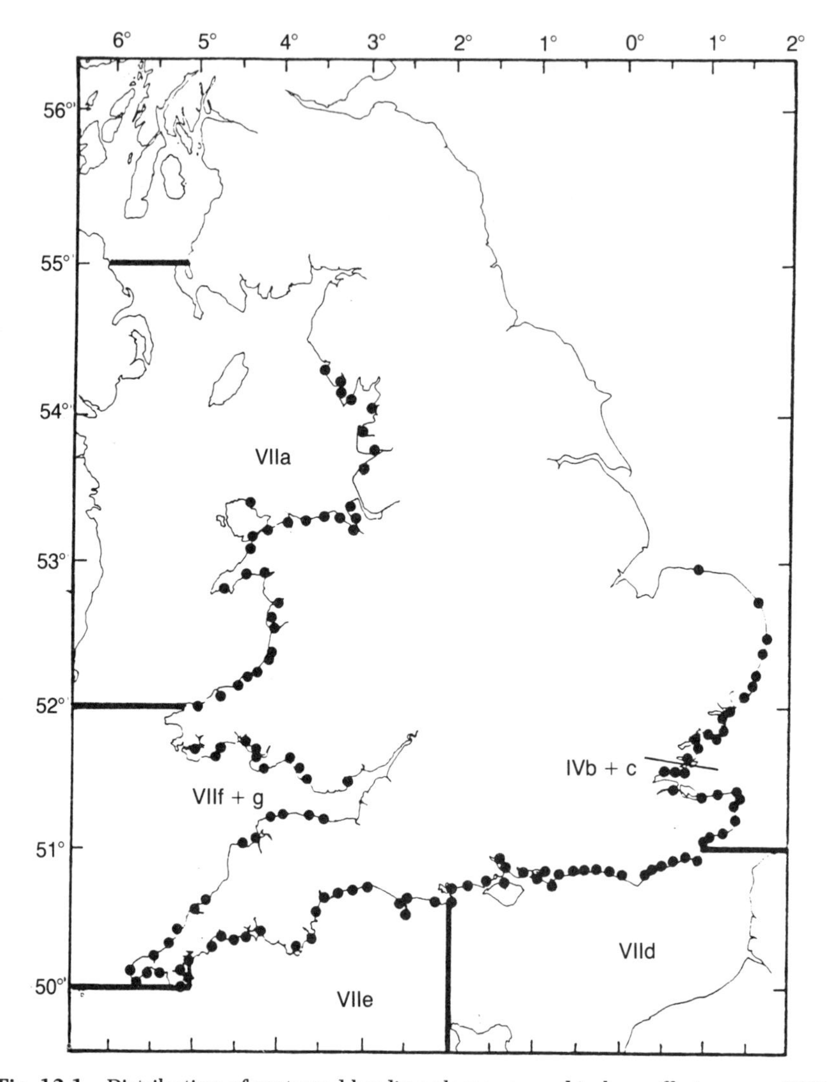

Fig. 12.1 Distribution of ports and landing places covered in bass effort census. ICES divisions are indicated. From Pickett, 1990

leaving aside gear types, one sample of each of the six fishing power categories per region (i.e. 6 × 5 = 30 logbooks) is considered to be the absolute minimum level required to produce an assessment of national catch, and ten samples in each fishery region (50 logbooks) would be needed to produce satisfactory regional catch estimates.

Not surprisingly, there were initial problems in finding sufficient willing

Table 12.1 Summary of the distribution of active vessels fishing for bass in England and Wales, 1985–88, by ICES divisions (from Pickett, 1990). F, full-time vessels; P, part-time vessels

Year		ICES divisions					
		IVd	VIId	VIIe	VIIf+g	VIIa	Totals
1985	P	166	533	514	119	364	1696
	F	18	102	44	7	8	179
	Total	184	635	558	126	372	1875
1986	P	228	607	403	217	380	1835
	F	26	132	40	22	51	271
	Total	254	739	443	239	431	2106
1987	P	225	601	339	267	367	1808
	F	26	81	60	27	33	227
	Total	251	682	399	294	409	2035
1988	P	264	570	319	265	375	1793
	F	21	85	39	25	37	207
	Total	285	655	358	290	412	2000

participants in the sampling scheme, and it was necessary to recruit new logbook holders where a stratum was poorly sampled or where individual fishermen dropped out of the scheme. Nevertheless, at least ten logbooks have been completed in each of five regions in most of the years 1985–92, with a maximum reached in 1986, when 66 books were returned. Return rates averaged 50% of the books given out and, although there has been a higher proportional coverage of full-time fishermen (11%) than part-timers (2%), more (35:25 in 1987) part-timers actually returned logbooks (Table 12.2). This is largely due to the chances of making contact with and identifying bass fishermen in each category.

A large proportion of the part-timers fished only on weekends or at night, and it is a much simpler task to identify vessels which are actively fishing than to contact a representative sample of cooperative individuals in each category. There may therefore be some positive bias in the selection of logbook holders, in that the more regular boats with conscientious skippers are more likely to be encountered and recruited. The system used to stratify vessels was chosen in an attempt to overcome this problem. Because any bias has probably been consistent, selection of logbook holders and the overall distribution of catch and effort data for the fishery is considered to be random.

Over the years, annual sampling levels have varied between 1.6 and 3.3% of the actively fishing boats, and this has made it necessary to devise a system for weighting sampled catches in order to interpolate through unsampled strata and to help correct any bias in sample distributions through the fleet.

Table 12.2 Logbook coverage fo fleet by 'fishing power' strata: regions and gears combined in 1987 and 1988 (from Pickett, 1990)

	Charter/casual angling vessel	*Single-handed vessel*	*2+ handed vessel*	*Total*
1987				
Full-time vessels	44	80	103	227
Logbooks	2	14	9	25
% Coverage	4.5	17.5	8.7	11.0
Part-time vessels	792	546	470	1808
Logbooks	11	19	5	35
% Coverage	1.4	3.5	1.1	1.9
Totals				
Vessels	836	626	573	2035
Logbooks	13	33	14	60
% Coverage	1.6	5.3	2.4	2.9
1988				
Full-time vessels	41	67	99	207
Logbooks	1	12	9	22
% Coverage	2.4	17.9	9.0	10.6
Part-time vessels	754	617	422	1793
Logbooks	6	14	9	29
% Coverage	0.8	2.3	2.1	1.6
Totals				
Vessels	795	648	521	2000
Logbooks	7	26	18	51
% Coverage	0.9	3.8	3.4	2.6

Catch of individual boats

Each logbook contains details of the type of boat and its fishing power category, its full- or part-time designation, port of landing and main fishing method, with which the vessel is allocated a position in the effort census matrix. The information requested from each fisherman is: day/ date, grounds fished, gear used (including mesh size etc.), weight and numbers of bass (and grey mullet, which are frequently taken in the same fishery) caught in three weight ranges, and numbers and total weight of other species caught. A simple weekly form (Fig. 12.2) is compiled into books of 25–50 sheets, depending on the expected length of the local bass fishing season.

When these books are returned to the Lowestoft Laboratory, daily data

Skipper code: 139				Week: 5				Port code: 420		
Date	Grounds fished	Gear used	Species	Enter numbers of fish in size range			Total weight (lb)	Other species (numbers and weight of each type)	For Fish. Lab. use	
				Under 1.5 (lb)	1.5 - 6 (lb)	Over 6 (lb)			Rect-angle	Gear
Sun 1st MAY	Outer Sand Bank	Drift Net 4" mono	Bass	5	12	3	57	4 stn Codling		
			Mullet	2	30	1	82			
Mon			Bass							
			Mullet							
Tue 3rd MAY	Main Channel	Long Lines	Bass	3	18	2	62	10 codling - 2 stn 5 roker - 3 stn		
			Mullet							
Wed			Bass							
			Mullet							
Thurs			Bass							
			Mullet							
Fri 6th MAY	Offshore Wreck mark	Rod & Line - lures	Bass		3	9	71	4 cod - 25 lb 20 pollack - 80 lb		
			Mullet							
Sat			Bass							
			Mullet							

Fig. 12.2 Logbook weekly form.

are extracted and input to a menu-driven computer program written specifically to compile and analyse the logbook and effort census data (Pickett, 1990). It is then a simple matter to calculate total catch (weekly, monthly or seasonally) or catch per unit of effort (weight or number of bass per boat day) for individual boats. These raw data are confidential to the individual originating fishermen and DFR.

Calculation of total catch and effort

The total catch of the bass fishery is estimated from a weighted mean of all daily catches, raised by total effort expressed in boat days. The monthly catch weight of each sampled boat is obtained and allocated to the appropriate effort stratum, and the mean monthly catch by sampled boats is raised by the total number of boats in each effort stratum (for details of data matrix see Pickett, 1990). Where there are significant differences in mean catch between part-time and full-time boats and between the various fishing powers, catches were weighted by multiplying the sampled catch with the appropriate fishing power ratio. These were originally set arbitrarily according to the number of crew on individual boats, i.e 1:2:3, but were later derived from observed catch rates in the various fishing power strata. Weighting values are only used when a particular stratum within one region is not sampled.

In effect, this system aggregates the monthly catches estimated for each gear group in each region. As the effort census is carried out port by port, it is possible to estimate monthly and annual catches at each port, even though only around 40% of the ports around England and Wales are sampled by logbooks. Because the weighting procedure will tend to smooth data between adjacent sampled and unsampled strata (but not between gears), a main source of variance is the distribution of logbooks through the fishery. This is particularly a problem if only one, unrepresentative, vessel is used as a sample for a particular regional catching power stratum. The total fishing effort for bass each year is calculated by raising the effort recorded in logbooks (i.e. the number of days that the sampled boats spent fishing for bass) by the number of boats in each region, fishing power, gear and full/part-time stratum.

Catch and effort

Estimates of annual bass catch and effort for the years 1985–91 are shown in Table 12.3 by gear-type for all regions combined.

During these 7 years, between 1875 and 2039 vessels were used in the fishery, and effort was estimated to have risen from 98 000 to 121 000 boat days annually. To obtain the final catch figures, those given in Table 12.3 are adjusted by adding data from non-sampled sources, e.g. landings by offshore trawlers and at some of the major ports covered by SFI statistics. Estimates of total UK bass catches obtained in this way are given in Table 12.4 (Pickett, 1990; MAFF, unpublished data).

These figures show that SFI's landings records have consistently under-reported the UK bass catch. A higher proportion of the total bass catch was recorded by SFI in 1988 to 1991, probably owing to a much higher catch being taken by trawlers in those years. It should be noted that with bass, landings are equivalent to catches; very few bass were discarded in the commercial fishery prior to 1990, when much tougher controls on landing size were introduced.

Table 12.3 Estimated bass catch and effort by gear type in inshore boat fisheries in England and Wales as derived from logbooks and effort census (adapted from Pickett, 1990)

Gear	*1985*		*1986*		*1987*		*1988*		*1989*		*1990*		*1991*	
	Catch (t)	*Boat days*	*Catch (t)*	*Boat days*	*Catch (t)*	*Boat days*	*Catch (t)*	*Boat days*	*Catch (t)*	*Boat days*	*Catch (t)*	*Boat days*	*Catch (t)*	*Boat days*
Otter trawl	53.6	11 370	13.8	4 159	17.6	5 732	37.0	4 494	92 5	–	121.5*	15 831 *	101.1	19 342
Drift net	19.6	866	33.5	3 891	119.0	7 178	30.0	3 682	46.3	3 874	32.8	4 930		
Gill net	77.6	31 502	141.1	32 590	293.8	28 543	195.9	38 886	70.0	18 195	127.6	25 482	193 1	43 896*
Trammel net	19.0	4 109	22.7	12 889	28.8	8 995	2.0	13 902	27.4	10 963	17.1	4 765		
Longlines	0.5	110	60.6	5 183	13.6	1 873	20.6	1 933	44.3	5 172	28.0	3 516	26.2	2 839
Angling/ handlines	408.8	50 229	326.2	47 545	287.1	54 679	185.9	55 011	243.5	56 678	228.4	41 500	396.3	63 327
Total catch (t)	579.2		598.0		760.0		512.1		524.1		555.5		716.6	

*Best estimate using logbook and SFI data combined.

Table12.4 Comparison of annual bass catch estimates for England and Wales (from Pickett, 1990, extended)

Year	*Sampled catch (t)*	*Estimated total catch (t)*	*Adjusted total catch (t)*	*SFI total catch (t)*
1985	12.9	565.8	579.2	105.8
1986	33.1	597.9	616.6	103.0
1987	29.6	759.9	784.7	125.1
1988	31.2	512.6	570.3	176.9
1989	27.4	469.7	524.1	173.6
1990	20.6	467.0	555.6	188.7
1991	20.7	645.1	716.7	236.5

Calculation of total effort and catch per unit of effort (CPUE)

Three forms of CPUE can be derived:

1. monthly or annual CPUE values for particular, individual boats, which are directly comparable between years and could therefore be used as indices of the local stock abundance of bass;
2. mean CPUE (unraised) of the sampled fleet, split by gear type if required. Although the values could be biased in relation to the whole fleet, these estimates would also provide a stock abundance index where the same boats have been used from year to year;
3. estimated mean CPUE of the total fleet (weighted and split by gear type), which is usually lower than the unweighted mean because catch and effort sampling is biased towards full-time boats. Estimates of CPUE by region and gear has enabled the relative catching power of different gears in different areas to be evaluated.

It can be seen from Table 12.5 that CPUE has varied much more between gears in the same year than with the same gear through time, an interpretation which assumes that the same part of the fleet is sampled each year. This highlights the influence that differences in the catching efficiency of gears have on catch rates. Moreover, trends in the annual mean catch per boat day, for the years 1985–91, differ widely between the major gear groups. Prior to 1989, the daily catches by drift net were consistently higher than those obtained by any other gear, although a downward trend against increasing effort is apparent. Complaints by some anglers and professional handliners, that their catch rates of bass have declined, seem justified by these CPUE figures, though this trend is observed clearly only in some parts of the fishery. It will surprise some of

Table 12.5 Mean bass catch (kg) per boat day for sampled boats (sampled effort in boat days in parentheses) (from Pickett, 1990, extended)

Gear	1985	1986	1987	*Year* 1988	1989	1990	1991
Trawls	5.1 (93)	3.0 (256)	3.0 (149)	15.5 (121)	3.7 (148)	7.2 (185)	6.2 (154)
Drift nets	34.1 (57)	26.1 (180)	20.4 (310)	18.0 (346)	14.0 (354)	5.1 (283)	8.8 (295)
Gill nets	4.0 (653)	8.4 (1093)	8.8 (904)	8.8 (846)	5.6 (1144)	6.0 (737)	5.5 (667)
Trammel nets	2.4 (106)	2.2 (569)	3.3 (453)	3.0 (305)	2.4 (386)	2.7 (186)	3.3 (246)
Longlines	5.3 (4)	12.1 (222)	9.1 (271)	9.4 (164)	8.7 (297)	9.2 (207)	13.3 (215)
Angling			8.4 (968)	9.8 (930)	6.0 (1524)	7.5 (1143)	8.2 (971)
	11.1 (690)	11.7 (1239)					
Handlines			8.9 (304)	7.8 (367)	12.4 (225)	10.5 (239)	8.9 (207)

Table 12.6 Distribution of average catches (kg year^{-1}) per sampled boat in the 'fishing power' strata – 1987 and 1988 data combined (from Pickett, 1990)

	Charter/casual angling	*Single-handed*	*2+ Handed*	*All*
Full-time boats	3	28	18	49
Total catch	1950	19 047	26 099	47 096
Catch per boat	650	680	1 450	961
SD	786.0	480.5	1 578.9	
Part-time boats	16	33	13	62
Total catch	2134	5890	6037	14 061
Catch per boat	147	179	464	227
SD	128.5	208.9	448.3	
All boats	19	61	31	
Total catch	4084	24 937	32 136	
Catch per boat	215	409	1 037	

these line fishermen that fixed gill net catch rates were generally much lower than those for handline or angling.

Whilst the CPUE values given in Table 12.5 provide a useful comparison of catch rates by gear for the sampled part of the fleet, they are not necessarily representative of the total fleet covered by the effort census. For example, the gross mean catch from vessels sampled in 1988 was 9.7 kg per boat day, whereas the weighted mean across the whole bass-catching fleet was 4.4 kg per boat day. The total annual catches of individual boats depend a great deal on the frequency of fishing – hence, the full-time/part-time designation – and may range from a few kg each season for some small, part-time boats for which bass is a bycatch species, to over 6000 kg per year for some of the larger, full-time specialist bass boats.

Mean annual catches for the combined years 1987 and 1988 were 961 kg for full-time and 227 kg for part-time boats, for the sampled part of the fleet. Inevitably, there was a wide variability of catches within each fishing power stratum, though Table 12.6 shows that mean catches by charter/casual angling and single-handed vessels were similar, for both full-time and part-time vessels.

Length and age composition of catches

Estimates of the numbers of fish caught in each age or size group are fundamental to the assessment of marine fish stocks. The quality of assessments thus depends on how well catches are estimated and on the

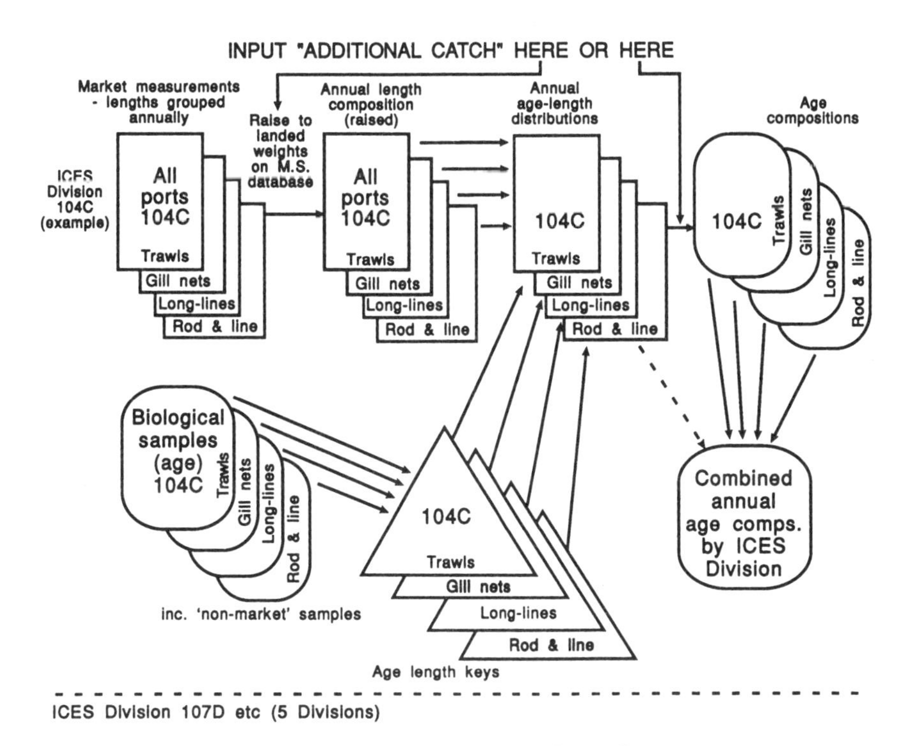

Fig. 12.3 Processing stages in bass assessment. After Pickett, 1990

adequacy of sampling of catches for length and age throughout the fishery. The catch weights of bass, estimated for each of the four main gear groups (gill nets, trawls, handlines and longlines) in each region, were used in processing biological sampling data on the DFR market sampling system at Lowestoft. This produces length distributions, age–length keys (ALKs) and age compositions in the manner developed for fish stocks which are assessed by ICES (Gulland, 1969).

Sampling bass for length and age within the various effort strata has not been comprehensive and, out of 20 ALKs produced each year, only 12–14 were usable by themselves. This is a normal situation in many assessments of the major commercial species' stocks and requires some ALKs to be combined across gears or regions. ALKs are applied to sampled length distributions, which have been raised by the ratio of total to sampled catch weight for each stratum. The processing for bass is illustrated in Fig. 12.3.

The products of this system were regional, gear-grouped and all gears combined age and length compositions, and examples of the length compositions are shown in Fig. 12.4. Sufficient catch at age data for bass

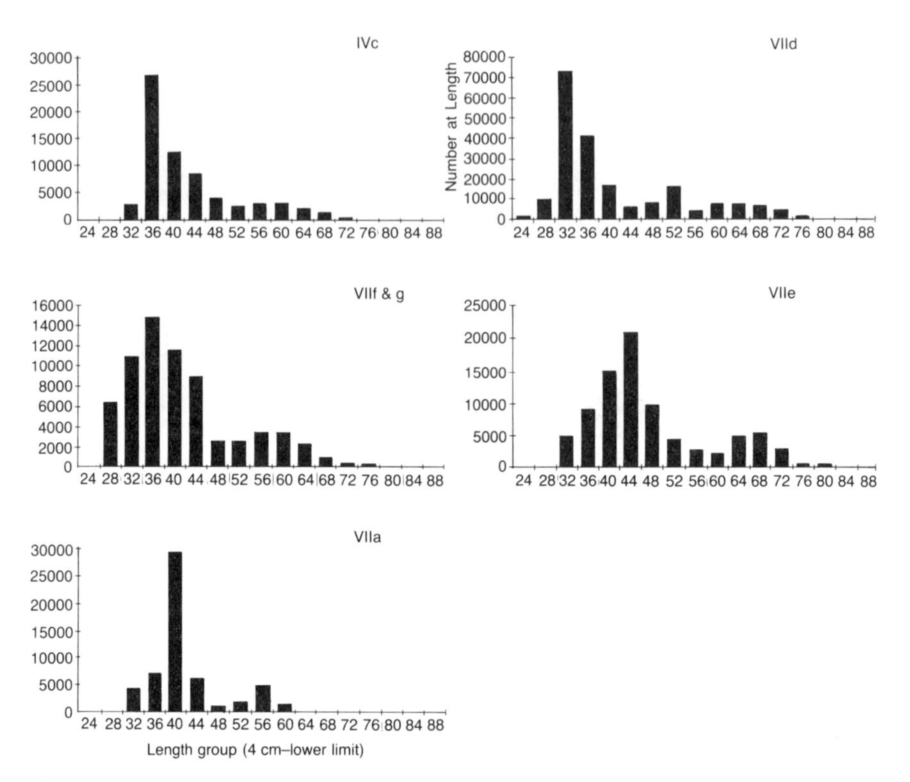

Fig. 12.4 Length composition (4 cm groups) of bass catch in ICES divisions in 1988 by all gears. Note differing vertical scales. (MAFF unpublished data)

have been available to enable the year classes from 1976 onwards to be monitored as they progressed through the fishery, to identify the characteristic exploitation patterns of the various catching gears, and to enable year-class abundance and yields to be determined (Chapter 15).

12.2 VALIDITY OF THE LOGBOOK SYSTEM

Compared with many marine fisheries, there are few independent catch data for bass with which to validate the results of DFR's logbook scheme. The system was used to obtain catch and fishing effort data in a relatively large and dispersed fishery, which was not well sampled by the established method of collecting statistics. The main shortcomings of the fishermen's voluntary logbook sampling method are the inevitable low level of sampling attained and the difficulty in obtaining a complete effort census each year. Because individual effort strata were seldom sampled by more than one logbook (in each region), it was not possible to determine between-boat variability. Unlike the output of many other catch recording systems, however, these logbook data did represent actual catches, because there was little incentive for the scheme's participants to misreport, and any discards were recorded.

As data accumulated over the years, it became evident that both the commercial and recreational boat fisheries for bass could be assessed in this way. The total annual catch of the UK fishery appeared to be relatively stable, and fluctuations between regions were due as much to the varying accessibility of bass to the inshore fisheries there as to errors in the estimating procedure. Whilst rough weather limits the fishing activity of small boats, it does not usually affect all regions similarly, and changes in the weather pattern will also affect the distribution of bass.

The main sources of variation apparent in the fish population were the overall abundance of the stock, the seasonal movements of the fish and their distribution in the fishery and, for those fleets which targeted recruiting 3–5-year-olds, the relative strength of successive year classes. The consistency in the catch-at-age trends, during the period 1985–91 (see Fig. 15.1), suggests that the system's output was satisfactory for the purpose of assessing population trends and the impact of exploitation.

There are dangers in using indices of stock abundance based on catch per effort data derived in the manner described, particularly as a common unit of effort (boat days) was used which did not necessarily reflect the relative catching powers of the various gears or the manner in which they were deployed. The prime aim was to estimate catch for biologically sampled gear groupings, and thus to arrive at numbers of bass caught at each age.

12.3 ECONOMIC APPRAISAL

Scope of the study

Economic information for the commercial sector of the bass fishery in England and Wales has been based, in part, on information obtained through DFR's effort census and logbook scheme. In addition, specific surveys by CEMARE provided a basis for forecasting the changes that might ensue from tighter management of the fishery. These surveys had three main objectives (Dunn *et al.*, 1989): (a) to collect descriptive data for the whole commercial fishery for bass, (b) to provide a profile of the activities of bass fishermen, and to examine the way in which bass fishing related to other fishing activities, and (c) to evaluate the perception of bass fishing as a commercial activity and to assess the response of fishermen to the prospect of changing catching or market opportunities.

Methodology

Data on fleet size, effort and catch, derived from the DFR effort census and logbook scheme, were used to describe the bass fleet's activities. The most practical approach to obtaining new, independent data was considered to be through a questionnaire distributed to fishermen. A prototype questionnaire, tested on a small group of volunteers, forewarned CEMARE of the difficulties which might be encountered in gaining the cooperation of fishermen in general, and it was necessary to modify the questionnaire considerably. The survey proper covered those skippers already involved in the DFR logbook scheme, charter skippers contacted through local SFCs, commercial fishermen contacted by port-based fishery officers, and other known bass fishermen contacted through a postal survey directly and through commercial fisherman's organizations.

Each respondent was asked questions relating to boat and gear specifications, past catches of bass and other species, fishing methods, fish prices, level of effort and reasons for taking up bass fishing. The data collected were rather disparate, and several analytical techniques were used to produce meaningful results.

Economic characteristics of the commercial fishery

On the basis of the fleet census data already described, and taking the DFR bass landings estimates as being a reasonable, if conservative, estimate of catches, the study indicated that the true commercial landings in 1987 were around 630 t. This was less than the DFR estimate of 785 t, because an estimated 155 tonnes of the latter was landed by recreational boat fishermen. At the prevailing average price of bass (£6.19 per kg to the fisherman) this gave a first sale gross value for the commercial fishery in 1987 of £3.9 million. A typical commercial boat, fishing full-time for bass in season, would be expected to have earned £10 850 from bass alone, obtained for an average catch of 1.75 t. A typical part-time boat would have earned perhaps £1800 from an average bass catch of 292 kg.

The questionnaire survey was heavily weighted towards full-time professional fishermen, who provided 77% of responses. The data suggested that a typical bass fishing operation in 1987 involved a boat 7.5 m long, of 60 horse power and 11–12 years old, worth between £8000 and £12 000 at resale value and carrying bass gear worth between £1400 and £1500. With an average of two crew members, the boat would have made a total of 124 bass trips in seven months (full time) or 72 bass trips in 5–6 months (part time). The total resale value of the 192 boats fishing full time for bass in England and Wales was £1.65 million, and they carried bass gear worth £0.26 million. The fleet of some 1000 boats fishing part time for bass would have had a resale value of around £12–22 million and carried bass gear worth £1.5 million. The most striking feature of the bass fishery was the apparent ease with which a casual operator could move up to a commercial enterprise with very little investment in gear, compared with a trawl fishery, for example.

The survey was able to give a breakdown of the average earnings in 1987 for the full-time bass fishermen, according to number of crew:

single-handed – landed 556 kg of bass worth £3450
double-handed – landed 1033 kg, worth £6400
3 plus handed – landed 2087 kg, worth £12 920

This pattern tends to reinforce our contention that fishing for bass and other local fish stocks is a self-limiting activity, which responds to a certain level of economic necessity depending on how many wages have to be paid from selling the catch. The proportional dependency on the bass fishery for total gross earnings (from fishing) was 56%, 51% and 68% respectively for the three groups given above.

According to the survey, fixed gill netting was the most widely used technique in 1987 and accounted for 47% of bass landed, with drift nets taking a further 19% and commercial rod-and-line effort accounting for 17% of landings. These figures are broadly in line with the DFR estimates although, as previously mentioned, recreational and commercial rod-and-line catches from boats are not distinguished in the DFR system.

As might be expected, entry into the bass fishery has a strong commercial motivation, and around 40% of those participating in the bass fishery in 1987 had entered during the period 1976–81. This influx may have been associated with events in other fisheries, such as local resource depletion or a tightening of controls, which suggests that bass fishing may be regarded as a displacement fishing activity. The study also reflected a high level of worker satisfaction in bass fishing – a bonus to the value of the catch that is landed and sold. The strongest evidence, however, was that the level of effort in the bass fishery is a function of bass prices, and perhaps the availability of bass. A sustained rise in real bass prices (from whatever cause) may result in an intensification of effort by those already in the fishery, and more people are likely to move into bass fishing if the relative value of other commercial resource species falls, or they become locally scarce.

These are important findings for fisheries managers to consider, because the aim of most regulation is to increase or maintain the supply (yield) of fish to the fishery. With many species, larger landings may result in a considerable drop in unit price and a consequent increase in effort in an attempt to maintain incomes, a combination which is undesirable to any of those involved in the catching side of the fishery.

Compatibility of data

The economic assessment covered a much wider spectrum of fishermen than those who filled in catch logbooks for DFR, though both sets of samples were biased towards the full-time fisherman. In the logbook system, allowance had been made for this when calculating total catch, by relating full-time and part-time boats' catch rates to the respective levels of effort in the fishery. Nevertheless, it is important to know how representative these samples were of the population of bass fishermen. Catch data obtained from the questionnaires for non-logbook holders were compared with those derived from logbooks for the same period. The range and average catch weights for the two groups were similar to within 10%, which suggests that, despite the relatively low sampling levels, the catching capacity of the fleet was being truly represented in the logbook returns.

The international context

To our knowledge, no other study has been carried out on the economics of a bass fishery, although some local fisheries investigations have been quite detailed and have included information on local and national prices, e.g. Bertignac (1986). The characteristics of other bass fisheries, particularly those in France, suggest that similar surveys could be carried out with a reasonable chance of success. Much of the landings of bass in France goes through the main markets, and landings statistics are available and published. As international markets and, to some extent, fisheries become more closely integrated (particularly for EC countries), it would assist international management of the resource if it were possible to predict the effect of the supply of bass on international prices, especially as the cultivation of bass is expanding in southern Europe.

Only since 1990 has the FAO (FAO yearbooks) statistics listed the UK as a bass-producing nation. This chapter provides evidence that credible statistics now exist for this fishery and that for England and Wales at least, details of the social and economic framework of bass fishing operations are available.

Chapter thirteen

Recreational fishery evaluation

13.1 METHODOLOGY

It is difficult to obtain accurate statistics for large-scale recreational (sport) fisheries. This is due to several factors, the most important being the large numbers of anglers involved, the varied timing and frequencies of fishing trips and the lack of systematic records of catches. The geographical distribution of the recreational bass fishery in the UK is well known, and numerous books and articles in the angling press have provided relevant information. In addition, Donovan Kelley, who has had a lifetime's experience of this fishery, has produced a pictorial description (Fig. 8.6) of the fishery's distribution.

Early in DFR's study on the UK bass fishery, it was realized that the number of participants in the sport of fishing for bass amounted to some tens of thousands of anglers. In view of their numbers, it was important to estimate the shore-anglers' catch, to assess its possible impact on the bass stock, and to evaluate the social and economic importance of this part of the bass fishery. However, the logbook system described in the previous chapter could not be used to obtain a representative sample of shore angling catches, owing to the large sample size required and the lack of information on categories of anglers. Fortunately, several charter-boat skippers and a few sport anglers who used boats were involved in the logbook scheme. From these samples, estimates could be made of catches in the recreational boat fishery, though in view of the variability of effort in the 'casual' sector, some independent means of validating these estimates was required. It was later possible to obtain catch records for some boat and shore anglers, which suggested that the total retained sport catch of bass was probably of a similar order of magnitude to that of the commercial fishery (i.e. hundreds of tonnes).

The economic appraisal of the UK bass fishery carried out by CEMARE in 1987 included in its brief a full evaluation of the recreational fishery for bass. It was designed so that there were areas of overlap with the DFR logbook system in the boat angling sector, and this helped with validation of catch data in both directions. The CEMARE estimates of shore anglers' catches of bass are unique, but these are just one aspect of the recreational fishery included in this study, which investigated:

1. the population of bass anglers and those sea anglers fishing for other species;
2. the motivation for bass angling;
3. the relative importance of bass and alternative species;
4. sea angling effort and catch, in general;
5. bass angling effort and catch;
6. the characteristics of bass and non-bass sea angling trips;
7. the expenditure by bass anglers;
8. anglers' evaluation of their right to fish for bass;
9. angling charters.

The survey thus combined statistical, social and economic elements, and was aimed at helping our understanding of the recreational usage, and importance, of the bass resource. The methods employed to obtain data differed from those used to obtain basic statistics of the commercial fishery, primarily because the main output of the latter is simply fish. In contrast, the main product of a recreational fishery is an enjoyable fishing experience, i.e. fishing rather than fish. In comparing commercial and recreational fisheries, therefore, it is inappropriate to use such a straightforward measure as the quantity of bass landings from each sector. Recreational fishing is a complex economic activity for which special value designations are necessary. Those requiring details of the methods used by economists are referred to Dunn *et al.* (1989). The CEMARE economic appraisal of the bass fishery was the first study of a UK marine fishery to employ these methods, which were originally developed for recreational fisheries in the United States and have been used to evaluate the sport fishery for salmon in the UK (Radford, 1985).

13.2 RESULTS OF THE SURVEY

The population of bass anglers and of sea anglers who fish for other species

There is little published information which can be used as a basis for estimating the number of sea anglers who fish for bass. An on-site census, as employed by DFR to determine the size of the commercial and charter boat fleets, is inappropriate where shore angling is concerned, owing to the wide variability in anglers' presence between sites and the influences on their outings of such factors as weather, tides and time of day. National angling surveys in 1970 and 1980 (NOP, 1970; 1980), and a subsequent survey by The Angling Foundation, provided some data on the total number of sea anglers in the UK; the answer was more than one and a half million people! Thus, the number of bass anglers could be estimated if the ratio of bass anglers to total sea anglers was known. This statistic was determined by on-site surveys during 1987, when a total of 406 sea anglers were interviewed at 76 sites in all coastal counties of England and Wales from Norfolk clockwise to Cumbria. The questionnaire was structured to provide information on many of the aspects of anglers' fishing activity listed above. In addition to the on-site survey, postal questionnaires were mailed to three groups of anglers: BASS members, anglers interviewed during on-site surveys and other anglers contacted through the angling press.

A total of 585 sea anglers participated in on-site and postal surveys, of whom 348 claimed to be bass anglers. Using data from the 1980 national angling survey, and after accounting for sampling bias, the total population of anglers who fished for bass in 1987 was estimated at 490 thousand, with a ratio of 1 in 825 anglers being sampled. Not all these bass anglers were actually fishing for bass when interviewed in 1987, but around 42% of interviewees said that they intended or hoped to catch bass during the course of 1987. Not unexpectedly, the numbers of bass anglers varied regionally, with fewer in the north and north-east and most in the south-west and South Wales. There was evidence that anglers travelled between regions for their fishing and the authors of the study noted that this could be a source of error when estimating the total number of bass anglers in each region.

Types of bass angler

Four (not necessarily exclusive) categories of bass anglers were identified in the survey:

1. sea anglers who were fishing in 1987 with the intention of catching bass on one or more of their sea-angling trips. There were nearly half a million of these, equivalent to the total population of bass anglers as given above;
2. sea anglers who did not intend to catch species other than bass on bass fishing trips, 220 600;
3. sea anglers who intended to catch bass along with other species on all of their angling trips, 88 900;

4. sea anglers who intended to catch only bass on all their sea-angling trips and fished for no other species in 1987, 24 500.

This last category represents the specialist bass angler, and these are equivalent to 5% of the sea-angling population who fish for bass. Interestingly, BASS, whose members contributed to the survey, had a membership of only around 350 in 1987.

The authors of this study point out that, because these figures are based on both sample and extra-sample data, they are subject to both the size of the sample and to the accuracy and relevance of the 1980 national angling survey, which was used to 'raise' the 1987 survey data. A second, repeat, survey is being conducted in 1992/93 and will attempt to produce estimates that are less likely to be biased in this way.

Motivation for bass angling

Anglers were asked what feature it was that made bass fishing attractive to them, and why they did or did not fish for bass in the current year. Although they were generally pessimistic about their chances of catching bass, they nevertheless valued the bass fishing experience highly. The total number of bass anglers does not seem likely to change much as a result of fluctuations in catch rates, although poor catches have caused some anglers to give up bass fishing, only to be replaced by new entrants to the sport; over a quarter of sampled anglers were new to bass fishing in 1987. It is clear that the sporting quality of bass was the single most important factor which made people take up or stay involved with bass fishing. It was claimed that the species' monetary value or eating qualities were generally regarded as being unimportant to anglers.

Relative importance of bass and alternative species

Anglers were asked to indicate which species they fished for and whether they directed their efforts at bass. The most popular species pursued by shore and boat anglers around the coasts of England and Wales are shown in Table 13.1, according to the proportion of sampled anglers that fished specifically for them in 1987.

Clearly, cod is the most important species to anglers in general, ahead of mackerel, bass and whiting.

General sea angling effort and catch

Anglers were asked to give details about the frequency of their fishing trips, the main species they caught, and how they disposed of their catch. Those who fished from the shore for species other than bass had made an

Table 13.1 The most popular species pursued by shore and boat anglers around the coasts of England and Wales, according to the proportion of sampled anglers that fished specifically for them in 1987. From Dunn *et al.*, 1989

Species	*% of shore anglers fishing for*	*% of boat anglers fishing for*
Cod	52.2	69.3
Flounder	47.1	–
Bass	45.1	14.9
Mackerel	43.4	26.3
Whiting	43.4	13.2
Ray/skate	–	29.8
Eel	27.6	–
Plaice	27.6	–
Pouting	25.3	–
Dab	22.2	–
Conger eel	–	20.2

unweighted average of 38–40 trips in 1986, and 28% of the anglers surveyed went shore-fishing more than once a week. After adjusting for sampling bias, an annual mean value of 21.2 trips was estimated for the shore-angling population in general, and 3.5 trips for boat anglers.

It was not possible to determine the total catch of each angler, but an impression of which species were caught was gained by asking about the catch taken on their most recent angling trip. Although 22% of shore anglers had caught nothing, a total of 17 species were recorded and bass did not feature in the 10 most commonly caught. Boat anglers had caught a total of 14 species, and bass was not recorded in the six species caught most frequently.

Bass angling effort and catch

Anglers in groups ranging from specialist catchers of bass to those who had never fished for bass, were asked how many trips they made over the whole of 1986, and how many bass were caught on these trips. Bias owing to the over-representation of anglers who fished most frequently was corrected in order to calculate true trip frequency, and the averages were raised to the total population of bass-specialist and general sea anglers. Numbers of bass caught overall were calculated in a similar way.

Shore anglers who fished specifically for bass at some time made an average of 31.9 trips in 1986, and although 44% of these trips were made with the intention of catching bass, the species was caught in only one in five trips. This same sample of anglers anticipated that 39% of their shore trips in 1987 would be aimed at catching bass, which suggests that poor success in catching bass is no deterrent to a bass angler's optimism! Around 60% of bass caught were said to have been returned to the water, largely because they were undersized. On this basis, and including an estimate of the bycatch of bass by other shore anglers, the total retained catch of bass taken from the shore in 1987 was estimated to have been between 570 and 645 t.

Boat anglers fished for bass on 27% of their trips in 1987, compared with 25% in 1986, when they made a weighted average of 4.6 trips for bass. Around 52% of bass caught from boats in 1986 were returned to the water. The landed boat angling catch in 1987 was estimated at around 153 t, of which 146 t was thought to have been taken by those fishing specifically for bass.

The average weight of each bass retained in these catches was approximately 0.8 kg for shore-caught fish (from CEMARE sample data) and 0.94 kg for boat-caught fish (from DFR logbook data). With the higher proportion of bass returned to the water, this suggests that shore anglers usually catch smaller bass than those fishing from boats. Over the same period, DFR's logbook-derived estimate of total boat angling catches (which includes commercial landings) was 257 t. It was not possible to identify the recreational component in this total, but the CEMARE estimate of 153 t for this seems reasonable when viewed alongside a commercial angling catch of 104 t (257 − 153).

Features of bass and non-bass sea-angling trips

Bass specialists and other sea anglers provided information on the hours they spent fishing, their mode of travel to the fishing venue, the miles travelled and journey time for their last fishing trip or a typical outing in the previous season. Boat anglers were treated separately to shore anglers, to enable comparisons to be made between bass and non-bass anglers in both the shore and boat fishing categories.

It was found that the typical shore fishing trip for bass involved 20 minutes of travel and 5 hours fishing, with 78% of anglers travelling by car for an average of 14.5 km. The typical boat fishing trip for bass involved a journey of 30 minutes and 8.5 hours fishing, with 87% of anglers travelling by car for an average of 38.5 km. It appears that shore angling for bass is more likely to involve local anglers than is boat fishing, in which there is a greater investment of people's time and money.

Angler expenditure

Anglers interviewed on site were asked to give details of their expenditure on travel to and from the fishing venue, on bait, tackle, food and drink, and upon accommodation and other incidental expenses. Boat anglers were asked similar questions relating to their most recent bass boat trip in 1987, including the cost of using the boat or charter fees. These data were used to calculate gross expenditure in the recreational fishery, although the resulting figures have little relevance in terms of the economic significance of the recreational fishery. For example, the valuation in terms of gross expenditure is not the net value of the fishery; none of the expenditure relates to purchasing the actual bass recreational fishing experience. If the bass sport fishery did not exist, the attendant expenditure would probably have been transferred elsewhere, very likely into other angling activities.

The average expenditure made by each bass angler in 1987 was estimated to be £3.27 per shore trip and £12.60 per boat trip. Total expenditure on UK bass angling during the year was estimated to have been at least £15.8 million, of which £12.1 million was spent on shore fishing. Of this, an estimated £4.3 million was spent on bait to catch bass, which represents a minor industry in its own right, supporting a trawl and seine fishery for live sandeels and hand gathering of moulting shore crabs. These values are interesting from a descriptive point of view, but by themselves they do not enable us to predict any changes in expenditure that might arise from changes in the fishery, whether through stock fluctuations or brought about by new legislation. Clearly, some other form of valuation of unlicensed recreational fisheries based on wild stocks, such as bass, is needed.

Willingness to pay and to sell

In a recreational fishery, where there are no market data on commodity production and prices which can be used to evaluate its economic significance, economists may try to determine the fishery users' willingness to pay for, or to sell, the right to participate in their sport. The method chosen for this study was the contingent valuation method, which has rarely, if ever, been used before in the UK for fisheries evaluation. This method is gaining widespread acceptance in the USA in this type of study and for environmental evaluations, such as quality improvements in air and water pollution (Daubert and Young, 1981; Loomis *et al.*, 1986). It can also be used to evaluate changes in amenity quality or levels that would follow the introduction of a regulatory measure. Bass anglers were asked about their willingness to pay for being able to continue to fish for bass for one year, and for a beneficial change in their prospects of catching bass. They were also asked about their willingness to sell their right to fish for bass. This was assessed as the minimum compensation they would require if, for reasons beyond their control, they were unable to fish for bass for one year. Although this approach has produced results in the USA which were consistent with the results using other techniques, care must be taken in interpreting the true meaning of the figures derived.

Probably the more reliable estimate of the net value of the sport fishery is the £8.7 million that UK sea anglers, as a group, would theoretically have been willing to pay for the right to fish for bass in 1987. To this would need to be added the amount spent on travel, tackle, bait, etc., during 1987, to obtain the gross value of the UK recreational bass fishery in that year. This amounts to a total of around £24.5 million. British anglers had produced their own estimate of the value of the recreational bass fishery which, although couched in different terms to those used in the economic survey, was quite similar, when considering values for tackle, bait, boat hire, etc. (Section 17.3). The same anglers' aggregate willingness to sell their right to fish for bass in 1987 was estimated at £111.6 million. Such a large difference in these respective evaluations is not unusual when using this approach.

Angling charters

The study also examined the charter boat sector, which is a commercial operation that is generally carried out as a part-time activity by licensed skippers of registered fishing boats. However, the catches are taken by recreational anglers and thus contribute to the recreational catch. These catches were covered by CEMARE's survey of boat anglers and also to some extent by the DFR logbook scheme. In order to assess this sector effectively, charter skippers had to be contacted directly, and questions relating to chartering for angling were also included in the survey of the commercial fishery. A response was obtained from 120 charter skippers. The DFR effort census data were used to estimate the total number of boats chartering for bass angling, on both a part-time and a full-time basis.

The commercial fishery survey's results suggested that 10% of fishermen in the England and Wales inshore fleet took out charter parties of anglers at some time. Of those who ran charter vessels as a full-time occupation, DFR's effort census revealed that 35 were heavily dependent on bass angling charters in 1987, with another 161 boats being involved part-time.

Only 15% of all charter skippers earned more than 90% of their income from chartering, with full-timers earning an average of 15% of their fishing-derived income from bass-directed trips, and part-timers earning 50%. Full-time charter boats worked a bass season of 3.5 months, on average, whereas part-timers worked an average of 5.5 months on bass. The latter, therefore, appeared to rely more heavily on bass than those catering full-time for angling parties, who had a relatively short bass season. On average, eight anglers were taken on each charter trip, which lasted about 7.5 hours.

Total catches by this sector were estimated at 75–95 t in 1987; this figure is included in the total recreational catch given earlier. There was little evidence, with bass chartering, of the skipper taking a share of the catch to be sold on the market as has occurred with some wreck fishing parties catching cod, ling, conger eel (*Conger conger*) or coalfish (saithe).

The economic data relating to chartering are not very robust, but, in 1987, those involved full-time in bass charter fishing (in season) made some 175 trips, and earned £15 750 per boat, of which perhaps £10 000 came from an average of 125 bass trips. Those involved only part-time in bass fishing (in season) made a total of 98 trips, and earned some £9000, of which 12% came from an average of 12 bass trips.

A large proportion of modern charter boats are not involved in bass fishing, and concentrate instead on trips to fish around wrecks or taking out holiday makers to 'feather' for mackerel. In some areas, e.g. Bradwell in Essex, charter skippers ceased to take out specific bass angling trips during the late 1980s because catches had declined to unsatisfactory levels. It seems, moreover, that this sector may be declining overall, although in 1987 some 300 people were still earning a living from charter boat fishing involving bass.

13.3 CONCLUSIONS AND IMPLICATIONS FOR MANAGEMENT

Despite data limitations inherent in the techniques described above, and the uncertain quality of responses obtained during this, the first economic survey of the recreational bass fishery in the UK, it is evident that the recreational catch of bass and the net value of its fishery are both substantial. Anglers have been the main lobbyists in the cause for stronger protective legislation for bass populations around the British Isles, but their campaigns have often been fragmented or at best engineered by small groups of people, who might not have represented the general angling public's views. Prior to this survey, the size of the bass angling population, its share of the total landed catch and its role in the national fisheries economy could only be guessed. The total UK landings of bass now appear to be more than double those estimated previously, because these had excluded the catch taken by shore anglers. The study concluded that, in total, some 1325–1360 t of bass were killed in England and Wales in 1987, less than half of which was sold. The potential impact of this previously unaccounted fishing mortality could be significant in future management of the bass fishery.

During various bass tagging schemes, reports of recaptures of tagged bass by anglers during the 1970s and 1980s exceeded those from the commercial fishery by a wide margin (Pawson *et al.*, 1987). Although anglers may have had an incentive to return tags, and some commercial fishermen may have deliberately not returned them, it was still evident that anglers were taking a high proportion of the UK catch of bass. In the meantime, this has created a problem – how to estimate and include the recreational catch in future assessments of the bass stock and its fishing mortality? It would be expensive and difficult to adequately sample catch and effort in the sport fishery, though the lack of data does not preclude our making good estimates of abundance trends and effects of exploitation by the commercial fishery on the bass population.

The economic assessment of the bass fishery lends support to the British Government's policy of regarding the maintenance of bass stocks around the UK as being in the interest of both the commercial and recreational fisheries. Consequently, proposals for management of the bass fishery in England and Wales have attempted to take account of the views and requirements of both sectors. It has been difficult to substantiate or refute claims by either sector as to the importance of bass, but improvements in catch and effort data quality, coming from the DFR logbook scheme and the CEMARE economic appraisal, have permitted sounder judgements to be made on the effects of the various measures which could have been introduced.

When survey data are made public, there will always be those who wish to use the information to justify their own position. For example, late in 1990, a small group of specialist bass anglers renewed its campaign for bass fishing to be assigned a sport-only status, using the above published values for the sport fishery to justify their case. This concept, when applied to a commercially exploited marine species in other countries, has raised some interesting issues. This matter is discussed further in Chapter 17.

Chapter fourteen

Assessment of bass populations and state of pre-recruit stock

14.1 INTRODUCTION

The welfare of the bass stocks around England and Wales has been monitored by two parallel but separate courses of action. An annual assessment of the age and size composition of the fishery's catch is made by DFR, to determine yields and exploitation patterns (Chapter 15). From this information, it is possible to determine whether growth overfishing is occurring, that is, are yields decreasing because there is too much fishing effort on small (young) fish? It is also important to know whether the spawning stock continues to be able to produce sufficient recruitment to the fishery to sustain yields, even if the exploitation pattern is not leading to growth overfishing. The approach has been to assess the abundance of succeeding year classes before they recruit to the fishery, and to determine the level of their survival and eventual contribution to the fishery.

Estimates of the relative strengths of successive year classes can only be obtained by establishing time series of data collected in a consistent manner. In the absence of absolute abundance values, short-term studies cannot produce truly comparative estimates, unless they have quantitatively evaluated reference points. It is also important to examine the causes of any trends in year-class strength, so that natural effects can be distinguished from any due to human influence. The following sections describe the methods that have been used to monitor recruitment to both the bass stock and its fishery.

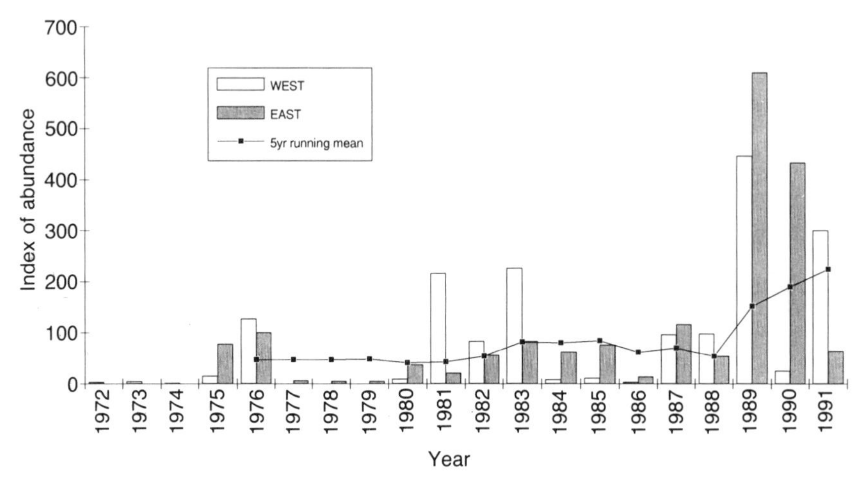

Fig. 14.1 Abundance indices of 0-group bass – numbers per standard volume of intake water at West Thurrock (east) and Oldbury (west) power stations showing 5-year running mean. (MAFF unpublished data)

14.2 0-GROUP ABUNDANCE

Power station intake screen samples

0-Group bass are regularly trapped on the cooling-water intake screens of coastal power stations in the autumn and winter in several localities, and regular samples from some of these sites have been used to estimate the relative strength of successive year classes (Riley *et al.*, 1986). The numbers of 0-group bass taken each year from the screens at West Thurrock (River Thames) and Oldbury (River Severn), for a standard volume of intake water over the period 1972 to 1991, are shown in Figure 14.1.

It is apparent that the relative abundance of each year class varies between sites, and this will depend on the size of the local population, differences in the susceptibility of 0-group bass to the particular power station's cooling-water intakes, and the regional variability of year class strength. Nevertheless, the high abundance of the 1976 year class could have been predicted from these data, 3 or 4 years before it was seen that these fish began to dominate bass catches in estuarine and coastal waters in 1980 and 1981. More recently, the 1981–83 spawnings appear to have been quite successful on both coasts, and the fishery generally benefited from recruitment of the the resulting year classes (Chapter 15). Recruitment on the west coast was weak between 1984 and 1987, but successful spawning and better survival in the warmer period from 1987 to 1991 seems to have produced some of the biggest broods recorded to date on both the south-east and west coasts.

14.3 TWO- TO FIVE-YEAR-OLD BASS ABUNDANCE

Fine mesh trawl surveys

Each year since 1981, high-headline bottom trawls have been used to monitor the abundance of 2–5-year-old bass in the Solent. This area contains a number of natural harbours, which are considered to be the most important bass nursery area in southern England.

DFR's surveys usually take place in May and September, when a standard grid of 35 trawling stations is worked over a 4 day period. Consistency has been acheived since 1983, by chartering the same inshore fishing vessel to use a trawl of standard pattern and size, and by fishing at the same part of the tidal cycle on each visit (falling springs). Likewise, tows at each station are of similar length, though their individual duration ranges between 10 and 20 minutes, at a consistent towing speed equivalent to 3–4 knots over the ground, depending on the strength of the tide.

The trawls used have mesh sizes of 80–85 mm in the wings and 60 mm in the cod end, although the latter is always fitted with a shrimp mesh (10–12 mm) liner. It has been possible to determine the selectivity of these trawls for young bass and, by comparing the numbers of 2–5 group fish between successive surveys, to derive values for the relative abundance of year classes. The results (Table 14.1) indicate that the 1976, 1979, 1981–83 and 1987–89 year classes were particularly strong in the Solent nursery areas (MAFF, unpublished data).

Considered together with the indices of 0-group abundance obtained from power station intake screen samples (Fig. 14.1), it appears that at least 8 of the last 15 (1976–90) year classes of bass have been of above-average strength, and that 6 have been relative failures. There appears to have been an upward trend in the index of year class strength over the period of sampling.

It has also been possible to demonstrate the variation of growth rate between different year classes (Section 6.10). The samples obtained by the trawl survey represent an important source of growth data for young bass, as most of the fish caught are under the minimum landing size. Such fish are difficult to obtain from commercial landings, and are then likely to be biased towards larger or faster-growing fish owing to gear selectivity.

Table 14.1 Year-group abundance indices from Solent trawl surveys, 1981–90 (MAFF unpublished data)

Age	*Survey time*	*Year class*													
		1976	1977	1978	1979	1980	1981	1982	1983	1984	1985	1986	1987	1988	1989
1	May						0		0			0	0		0
	June/July														
	Sept						0	2.12	1.23		0.03	0.02		6.81	23.63
2	May					0.51		0.95			0	0		2.84	
	June/July				0				24.33						
	Sept					3.25	9.87	1.38		0.27	0.05		6.68	2.81	
3	May				2.17		2.66			0.42	0.02		2.48		
	June/July			0.3				10.33							
	Sept				10.10	0.91	0.65		4.26	0.28		0.37	1.15		
4	May			0.16		0.43			3.18	0.47		0			
	June/July		0.25				2.56								
	Sept			0.38	1.88	0.09		1.31	2.27		0	0.02			
5	May		0.05		0.27			0.93	2.04		0				
	June/July	2.51				0.19									
	Sept		0.20	0	0.02		0.37	0.90		0.19	0				
Recruit index		1.86	0.40	0.18	1.35	0.44	1.40	1.59	2.25	0.38	<0.01	0.04	1.33	1.84	4.89

Table 14.2 Relative densities and indices of abundance of O-group bass in west coast estuaries (source, Don Kelley, unpublished data)

Year class	*River Camel*		*River Tamar*		*River Taw*	
	No. per m²	*Index*	*No. per m²*	*Index*	*No. per m²*	*Index*
1982	1.84	78 (base)	–	–	–	–
1983	0.32	14	–	–	–	–
1984	1.33	56	1.50	35	–	–
1985	0.22	9	2.47	58	–	–
1986	0.01	1	0.03	1	–	–
1987	0.19	8	0.10	2	0.39	12
1988	0.30	13	4.77	111	1.36	43
1989	1.01	43	4.52	105	1.31	41
1990	1.69	72	3.09	72 (base)	2.28	72 (base)
1991	0.26	11	0.24	6	2.28	72

Estuarine seine net surveys and rod-and-line catch rate

Independent estimates of bass year class strength have been made by Donovan Kelley (pers. comm.), who initially based these on 'standardized' angling catch rates of 4–5-year-old fish, but more latterly used seine net surveys in estuaries to monitor 0-group and 1-group bass. Table 14.2 shows Kelley's year class abundance indices of bass for South-west England, which are not published elsewhere. The trends in relative year-class strength are similar to those based on power station intake screen samples (Fig. 14.1). Kelley's data also indicate that there is frequently a discrepancy in the abundance of recruits between the north side (Bristol Channel) and the south side (western English Channel) of the Cornish Peninsula.

14.4 POSSIBLE CAUSES OF STRONG AND WEAK YEAR CLASSES

The bass population which is exploited in the eastern English Channel and southern North Sea spawns during February–May in the English Channel (Thompson and Harrop, 1987), and 0-groups recruit to the sheltered waters of the Solent (Fig. 14.2) in July and August.

They remain inshore along the central southern English coast for 4–6 years, before recruiting to the adult population (Pawson *et al.*, 1987). Because Britain is near the northern limit of the geographical range of bass, many authors have suggested that sea temperature will affect the distribution, reproduction, growth and mortality of bass there (e.g. Holden

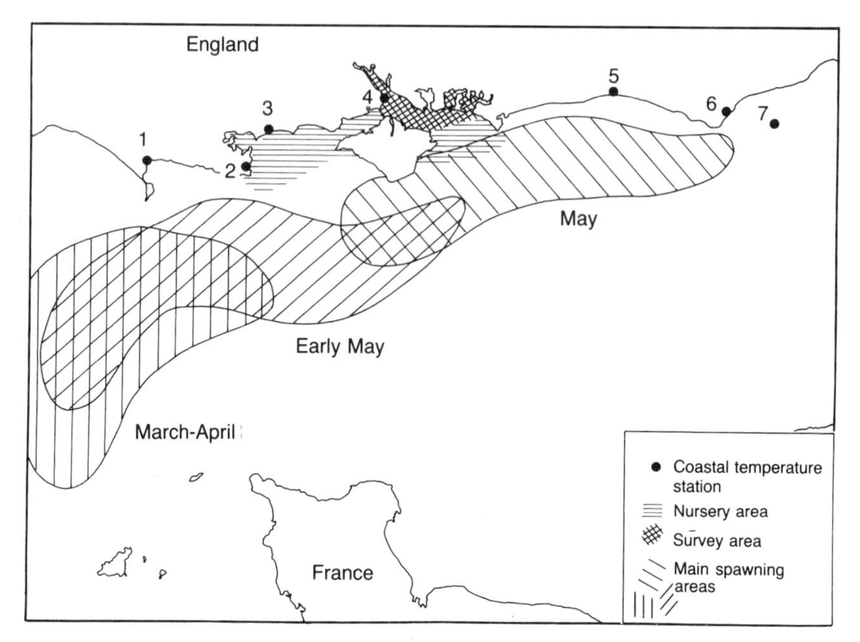

Fig. 14.2 The central English Channel, showing bass spawning areas, nursery area, trawl survey area and coastal temperature recording stations. From Pawson, 1992.

and Williams, 1974; Devauchelle, 1984; Kelley, 1987; Pawson *et al.*, 1987). It is also thought that unusually low temperatures cause increased mortality of bass during their first winter (Chapter 4).

Sea surface temperatures recorded at several coastal stations near the spawning and nursery areas in the central English Channel have been used to calculate the annual deviations of monthly means from the 10-year (1980–89) mean for the periods November–March, March–May and May–November (Fig. 6.10; Pawson, 1992). There is a pattern of warming through 1981–83, cooling in 1984–87, and a rapid warming through 1988 and 1989, which correlates positively with the abundance indices of bass year classes (Fig. 14.3). It is possible that the distribution of adult bass prior to and during spawning and the timing of gonad maturation, and consequently the level of recruitment of 0-groups to nursery areas, are affected by temperature. However, temperature trends prior to and during the spawning season (Mar.–May) did not appear to be as closely linked with the subsequent abundance of each year class of juvenile bass as were those during the following summer and first winter. It seems, therefore, that growth opportunities during the first year and the subsequent overwintering survival of 0-groups, have the greatest influence on year class strength.

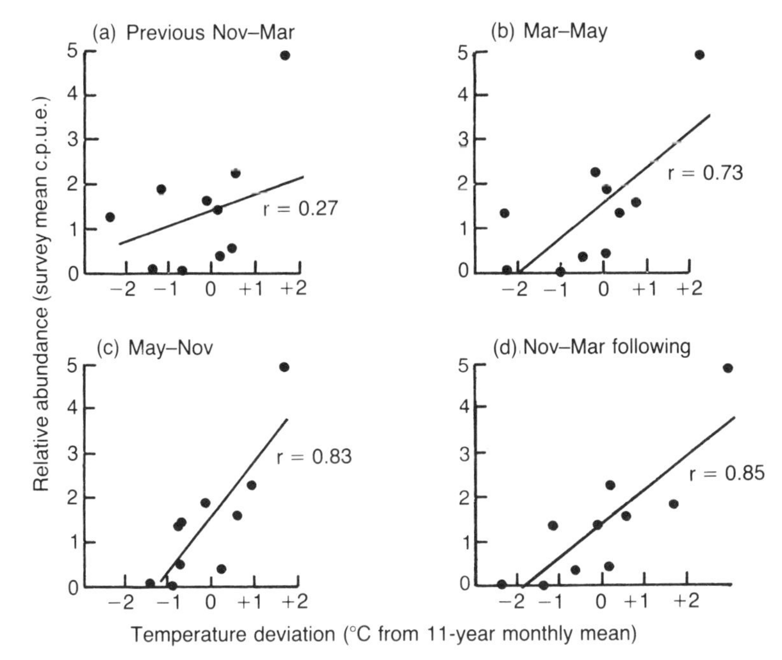

Fig. 14.3 Relationship between relative year class strength of bass and seasonal water temperature deviations (Bournemouth) from 1979–1990. Lines fitted by linear regression. From Pawson, 1992.

The growth of juvenile bass in the Solent nursery area also appears to have been affected by this climatic pattern (Chapter 6). The mean lengths reached by successive year classes of bass at the end of their third year are positively related to temperatures in the nursery area during the two preceding growth periods (May–Nov.). There is also evidence of a positive relationship between growth and year class strength, in other words, above-average growth of 0-, 1- and 2-group bass is usually found in year classes of above-average abundance. This supports the suggestion that interspecific competition and food availability are not usually limiting factors in the growth of juvenile bass, but that temperature is more important (Kelley, 1988b; Lancaster, 1991).

The comparative failure of the 1984–86 year classes caused some concern as to the state of the bass spawning stock. It is now evident that changes in the climate during the 1980s – as represented by temperatures of coastal seawater – had a considerable influence on recruitment of bass to the population in the English Channel, and on the subsequent growth and survival of juvenile fish (Pawson, 1992). Prior to 1983, bass in the Solent reached 36 cm (now the EC minimum landing size) as 5-groups, on average, but warming during 1981–83 resulted in faster growth of the strong 1981 and 1982 year classes, which reached 36 cm almost one year earlier. Growth of the locally strong 1983 year class was depressed during the cooler period from 1984 to 1987, but warming in 1988–90 once again enabled the above-average 1987–89 year classes to recruit to the fishery as 4-groups. Thus, not only has climate influenced the strength of recruitment to bass nursery areas, but it also has had additional effects on recruitment to the fishery, and appears to have a stronger effect on growth and survival than do changes in relative abundance.

Chapter fifteen

Assessment of the impact of fishing on the fishable stock

15.1 TOWARDS A STRATEGY FOR MANAGEMENT OF THE BASS FISHERY

We have already described how the level of recruitment of bass to the population is subject to natural fluctuations in spawning success and 0-group growth and survival. Around the UK, the abundance of 0-group bass appears to be related more to climate and weather conditions than to changes in the abundance of the adult stock through exploitation; nevertheless, Man's activities increasingly become the major cause of mortality of bass after their first year. Some of this mortality may be caused by pollution or impingement on power station intake screens, but most will be directly related to fishing activities. Some bass will be killed unintentionally, for example by anglers who may hook a small bass deeply and would otherwise have returned the fish alive, or by fishermen using small-meshed nets for other species. Nevertheless, the main impact on bass stocks, around the UK at least, has resulted from fishing directed at bass from around 25 cm upwards. Different parts of the fishery have targeted different sizes of bass to meet a particular need, whether to supply the local restaurant trade, by using gill nets which are highly selective for size, or when hunting large fish with rod and line for sport. The availability of fish to each sector of exploiters will vary from season to season and according to year class strength, and the exploitation pattern of different parts of the fishery will also, to a certain extent, affect who catches what.

Control measures that are designed to modify exploitation patterns (age-related mortality) will also change the level of exploitation of a fish population, if those fish which continue to be caught are more or less vulnerable than those which were previously exploited. So, to decide how best to conserve bass, and to understand how control measures will affect the fishery, we need to know how the bass population is affected by fishing activity. Most of the data presented in Part 2 were collected with this objective in mind, and this chapter explains how they have been used to investigate the problems facing the bass fishery and to indicate how a solution may be found.

15.2 STATE OF THE FISHABLE STOCK

DFR's estimates of the total annual catches of bass in England and Wales, in numbers at each age for the years 1985–91, are shown in Fig. 15.1.

These are based upon landings data obtained by port-based fishery officers and through the DFR logbook scheme (Chapter 12), which have been used to raise the numbers at age caught in each of the ICES divisions bordering England and Wales, using age–length keys and length distributions obtained from samples of catches taken each year (Chapter 6). The age-structure of the catch on a national basis has changed as a result of the different strengths of recruiting year-classes. The fishery from 1985 to 1988 was dominated by young fish of 3–5 years old. In recent years a higher proportion of older (7–9 years) fish has been taken, although the good 1987 year class made an impact on the fishery at 4 years old in 1991.

An examination of logbooks which were provided by the same group of fishermen over a period of 8 years (1985–91), revealed fluctuations in the annual catches of adult bass taken by a standard level of fishing effort (Chapter 12). There has also been considerable variation in the average daily catch rate by the regional fleets over the period, with declines in VII e (W. channel) and IV c (North Sea), for example (Fig. 15.2). In some local fisheries (e.g. West Cornwall), catch rates declined during the second half of the 1980s, when they were increasing in other regions (e.g. South-west Wales). On the basis of these catch and catch at age data, there is no positive evidence of any overall change in the abundance of adult bass in the fishery as a whole during the study period, though the CPUE data suggest that stocks may have declined in some areas as fishing pressure on them has been increased.

To reveal the age structure in the catch, which is characteristic for each region and depends largely on fishing patterns and gear used, the age frequency distributions of bass caught in the various regions of the English and Welsh fisheries were averaged over the years 1984–1987 (Fig. 15.3).

It can be seen that bass then generally recruited to the fishery between ages 3 and 5, but Fig. 15.1 shows that these patterns were distorted by the differences in relative abundance of each year class. Since 1981, the strong 1976 year class has been well represented throughout the fishery: even in 1991 when, at 15 years old, this cohort accounted for around 7% of the total UK catch in numbers. This was despite the presence of other subsequent large year classes, including that of 1983, which accounted for 26% (in numbers) of the catch in 1991. In contrast, the 1977 and 1978 fish have been relatively scarce, particularly to the east of Devon. The 1979 year class contributed strongly to catches in the central English Channel in 1983 and 1984, although it did not recruit significantly to the commercial fishery outside this region until 1985. The abundant 1982 and 1983 year classes were exploited from age 3, with peak catch rates (in numbers) occurring at age 5 and 6, respectively. This has resulted in the fishing pattern in many regions being directed towards the younger age groups, and it was apparent throughout the 1980s that these small fish were the main target of a large part of the fishery for bass. This was a major cause of concern for the future of the bass and its fishery.

Thames Estuary and southern North Sea

The fishery here appears to have been directed at bass mainly within the 7–11 age range and, of recent strong year classes, only the 1976 brood featured in landings until 1986, when the 1979 year class first made an appearance. The age structures of catches taken offshore and inshore and at different times of the year were similar, though most of the very large bass tended to be taken near sand banks and around wrecks situated well offshore.

Eastern English Channel

Small bass of 3–5 years old have been the mainstay of the fishery in the eastern English Channel throughout the 1980s, but adults became more important as fish of the abundant 1976 year class grew and left nursery areas. At times when there was a preponderance of juvenile bass of a particular age group in the Solent, they attracted much of the fishing effort, and the pattern of exploitation has been markedly influenced by the strength of recruiting year classes. Conversely, when two or three poor year classes followed each other, effort moved onto larger fish. This phenomenon was seen clearly in 1989 and 1990, when the weak 1984–86 year classes were noticeably absent as 3–5-year-olds. Many boats were forced to fish further offshore, often changing from gill netting to angling or longlining in the less sheltered water where larger bass were found.

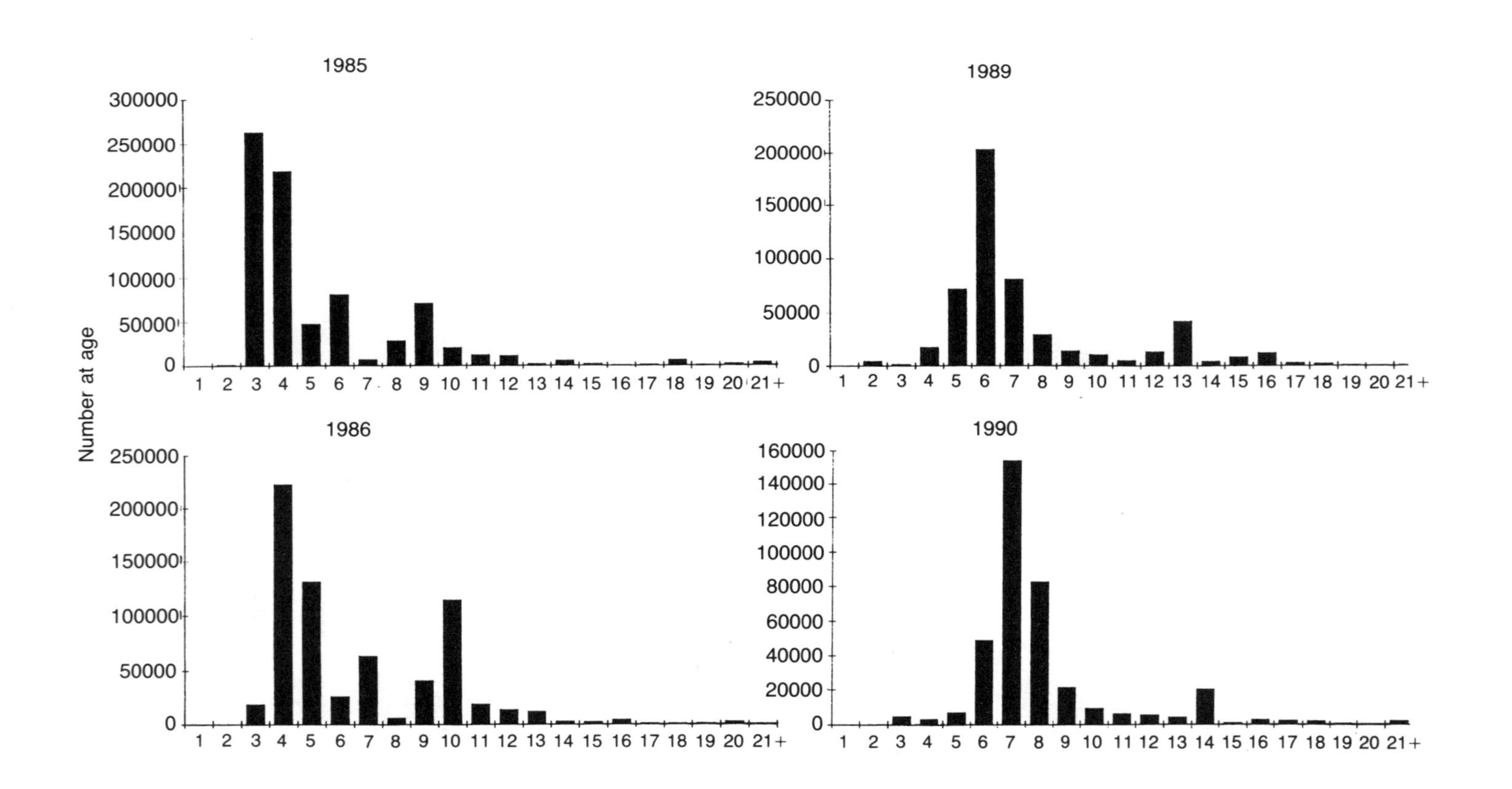
1985
1989
1986
1990
Number at age
300000
250000
200000
150000
100000
50000
0
160000
140000
120000
100000
80000
60000
40000
20000
1 2 3 4 5 6 7 8 9 10 11 12 13 14 15 16 17 18 19 20 21+

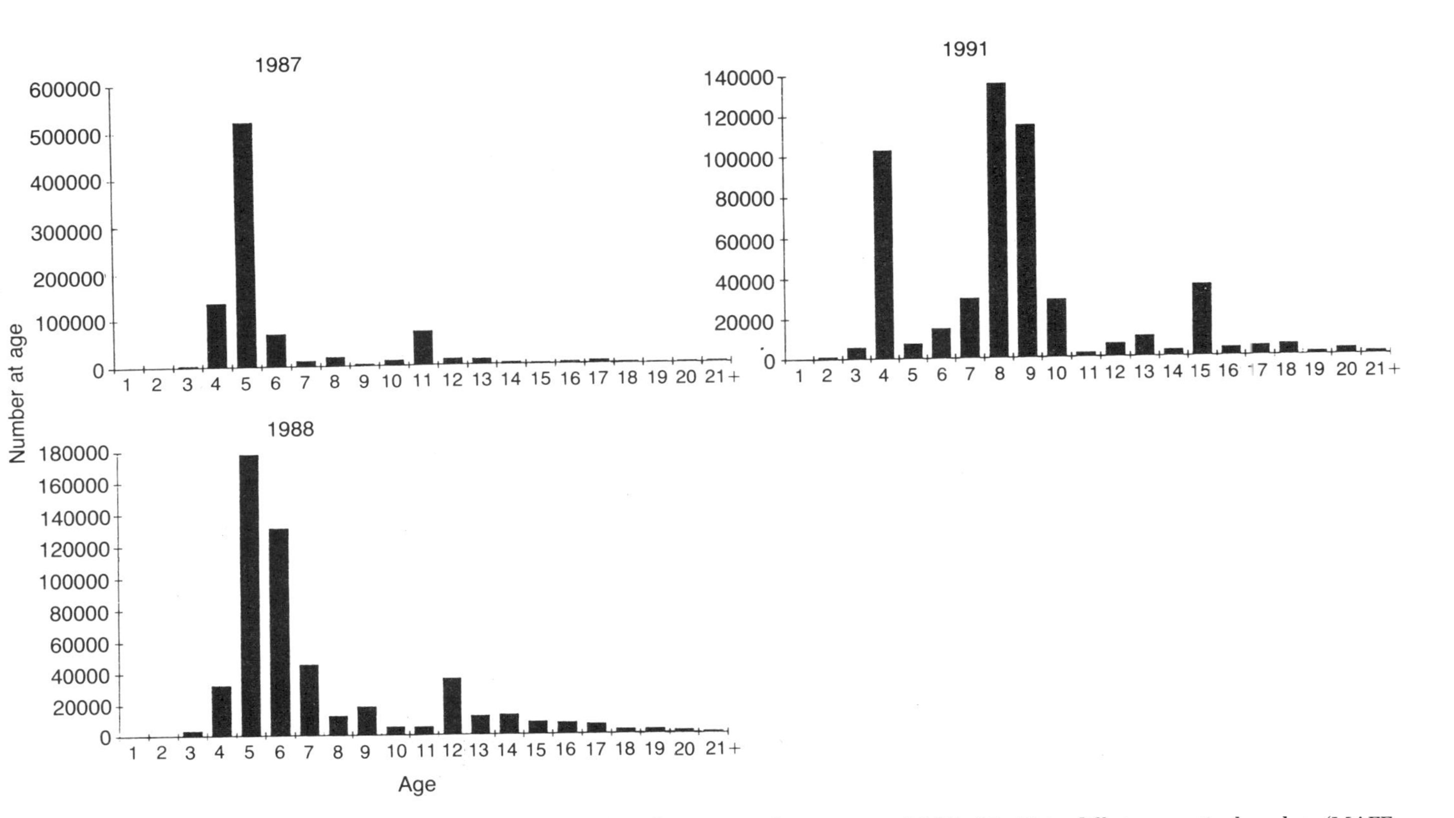

Fig. 15.1 Total annual catches of bass in England and Wales, in numbers at age, 1985–91. Note differing vertical scales. (MAFF unpublished data.)

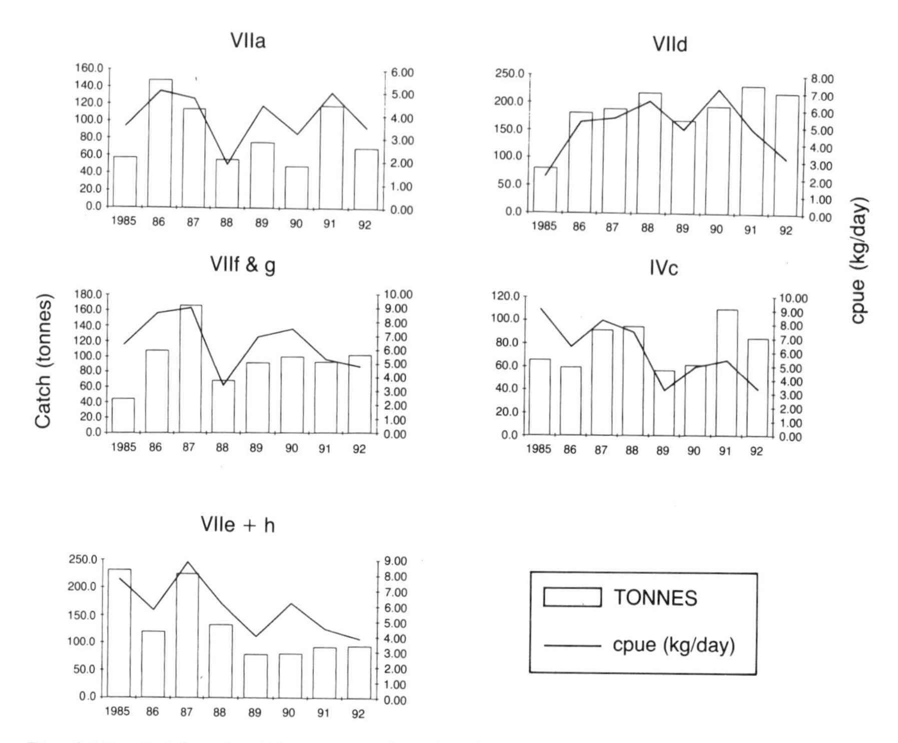

Fig. 15.2 Catch rate of bass in various ICES regions. Columns, catch (tonnes, left axis); curve, CPUE (right axis). Note differing vertical scales. (MAFF unpublished data.)

Western English Channel

This fishery has exploited the full available age range of bass and appears to have been least affected by year class strength fluctuations during the 1980s. There was, however, a clear distinction between the fishery for juvenile bass (mainly 3- and 4-year-old fish), which took place in the estuaries, and that for the adults around reefs and further offshore. Recently, however, the estuarine fisheries have been greatly restricted by controls on gill nets introduced as local authority bye-laws under the *Salmon Act 1986* (Great Britain–Parliament, 1986).

Bristol Channel and Celtic Sea

Five to seven-year-old bass predominated in catches here, partly because gill nets are used with mesh sizes too large to catch younger fish (the minimum mesh size is 100 mm in South Wales), and also because the

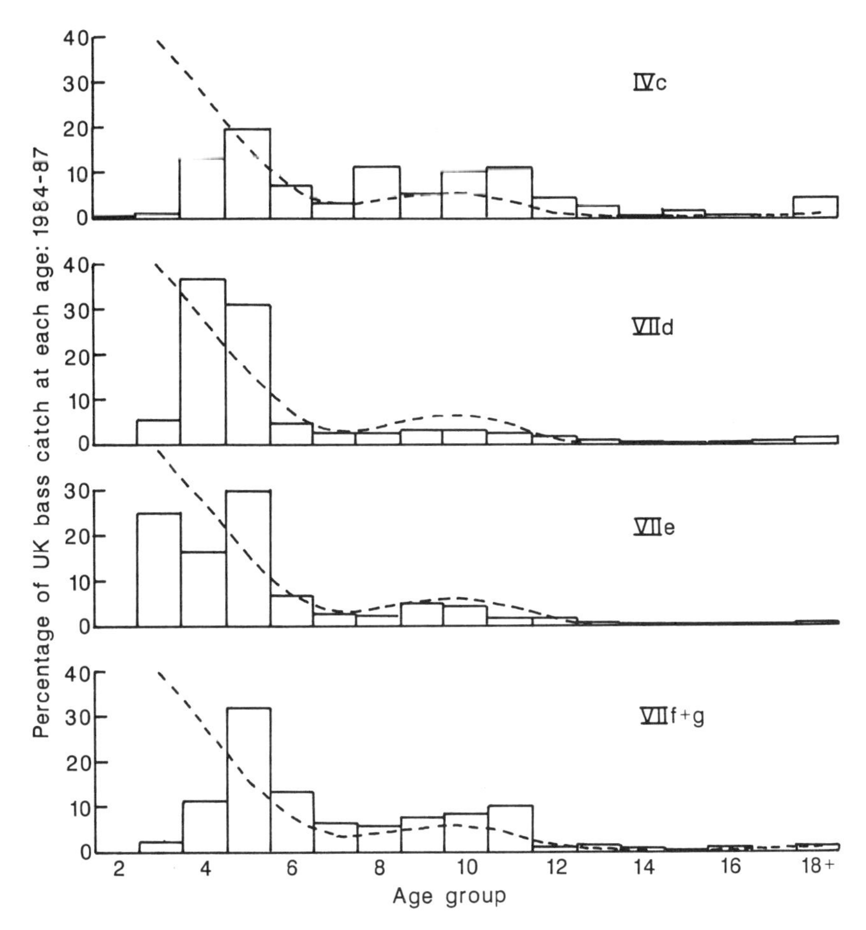

Fig. 15.3 Age frequency distributions of bass in various ICES regions, averaged over the years 1984–87. From Pawson and Pickett, 1988.

fishery takes place in a very exposed area and fishing boats tend to remain close inshore, where there is a reduced chance of taking older bass during much of the year.

Irish Sea

The commercial fishery in the Irish Sea, from West Wales northwards to Cumbria, has been restricted mainly to 4–8-year-old bass, because it tends to take place close inshore and there is a gill-net mesh size regulation (89 mm minimum under SFC bye-laws). There was a relative scarcity of older fish in landings in the early 1980s. More recently, rod-and-line fishing and longlining have increased owing to the restrictions on inshore netting, and this trend has resulted in more of the larger bass being caught. A contributory factor to this change is the more northwards dispersion of large bass from the 1976 year class which, since 1982, has supported some previously insignificant fisheries (e.g. in Morecambe Bay). There is good evidence that the 1959 year class dominated the west coast population in the same way until the early 1980s (Kelley, 1988b). It is possible that the adult bass stock, from Mid-Wales northwards, is composed mainly of such outstanding year classes.

15.3 DYNAMICS OF EXPLOITATION

For the purposes of assessing the effects of exploitation on fish stocks, it is necessary to determine the age distribution within the stock, rather than just in catches. For the major commercial species, this is achieved by using virtual population analysis (VPA, Pope, 1972), but adequate data are not yet available for bass. An approximation of the population structure of bass was obtained by taking an average of the numbers at each age in the total UK landings, over the years 1984–87, and extrapolating back to 2-year-old fish (as represented by the dashed lines in Fig. 15.3), i.e. crudely correcting for the effects of gear selectivity and regional variations in population structure (Pawson and Pickett, 1988). The exploitation pattern of the fishery in each region is then given by the ratios of the proportion at each age recorded in the catch there to that in the stock as a whole. These ratios have been used to model the changes in the fishery's catch for a nominal level of recruitment (i.e. yield per recruit), which would be expected with changes in the age at which the bass are first caught. The effects that different exploitation patterns have on yield per recruit at a constant level of fishing effort are illustrated in Fig. 15.4. These calculations can be used to estimate actual yields, or variations in yield, if the absolute or relative abundance of each year class is known.

There are four important assumptions in this approach which require examination to judge the confidence which can be placed in the results:

1. That the overall fishing mortality level on the stock affects each part of the fishery equally. Total annual mortality values for fish aged 7 and over appeared to be similar in fisheries throughout the English Channel and southern North Sea, and tagging has shown that bass populations there mix freely. Until the mid 1980s, the population on the west coast was exploited in the Cornish winter fishery, where the age structure of adult bass in catches suggested that mortality rates were no higher than

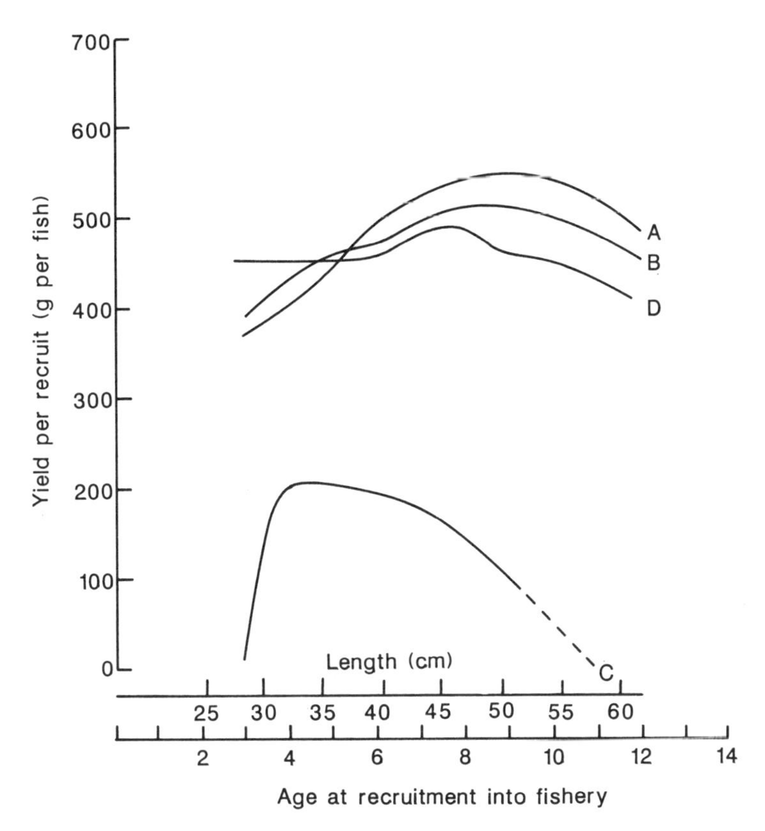

Fig. 15.4 Effect of increasing the age at which bass are first caught, on the yield to fisheries with the exploitation patterns of (a) the total UK fishery, (b) the Devon and Cornwall fishery, (c) the west coast fishery, (d) the Thames Estuary fishery. In each case, the total instantaneous mortality (Z) is 0.25 and the instantaneous mortality attributable to fishing (F) is 0.15. From Pickett and Pawson, 1988.

elsewhere. More recently, large bass have been well represented in the South Wales fishery and further north. This suggests that the low numbers of adults in earlier catches on the west coast were not necessarily the result of higher overall exploitation rates there, but were rather the result of a fishery which was directed chiefly at 4–8-year-old fish.

2. Mortality from causes other than fishing is constant, regionally and with age. Even in well-studied fish stocks, it is difficult to quantify 'natural' mortality other than to use it to explain the difference between estimates of the total mortality of the stock and that caused by the fishery. For bass, the assumption is that from the age of 2 or 3 years onwards, approximately 10% of the population dies each year as a result of non-fishing

causes. When applied to a hypothetical unexploited population age structure, this matches the observed longevity of the species in UK waters of 20 years or more. We conclude, therefore, that there is no reason to assume other than a constant level of mortality on the exploited part of the bass population throughout its UK range.

3. All year classes are represented in each part of the fishery in proportion to their overall abundance. Although there are geographical, seasonal and year-to-year variations in the relative proportions of adult and juvenile bass throughout the UK fishery, the age/size distributions within regions appear to have remained similar during the years 1984–91. Apart from environmental and climatic influences on seasonal movements, and the effect of the fishery itself, the most likely cause of a change in the bass distribution is when a very large year class spreads beyond the normal range of the juvenile bass population, which is associated with unusually warm years. It appears that variation in year class strength has had the greatest influence on the proportions of each age group in different parts of the bass population's geographical distribution. This effect is taken into account in the calculations by the use of appropriate regional exploitation patterns.
4. The strength of recruiting year classes has no effect on the exploitation pattern. After 1980, the unusually abundant 1976 year class attracted considerable fishing effort. During 1981 and 1982, when these fish were in the size range 35–45 cm, only a small proportion of the landings in some fisheries (e.g. Solent, parts of the west coast) contained bass over 45 cm. In 1983 and 1984, however, the increased size of the 1976 fish and the relative scarcity of smaller, legal-sized, bass led fishermen to use catching methods which could take larger bass (e.g. gill nets with meshes of 100 mm or larger and longlines) and much greater numbers of 9–16-year-old fish were caught. This trend was seen to reverse in some fisheries in 1985–88, when the above-average year classes of 1979, 1981, 1982 and 1983 recruited. This last assumption is therefore not upheld, but the use of catch-at-age data averaged over 1984–87 helps to minimize bias in the calculations.

The observation that fishing patterns change according to the relative abundance of recruiting age groups has an important implication for management of the fishery. It suggests that the amount of fishing effort aimed at bass would not necessarily diminish if the mean age of recruitment was to be raised by technical measures (within reason). Because the average size of fish caught will be larger, the yield to the fishery should increase, provided that the part of the bass population still available to the fishery is no more difficult to catch than that part which is then protected.

Effects of recruitment size on the fishery

It can be seen from Fig. 15.4 (curve A) that a maximum yield (in catch weight) for the UK bass fishery as a whole might theoretically be obtained if bass were first caught at an average length of 50 cm, provided that the fishing effort could be effectively aimed towards larger bass, and that fish below 50 cm suffered natural mortality only. It is encouraging to note that the yield curve for the fishery in Devon and Cornwall (Fig. 15.4, curve B) approximates to this ideal, possibly because the fishery in that region has the capability to exploit bass of all ages, from juveniles in the estuaries to adults well offshore. On most of the west coast, however, it appears that maximum yields would only be obtained with recruitment to the fishery at between 32 and 36 cm (curve C), and that a higher value would not enable yields to be maintained with the fishing pattern of the early 1980s. In the Thames Estuary, by contrast, bass under 40 cm have not been as predominant in commercial catches as elsewhere, and a relatively small increase in yield would be achieved by raising the recruitment size, up to a maximum at about 46 cm (curve D). The inshore fishery in and around the Solent is less easy to model, in view of the strong influence that the local abundance of recruiting year classes has on exploitation patterns. There appears to have been a peak of catchability in the size range 32–40 cm, and yields are likely to decrease with a larger recruitment size, unless the fishery could change its exploitation pattern by targeting larger fish, as was the case in 1989 and 1990. This flexibility in the fishing effort would probably enable it to adapt to changes in the size at recruitment up to at least 40 cm, in order to maintain yields.

From the above assessment, made in 1988, it was concluded that heavy exploitation of bass at an early age, and often whilst under the legal minimum landing size (then at 32 cm, 4 years old), had prevented the potential yields from being realized during the 1980s. Whilst it might be desirable to increase protection for the more vulnerable small fish, and to allow them to grow and maintain or raise yields generally and to safeguard recruitment to the spawning stock, this must be weighed against the need of fishermen in all regions to exploit bass of a size that would enable them to sustain their income.

These predictions, of the effects of changes in recruitment size on potential yields in the UK bass fishery, were based on the assumption that fishing patterns in the various regions around England and Wales are well established, having evolved to take maximal advantage of local resources (including species other than bass). If it was not possible to modify the pattern of local effort, to take account of an enforced change in the recruitment size for bass, then the benefits might not be obtained in the

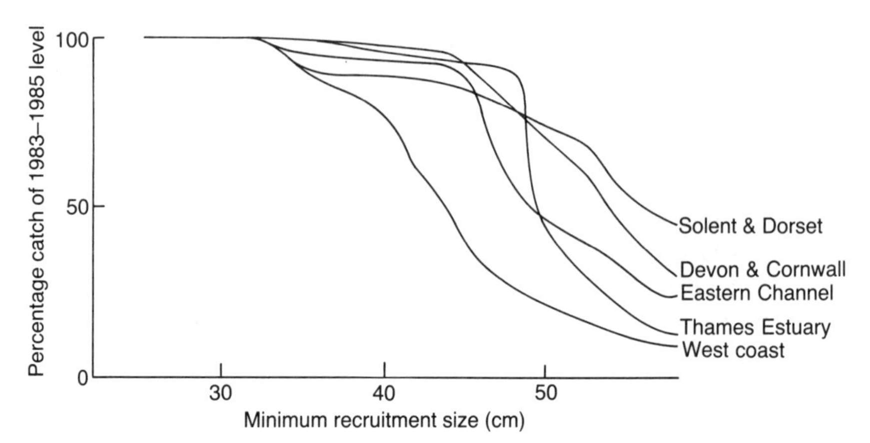

Fig. 15.5 Expected short-term losses to the bass catch (in weight) in five regions of England and Wales in relation to size of first capture. From Pickett and Pawson, 1987.

same fishery. Although this seemed unlikely, given the flexibility demonstrated by the bass fishery in the Solent, for example, the prospect of a decrease in effort on bass in some regions if the size at recruitment were raised to too high a level by technical measures required quantifying. Estimates of the magnitude of this potential loss in each region are illustrated in Fig. 15.5, which shows the proportional catch in weight that would occur in each region with the fishing pattern observed there in the early 1980s, at various recruitment sizes.

On the west coast and in the Solent, an increase above 32 cm in the size at which bass recruit, could have led to a progressive decrease in catches, but elsewhere it appeared that the short-term loss would be less than 10%, even with a recruitment size as high as 44 cm. At this level, the UK fishery as a whole would have probably suffered an immediate decrease in bass catches of less than 10%. It was expected, however, that the fishing pattern would soon be changed to compensate for this loss, although the expected longer-term gain would only materialize if the fishery could be restructured towards catching and marketing the larger bass. One important benefit, of course, would be a greater stability in the fishery as a whole, as it would not be so dependent on the strength of one or two successive recruiting year classes.

Effects of year class size on the fishery's yield

Because the strength of recruiting year classes may affect the exploitation pattern, yields in some parts of the bass fishery can vary considerably from year to year as strong or weak year classes pass through. The

Table 15.1 Contribution to UK fishery (number of fish caught) of bass at ages 4–6 for the year classes 1979–85. (MAFF unpublished data)

Year class	*Numbers caught*
1979	500 649
1980	231 007
1981	419 294
1982	875 332
1983	527 908
1984	154 149
1985	38 985

number of fish taken will also vary, depending on the size of fish being exploited. This effect is apparent in Table 15.1, which is derived from assessments of catch at age in the bass fishery between 1983 and 1991. It also shows the contribution to the fishery of succeeding year classes.

Effect of changes in growth and age at recruitment on the fishery

There may be considerable differences in growth between year classes of bass (Chapter 6) and, consequently, the mean age at a particular minimum length of recruitment to the fishery will not be constant. The 1976–79 year classes, for example, recruited to the fishery as 4-groups, at a length of around 32 cm. Faster-growing year classes from the early and late 1980s would have reached this size as 3-groups. This can only be to the advantage of the fishery, because (a) the yield from a particular cohort may be taken earlier, (b) there will have been 1 year less natural (and undersize fishing) mortality, and (c) escapement to the spawning stock is unlikely to be reduced, because first maturity appears to be more closely related to length than to age. As movements and migrations also appear to be linked to size and maturity, faster-growing bass may be maximally exploitable by some parts of the inshore fishery for only a short period, compared with slow-growing fish. The main effect of a decrease in growth rate will be that a year class will be exposed to the inshore fishery for a longer period than usual.

15.4 SUMMARY

From this assessment, it appeared that the potential yields from the bass fishery around England and Wales were not being realized in the early 1980s, and that the trend of increasing exploitation on each succeeding year class at an early age could be leading to a reduction in recruitment to the spawning stock. It was concluded that precautionary action was called for, and the most effective strategy would be to increase the size at which bass are first caught in the fishery. This could, in theory, be raised to around 45 cm in such regions as the Thames Estuary and around Devon and Cornwall. With the exploitation patterns observed in the mid-1980s, however, there would have been little benefit to the commercial fisheries in the Solent and on most of the west coast in attempting to achieve a minimum size at recruitment higher than about 36 cm. Nevertheless, evidence of a decline in catches of adult bass on some parts of the west coast suggested that the fishery in this region should be directed away from bass under 40 cm, in order to improve recruitment to the adult stock.

Even with the low levels of fishing observed in the period 1983–87, it was unlikely that any modification to the exploitation pattern that would be acceptable to the fishery would result in the number of adults in the stock approaching that which would exist in an unfished population. This is essentially what the participants of the recreational fishery would like to

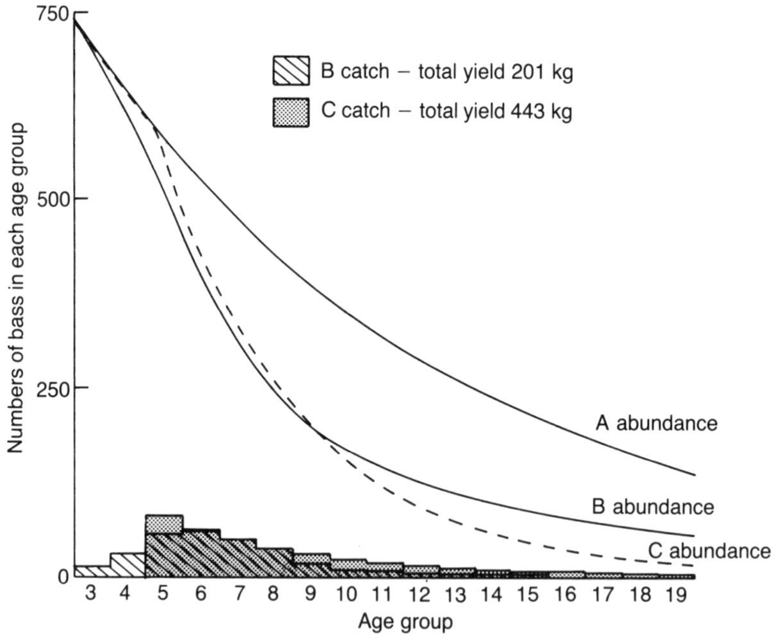

Fig. 15.6 The effect of exploitation pattern at a constant level of fishing activity on the abundance and yield (catch in numbers at each age) to the fishery of a standard year class (starting as 1000 0-group) of bass as it goes through life: (a) no exploitation; (b) maximum exploitation at 4–8 years old; (c) constant exploitation from age 5 onwards. From Pickett and Pawson, 1987.

see. The abundance of bass in each age group, which would be expected from a nominal 1000 0-group entrants each year, is shown in Fig. 15.6, for three different exploitation patterns.

It is apparent that the number of recruits entering the adult stock would be similar, whether the fish were not caught before the age of 5 (as a result, for example, of stricter protection of juveniles), or maximally exploited between the ages of 4 and 8 inclusive, as occurred in many areas during the 1980s. In the latter case, there would be a lower mortality of adults and, theoretically, more large bass in the population, but this would result in the fishery forfeiting a potential doubling of the yield from the bass stock. Recent increases in growth rates and the strength of recruiting year classes may result in a higher yield being taken in the UK bass fishery, despite the exploitation pattern accounting for younger fish.

Summary to Part Two

In this part of the book, we have described the ways in which bass are exploited in the European fisheries for which information is available. We have shown how the level and pattern of exploitation can be assessed, with particular reference to DFR's investigation of the bass fishery in England and Wales, and have discussed the effects of exploitation on the bass stock. The conlusion is that, although the bass stock around England and Wales has not to date been subject to excessive exploitation by local fisheries, there has been an undesirable trends towards targeting juveniles and some growth overfishing has occurred. Clearly, bass stocks could be further depleted if there was poor recruitment for three or four successive years, as occurred in the mid 1980s. It has been encouraging to note that there are signs of replenishment following good recruitment in the late 1980s and early 1990s and, although there is therefore no immediate concern about the state of the spawning stock, this might be sustained, and a better yield obtained from the bass fishery, by a change in the exploitation pattern.

Sufficient knowledge of the bass population, and its fishery, has been gained during the 1970s and 80s to enable the relative levels of recruitment to stocks around England and Wales to be assessed and the impact of exploitation to be evaluated. Advice can now be given on how the bass fishery may be managed so that stocks around England and Wales are not depleted to the extent that the spawning population is too small to maintain at least average recruitment, given favourable climatic and environmental circumstances. The management measure proposed by Holden and Williams (1974) – a 38 cm minimum size limit in the UK – might have sufficed to protect the resource itself, were it adequately enforced, but we are now in a much stronger position to advise a Government confronted with protests, proposals and counter-proposals from increasingly well-informed angling pressure groups and commercial fisheries interests.

Part Three

Conservation and Management

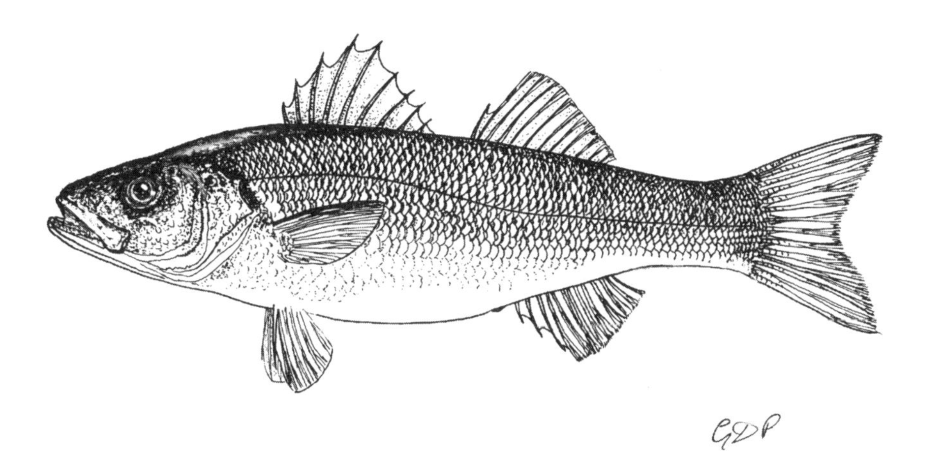

Sea Bass.

Introduction to Part Three

In the introductory chapter, we outlined the case for conservation of bass stocks around the UK as put to the government by sea anglers during the 1960s and 1970s. The gist of their complaint was that big fish in particular were no longer being caught in the numbers they had come to expect, and that 'conservation' action was required before the population of their favourite sport fish dwindled to a level at which it would no longer be able to replenish itself. The symptom was identified as overfishing – caused, initially, by anglers themselves – but latterly the accusing finger was pointed at the commercial fishery and, in particular, the increasing use of gill nets. Such a simple perception of the situation did not permit us to evaluate the real impact that exploitation was having on bass stocks, to identify who, if anyone, was to blame, and to determine what remedial measures should be considered.

In Part Two, we have described the fisheries for bass, historically and in the 1980s, and have come to the conclusion that, with the growth of that part of the commercial fishery that targets juveniles, bass stocks could be seriously depleted, and the fishery could collapse, if three or four successive years of poor recruitment occurred. In this, the final part of the book, we discuss the conservation needs of bass and outline the problems that have been faced in bringing about better management of the species' fishery. To illustrate the objectives of, and constraints on, regulatory actions, we present opposing views of those with an interest in the bass and its fishery, and describe the legislative framework within which controls on the exploitation of the bass resource have been implemented in the United Kingdom.

Chapter sixteen

Conservation requirements

16.1 TO EXPLOIT, OR NOT TO EXPLOIT

The question of whether bass should be exploited at all might appear academic, but recreational anglers could argue that by releasing alive all the fish they caught, there need be no exploitation as such, just enjoyment of an amenity. After all, catch and release is the accepted policy for all freshwater fisheries in the UK, excepting those for salmonids and eels, though it is not the norm in the rest of Europe. However, when a fish is good to eat and has a market value, it would be foolish of any government to attempt to preserve stocks in their pristine state, even though such a stock would contain many more of the larger (older) fish that sport anglers so desire, than if it was exploited. Fish populations produce a higher yield to a fishery as exploitation increases, up to a point (Beverton and Holt, 1957), and the optimization of the taking and allocation of this surplus production is the *raison d'être* of thousands of fisheries scientists, administrators and managers throughout the world. Even in a fishery with the social and economic (i.e. political) importance of that using rod and line to catch Atlantic salmon, it is acknowledged that a crop should be taken; the argument, as with bass, is about how much and, especially, by whom.

A policy of rational exploitation should ideally incorporate some biological objective against which the requirements of all interested parties can be measured; for example, a total allowable catch (TAC), a precautionary minimum level of spawning stock biomass, or the quotas required by the commercial fishery's various sectors to stay viable. The conservation objective could be to sustain the bass spawning stock at or above a level that is known to be capable of producing strong year classes of juveniles, whilst allowing the fishery, both commercial and recreational, reasonable access to take catches that at least satisfy some minimal criteria. In the case of the commercial fishery, this requirement could be the ability of an individual operator to make a reasonable living from bass fishing, in season. For the angler, a quality of sport that encourages further participation in this pastime might suffice. It is not possible to compute with any certainty the respective appropriate catch levels, in view of the variability in fishermen's skill and the running costs of their boats, and the enormous range of expectations that anglers have of what constitutes good fishing. Nevertheless, CEMARE's economic appraisal of the UK bass fishery (Dunn *et al.*, 1989) has shown that the attitudes of both sectors to taking part in the fishery can be evaluated against factors such as changes in catch rates, market prices, constraints on access to the resource, or even the probability of catching a bass above a certain size.

The most recent indices of juvenile bass abundance (Chapter 14) suggest that the spawning stock of bass around Britain is continuing to provide at least average recruitment to the population, and it could be argued that the stock might be sustained if enough protection were given to bass up to the time of first maturity. The question of whether to extend the protection given to juveniles to the spawning stock itself is also relevant, but recruitment has probably been better through the 1980s than for many years previously, when exploitation levels were certainly lower and adult bass were more abundant than nowadays. Unfortunately, catch and effort data have only recently been sufficiently comprehensive to enable an estimate to be made of the trends in the UK bass stock's size. It appears, however, that unless fishing effort for bass increases significantly, there is at present little danger of exploitation alone causing a stock collapse. The age structure in catches shows none of the classical signs of overfishing, such as a dramatic reduction in the number of adult age classes, the rapid decline of year classes of initial high abundance, or several successive poor recruiting year classes in otherwise favourable circumstances. The most consistent observation has been of increased targeting on juveniles by some parts of the fishery, which has underlined the need for precautionary management measures (Pawson and Pickett, 1992).

16.2 THREATS TO THE FISHES' ENVIRONMENT

Of more immediate interest is the impact of environmental change, whether as a direct result of human activities – barrages, marinas, power stations and pollution – or as the more indirect effects of possible global warming. In all these cases, it is the implications for the early stages in the life history of bass that matter most. The coastwards movement of bass larvae and their recruitment to inshore nursery areas brings them into a hazardous environment at the most sensitive stage of development. Their extended duration of stay in the juvenile habitat, which is probably longer than that of any other commercial fish species in the North Atlantic, has extended the species' vulnerability to human disturbance of the coastal zone (Jennings, 1992). Even episodal contamination of an estuary can wipe out a year class locally (Section 4.7), though the subsequent decrease in a district's juvenile bass population might be less significant than local annual variations in recruitment. A chronically polluted estuary, however, which would otherwise provide a productive nursery environment for juvenile bass, will lead to a permanent reduction in the potential recruitment to the fishery and to the adult stock.

The discharges of warmed water from power station condensers into coastal waters are known to affect the distribution and growth of juvenile bass, and these aspects have been studied in detail in the area of Kingsnorth power station (Medway Estuary) by Langford (1987). Mortality of bass owing to these elevated temperatures, or to the chlorine added to prevent the growth of fouling organisms in the cooling system, has not been demonstrated.

The entrainment of fish onto the cooling-water intake screens of coastal power stations (Fig. 16.1) is a well-known phenomenon, and bass have been recorded in large numbers at many such sites in the UK (Turnpenny,

Fig. 16.1 Band screens at a coastal power station.

Fig. 16.2 Sorting young bass from band screen washings.

1988, Van den Broek, 1979). Each year since 1970, CEGB* and MAFF have monitored the numbers of bass which become impinged on intake screens of some coastal power stations, e.g. Thurrock (River Thames) and Oldbury (Severn Estuary) and for shorter periods at Kingsnorth, Dungeness (Kent coast), Sizewell (Suffolk coast) and Fawley (Southampton Water). Until 1987, however, no attempt had been made to quantify the mortality of bass as a proportion of the local population.

A study was instigated by DFR and CEGB at the Kingsnorth power station site and locally in the Medway Estuary. Screen washings were sampled intensively for short periods during the autumn and winter in 1987/88 and 1988/89 to estimate the numbers of impinged bass (Fig. 16.2). The local distribution of bass was surveyed using a chartered vessel fishing with fine-mesh otter and beam trawls, and additional sampling was carried out from the shore with traps and seine nets. Around 4000 0-group bass were caught alive in trawls and dye-marked and released, and the numbers of marked and unmarked fish on the screens were monitored. In addition, the intakes were 'seeded' with dead, marked bass

*Central Electricity Generating Board – now privatized and split into separate companies – Power-gen, National Power and Nuclear Electric.

to determine detection rates by the sampling team. The local population of 0-group bass was also resampled by trawl to estimate the proportion of marked fish.

Standard Petersen indices (Petersen, 1891) and fractional survival rates (Rounsefell and Everhart, 1953) were used to estimate the size of the local 0-group bass population to be between 0.5 and 1.2 million fish (the 1987 year class was a relatively strong one), of which 2–5% were estimated to have been killed on the screens during the peak period of impingment during the 1987/88 winter. Thus, the level of mortality on 0-group bass, caused by a single power station such as that at Kingsnorth, is probably no greater than the variation in the annual natural mortality rate. At Kingsnorth, the outfall and intake are around 3 km apart. At power stations where the outfalls and intakes are closer together, impingement levels could be higher owing to the attraction for young bass of the warm-water discharge. Whilst there is no cause for complacency in this respect, given the current management aim of protecting the juvenile bass population from avoidable mortality, it is as well to be aware of another, less detrimental, side to the interaction between power stations and bass.

Bass of all sizes are attracted to the warm-water effluent from power stations (Langford, 1987) and, in estuaries such as the Medway and Thames, the survival of overwintering 0-group bass may be enhanced by the benign influence of this warm water. At Kingsnorth, for example, not only does the bulk of the 0-group bass population spend the period from November to March in and around the warm-water plume discharging from the power station, but the size of individual fish at the start of the normal second and third summer growth periods (May–June) can be nearly twice that of fish in similar, but unwarmed estuaries along the English south coast.

The implications for bass in nursery sites for which there are proposals to build tidal power barrages (e.g. the River Severn), are not so predictable. If the productivity of an estuary is not impaired by such schemes, it is possible that because of their small size, compared with migrating adult salmon, sea trout or potential predators, juvenile bass might actually benefit from the subsequent ecological changes. The tidal lagoon behind the barrage will become more like the *valli* used to rear bass extensively in Italy. The main drawback, of course, is the mortality of fish which pass through the generating turbines (Davies, 1988), and the potentially poor quality of the water which is impounded behind such barrages.

It is equally uncertain whether the proliferation of marinas and new port facilities around British coasts in recent years (Davidson *et al.*, 1991) is having any detrimental effects on bass populations. The silting of estuaries and encroachment of marsh vegetation in the absence of navigational management, reduces potential habitat area for juvenile bass, which are strongly associated with dockyards, piers and similar structures. With increased boating interest, however, comes a higher loading of sewage, petro-chemicals and antifoulants in coastal waters. The development of anti-fouling compounds containing tri-butyl-tin (TBT) has led to concern for the effects of this poison on oyster (*Ostrea edulis*) and other mollusc stocks in the 1980s (Waldock *et al.*, 1987), and the implementation of legislation to control the use of preparations containing TBT suggests that water quality in heavily used estuaries and harbours will be monitored more closely for other types of chemical contamination in future.

There is today a considerable lobby for better surveillance of water quality and habitat degradation in the coastal zone, coming from bird-watchers and conservation bodies (Davidson *et al.*, 1991) and from the fishermen themselves. Because bass are an integral part of the ecosystem which these people wish to protect, it is likely that any initiative along these lines will have a beneficial impact on bass stocks. We might suggest, moreover, that ongoing assessment of an estuary or harbour's status as a bass nursery area could provide a suitable biological benchmark against which to judge the severity of such perturbation against the known fluctuations in year class abundance, until each cohort has grown sufficiently to produce a 'reasonable' yield. In the meantime, the main emphasis of those who wish to conserve the bass fishery must now be placed on the maintenance of a spawning stock that is sufficiently extensive to provide good recruitment throughout the species' normal range, and the protection of the resultant juveniles from too high a level of exploitation.

Chapter seventeen

Management of the UK bass fishery

17.1 BACKGROUND

The habits of juvenile bass are well known to those who would exploit them and, in the south of England and Wales, they are accessible during the greater part of the year to fishermen and anglers operating from the shore and close to the shore in small boats. It is partly because of this vulnerability that the commercial bass fishery there has developed so rapidly in the last 15 years. Once bass mature, however, they adopt the habit of migrating between summer feeding grounds and overwintering areas to the south and west (Chapter 5). During this time, the larger bass are fleetingly accessible to the British inshore fishermen before they arrive at the offshore pre-spawning and spawning areas. Here, the main fishing effort has been by French midwater pair trawlers which, through the late 1970s and 1980s, have increasingly targeted this part of the bass population during winter and early spring. UK vessels fishing for other species sometimes take a bycatch of bass which, though relatively small in the offshore fishery context, would represent a significant tonnage if it was taken in the inshore fishery. Nevertheless, adult bass are still not subject to the high level of exploitation (annual fishing mortalities of between 0.5 and 1.0) that stocks of cod, sole, plaice and many other species experience, caused by mobile fleets which are not limited by the seasonal nature of local fisheries.

The consequence of this is that, during the 1980s, the UK Government was faced solely with the challenge of finding a means to direct the bass fishery away from juvenile fish, i.e. to improve the exploitation pattern. This strategy not only has the quantitative benefits of increasing yield per recruit (Section 15.3) and enhancing recruitment to the spawning stock, but also has a less tangible benefit of shifting fishing effort away from the vulnerable juvenile bass towards the relatively less catchable adults; i.e. the same level of fishing effort will give rise to lower overall mortality rates on the exploited stock. Put like this, fisheries management appears a straightforward task, but no-one should under-estimate the procedures involved.

It might seem to be unimportant for fisheries scientists to have a knowledge of the legislative framework within which management actions can be implemented, and some might profess that it is irrelevant to their declared task of providing information and forecasts based on accredited scientific data and an objective interpretation of the implications for the fisheries. In an applied scientific field, however, it might equally be argued that the implementation of management measures cannot be effected without appropriate regulations, which must be promulgated through enforceable legislation. It is important, therefore, to describe some of the legal technicalities relating to sea fisheries in the United Kingdom, in order to understand the constraints, and opportunities, that have led to the present management regime in the bass fishery.

17.2 MANAGEMENT CONSIDERATIONS

Since the mid 1970s, assessment of the impact of exploitation on British bass stocks has led to the conclusion that to reduce fishing mortality on juvenile bass, it is necessary to raise the size at which they first recruit to the fishery towards 38 cm (Holden and Williams, 1974). With the intention of protecting juvenile bass, by making it illegal to land them and by prohibiting the use of fishing methods that would catch or kill them, particularly in areas where they were most vulnerable, a package was proposed in July 1986 which combined a minimum landing size (MLS) of 36 cm, a minimum mesh size (MMS) of 100 mm for enmeshing (gill) nets and the prohibition of fishing in designated nursery areas (Pawson and Pickett, 1987).

In providing the scientific advice upon which these measures were based, the Fisheries Department was required by Ministers to make an assessment of the impact that any forthcoming regulations would have on the bass fishery. It was quite obvious that over the previous ten years there had been an increase in exploitation of bass by commercial fishermen, prompted particularly by a growing scarcity of other inshore fish resources and the high prices offered for bass. DFR's estimate of the commercial landings of bass in 1986 put their first sale value at £3–4 million, making it the seventh most valuable finfish species in England and Wales at that time. In the main bass fishing areas, it accounted for a much higher proportion of the earnings of many inshore fishermen than did any other single species. Around 400 full-time commercial fishermen using 270 boats were known to be directly involved in the bass fishery. About 2500 other fishermen using over 1800 boats took bass on a part-time basis or as a bycatch. Annual catches in the UK commercial fishery had fallen from a peak of over 1000 t in 1983–84 to around 600 t in 1985–86, and it was predicted that if the fishing effort directed at juvenile bass continued at this level, a further decline in the fishery's yield was likely. In the event, annual catches are estimated to have remained at around 500–600 t for the remainder of the 1980s (Chapter 12).

The degree to which those participating in the fishery had been affected by the declining availability of bass varied considerably from area to area. For example, by 1986, there were fewer than 20 commercial fishermen who, in season, directed their efforts at bass around rocky headlands and reefs in South-west England, whereas five years earlier, more than 150 people were involved in the fishery. Similarly, the eight charter skippers who took out angling parties especially to catch bass during summer in the Thames Estuary in the 1970s, had increasingly come to rely on other species to attract trade. In some other areas (e.g. the Solent), overall bass catches and the number of people in the fishery appeared not to have declined, partly because a high proportion of the catch consisted of small bass which continued to recruit to the fishery from a wide area, but also because the fishery had been developing and adjusting its catching methods and fishing grounds to enable landings to be maintained. There were even areas where bass landings increased in the mid 1980s, chiefly those where commercial exploitation was previously slight (e.g. South Wales and Cumbria), but it was feared that these fisheries would soon deteriorate if fishing effort increased further and the bass stock continued to be fished down.

CEMARE had estimated that some 24 500 sea anglers fished regularly for bass in the UK, and nearly half-a-million other anglers were thought to catch bass on an occasional basis (Dunn *et al.*, 1989). Because of its edibility and market value, most anglers retained some of their catch of bass, and many of the more successful sold them to finance their angling activities. Taking into account the wider benefits, such as tourism, the angling associations have pointed to an annual expenditure in this sector in excess of £20 million. More significant, perhaps, is the estimate of the anglers' total catch of bass (Chapter 13), which is not dissimilar to that recorded from the commercial fishery.

At the time that DFR was being asked for advice on the management of the bass fishery in England and Wales, in 1986–87, both national and European Community regulations for bass fishing prescribed only an MLS of 32 cm, although for the coastal waters of Cornwall the local SFC had a higher limit for bass, at 37.5 cm. Any additional measures for management of the bass fishery would need to be the minimum required to provide the most effective conservation of the bass resource, whilst being applicable to the wider interests of both commercial and recreational fishermen. Introduction of controls that were too stringent would not be enforceable, and might actually militate against better management of the fishery. Alternatives, such as a code of practice and voluntary agreements on fishing restrictions, had been considered, but these were not regarded as practicable because of the wide range of interests and the number of participants in the fishery. Voluntary agreements in the fisheries sector are not unknown (the demarcation of potting and mobile gear zones offshore in the western English Channel is an enduring example, where both sides derive benefits from the arrangement), but these often break down under commercial pressures. The prospects of commercial fishermen and recreational anglers reaching an accord over minimum landing sizes, controls on gill netting and the designation of restrictions applying in nursery areas, seemed extremely remote.

Quantitative approaches to regulating the UK bass fishery, such as licensing and quota management, had also been considered, but these were rejected, partly because of the difficulties in obtaining satisfactory assessment data on which to base recommendations for direct catch controls, and in view of doubts about the efficacy of these types of control in such a fragmented fishery that employs mainly small, inshore boats.

It was envisaged that the regulations for managing the bass fishery would be enforced by the UK Fisheries Departments, principally the Ministry's SFI, with assistance from local SFCs and the NRA, and through the self-policing activities of both commercial fishermen and anglers. For this to succeed, an understanding and respect for the need for these controls was important, and every effort was made to get the message across during the extensive consultations which took place during the investigation of the bass fishery and following the publication of DFR's findings and recommendations in 1986. Ministers also advised Members of Parliament with constituency interests in the bass fishery, that the industry and angling associations would have the opportunity to comment on any proposals for the future management of the fishery. Thus, account was to be taken of the legitimate concerns of the industry, in particular that measures to protect juvenile bass should not unduly disrupt important local fisheries for other species. The responses to MAFF's proposals for managing the bass fishery came from commercial fishermen, anglers and management authorities, i.e. district SFCs and the regional water authorities, now the NRA, and from a range of associated interests: tackle manufacturers, harbour authorities and local councils.

17.3 MANAGEMENT POLICY

The UK's strategy for managing its bass fishery gives no priority to either commercial or recreational use of the resource. This contrasts with that pursued in the Republic of Ireland, where it is implicitly recognized that bass angling is an important tourist attraction and takes priority over any commercial fishery for bass. On May 29 1990, the Department of Marine introduced regulations 'to conserve bass stocks in the Republic' (Maranuacht, 1990). These included a prohibition from 1 July 1990 on bass fishing from Irish sea fishing boats, and the stipulation that bass are not allowed to be carried on board or caught in nets from boat or shore within the territorial waters of the Republic of Ireland, out to the 12-mile limit, and that no person shall have in their possession bass of less than 40 cm total length. Thus, commercial fishing for bass from Irish fishing boats or from the Republic's shore is effectively prohibited. Additionally, no person must take, kill or have in their possession more than two bass in any 24 h period, nor must bass be offered for sale (unless imported). During the period 15 March to 15 June, bass fishing by any method is prohibited.

It therefore appears that, in the Republic of Ireland, only *bona fide* anglers are to be allowed to fish for or take bass, and that the two-fish bag limit and closing down of the commercial market were introduced in an attempt to diminish any trade in bass and cut fishing mortality on the stock. Clearly, none of these measures should inconvenience true sport anglers, but it remains to be seen whether the package as a whole has the desired effect of suppressing exploitation levels and rebuilding bass stocks around Ireland.

In the USA, those with the responsibility of managing striped bass fisheries continue to recognize the rights of both recreational and commercial interests. For example, by 1985, the Maryland stocks had become so depleted that a moratorium on fishing for them was implemented until 1990 (according to J. Valliant, in an article in National Fisherman, USA, in 1992). During the second year (1991) in which exploitation was cautiously resumed, commercial netsmen fished to a total quota of around 200 t, compared with average annual landings of 1782 t recorded over the peak period, 1958–1974 (FAO, 1980). In the event, they only caught a little over 100 t, partly because they were still learning how to fish under new, complex regulations, which limited each licensee to just 365 m of 125–175 mm stretched mesh gill net, of which not more than 1095 m was allowed per boat. Fishing was further restricted to weekdays and daylight hours, and catches had to be registered at one of 22 state-operated checking stations, before being sold. The commercial fishery was also handicapped by a lack of retail outlets, many of which had closed during the moratorium or, as with New York, had been declared out of bounds to Maryland's fishermen. All this had deterred many former operators from rigging new gear or buying the $250 annual licence, even though there was no shortage of fish. The sport fishery's season in 1991 was declared closed after ten days, even though it was admitted that policing it and monitoring catch statistics accurately were not simple tasks.

On the positive side, the collaboration between Maryland's Department of Natural Resources and Watermen's Association had given hope that, on the basis of this experiment, a relaxation of restrictions in future years would enable quotas to be more closely matched to catching opportunities. Whether these will be taken up depends, ultimately, on the cost of fishing. It was noted, however, that 'the success of the commercial season two years in a row has blunted attempts by sport fishermen's groups to reserve striped bass for themselves.'

Nevertheless, it took very little time for some anglers in England and Wales to draw the British Government's attention to Ireland's attempt in 1990 to curtail commercial exploitation of bass, and to reiterate their request that the bass should be declared to have game fish status, conferring, presumably, a priority for recreational rather than commercial use of the resource. Their argument is that, by any economic yardstick, a bass caught by an angler represents a greater financial activity than does the same fish landed by a commercial fisherman.

To give the UK angler's viewpoint, we present below an extract from a submission by the BASS to the Fisheries Minister, in response to the announcement of management proposals in 1988 (Harrison, pers. comm.).

Bass: the case for special status as an 'anglers' fish.'

"The MAFF report (i.e. Pawson and Pickett, 1987) misrepresents the current state of bass stocks. The short-term study has failed to observe the long-term decline, presented elsewhere in considerable detail (Kelley, 1979). Other evidence presented by D.F. Kelley clearly indicates that the supposedly strong 1976 and 1982 year-classes are already disappearing from the scene due to over-exploitation.

"MAFF recommends a weak management strategy which, even if implemented fully, is unlikely to meet the need for the protection of bass as an endangered species, and will certainly not protect the bass at a level compatible with sport-fishing. The tactics might indeed protect the population up to spawning size, but allow for unrestrained cropping beyond this size. In view of the growth of the commercial fishery and its exploitation practices, the consequence is likely to be the severe depredation of stocks at all sizes above MLS. The loss of larger fish is clearly contrary to the interests of the angling community, especially since inshore waters, by their very accessibility, inevitably bear the brunt of such population changes.

"There is a failure to discriminate between the sport-fishery and the routine use of rod & line methods to catch bass for commercial gain, whether by full or part-time fishermen. The report attaches considerable importance to the contribution of rod & line fishing to the current depleted state of bass stocks, but little evidence is presented to substantiate this figure.

"The potentially most promising suggestion was for the closure of nursery areas, though this 'needs further detailed consideration at local level and to be successful would need the full co-operation of the sport-fishing interests and local bodies such as sea fishery committees. . .' The response so far from anglers indicates that this suggestion would be welcomed, but, in view of the urgency of the situation, any delaying tactics must be set aside, and a blanket restriction imposed.

"Most seriously, the report fails to satisfy (the anglers) because, inevitably, it is pervaded by a MAFF philosophy which regards marine resources as pre-eminently the preserve of commercial fishermen. The relegation of sport-fishermen to a position of peripheral importance is an injustice and, with the growth of angling to its present position of economic importance, this represents a profound misjudgement of the basis on which resources are managed and allocated. A philosophy of resources management which held good in former times, has long ago been overtaken by social developments in leisure time and consumer spending power. Consequently, the report fails to identify and subsequently satisfy the fundamental requirements of a sport fishery, which are not the same as for a commercial fishery:

(a) The sport-fisherman conducts his activities primarily in inshore waters, and so requires a measure of protection specific to the needs of these areas.
(b) There is considerable evidence for the decline of certain localized (bass) fisheries to the point at which sport fishing is ceasing to be viable. In these areas there is an urgent need for measures designed to significantly increase the stock of bass.
(c) Sport-fishermen expect that they should be able to catch a reasonable proportion of fish of well above average size. This is contrary to the demands of commercial fishermen, whose primary aim is to capture the more marketable smaller fish – smaller than even the present MLS (*at 32 cm in 1987*).

"The bass in UK waters most certainly needs a Total Allowable Catch regulation. It further needs measures in the shape of protected nursery areas, serious limitations on gill-netting, agreements under the Common

Fisheries Policy to reduce offshore trawling, and a substantial rise in MLS which would guarantee survival well into adulthood for the majority of bass. Most important of all, we suggest, is a fundamental change in our approach to the species.

"The 1985, commercial bass landings are estimated to have a total value a little over £4.75M. If we were to add 50% for the incidental economic activity generated by the commercial pursuit of bass, we arrive at a turnover of little more than £7M. Even by a calculation which is deliberately cautious in arriving at the sport fishing turnover, we see that at £20M (*as estimated by BASS in 1987*), the bass sport fishery is of far greater economic importance to the country than is the commercial fishery.

"We believe these arguments and figures speak persuasively and that there is an unassailable case for administering the bass fishery as pre-eminently a sporting resource. We think further that the position of bass stocks in UK waters is marginal and has been severely depleted by virtually unrestricted commercial exploitation; that the conflicting demands of commercial and sport fishermen are mutually exclusive; that the commercial significance of the bass fishery is so small on a national level that this precludes a sharing-out of the resource even if it were administratively feasible.

"In view of the above, we argue strongly that immediate consideration be given to the implementation of measures which acknowledge the position of the bass as an angler's fish, and which lead to a discontinuation of the commercial pursuit and sale of bass in the UK."

From a stock manager's position, however, a rather different side of the argument can be seen, particularly when the long-term needs of the bass fishery are taken into account. First and foremost, it must be recognized that the European bass stock is an international property. Although one nation can control the fishing activity of its own people and, to a lesser extent, that of other nations' vessels fishing inside its territorial waters, agreement has to be reached at a multinational level before exploitation of the whole stock can be controlled. It is not just a matter of apprehending and prosecuting local poachers, who catch fish out of season or in closed areas, or who fish without a licence. There is a need to ensure that mobile fleets, displaced from other fisheries by stock declines or tight quota controls, do not begin to look to bass as a free-for-all, until it too is severely depressed. It is important, therefore, to show how an international bass fishery might be managed (Section 17.6). Declaring the bass to be a protected species, with a high minimum landing size and no commercial exploitation inside the UK 6 mile zone, would be a precarious stance from which to negotiate with the administrations of countries that at present have no management plans for their bass fisheries.

Furthermore, such negotiations are nowadays most gainfully held against a background of sound scientific assessment of the state of the stocks in question and advice on the biological and economic implications of the various options for management. There is no marine fishery in the North-east Atlantic where sufficient data for this purpose could be obtained from the recreational sector. The very act of commercial exploitation produces the information upon which assessments of fish stocks are made, and the more that act is harrassed and outlawed the less sound the resulting advice becomes. Even in the matter of the conservation of Atlantic salmon, a species with an undeniably greater claim to game fish status and with more tangible threats to its survival than the bass (Mills, 1990), a strong case is put for the continuation of some form of commercial fishing in or around estuaries, which will provide assessment data and an infrastructure upon which to base management and regulation outside the rivers.

It is not easy to prepare a proposal for managing a fishery which appeals equally to all interested parties. The problem is that each side has a unique requirement of the exploited resource, the demands of which lead the proponants to their own subjective examinations of the evidence and actions given in any published proposal.

It is enlightening to compare the angler's views with those of a typical (if more than usually erudite) inshore commercial fisherman, written in response to the February 1988 proposals (K. Matthews, pers. comm.).

"These proposals are clearly based on research by your scientists Pawson and Pickett (1987). Broadly, one must question whether the data are reliable, whether the stock really is in danger, and whether the proposed conservation measures are fair to all interested parties involved in the fishery.

"In considering bass conservation measures, we should bear in mind the views of the American biologist, Dr Daniel Merriman, who is quoted as saying 'It can perhaps be demonstrated that in certain areas it is sociologically and economically desirable to make the striped bass a game fish and hence eliminate commercial fishing in those places. If that can be done in a democratic fashion, then let the legislation be debated on that basis, but do not let that legislation masquerade under the cloak of conservation.' Is the current debate really about adequate stock protection, or really a conflict over who catches what? Ultimately, like most questions of resource management, this has to be seen as a political issue.

"Catch data on bass are unreliable and patchy. Estimated catches are given at 8 or 9 times the actual recorded figures and are therefore little more than 'guesstimates'. Whilst there is a brief mention of the mid-

Channel fishery by large French trawlers, little significance is given to it. Current *Fishing News* reports give some boats taking up to 40 tonnes of bass per trip; this represents the equivalent of about 40% of the total recorded catch in England and Wales for the whole of 1986. Tagging has relied almost exclusively on small bass caught inshore, and 'non-reporting of recaptured fish seriously devalues these data for assessment purposes'. Reliable angler-caught bass data are lacking and no distinction is made between commercial and sport rod and line-caught fish. There are also dark hints about amounts of undersize bass (down to 24 cm) being caught and marketed.

"One understands the difficulties faced by scientists trying to gather reliable and meaningful information from a 'fragmented and opportunistic' fishery – they are almost as bad as those faced by the fisherman trying to actually catch the fish itself. But this must not be allowed to cloud our judgement about measures that will adversely affect thousands of inshore fishermen.

"There is no firm scientific proof that the spawning stock is actually at risk at the moment. Strong year classes, resulting from good summers like 1976, have maintained numbers in spite of increased fishing effort. Even if there might be the possibility of a future stock collapse, what management controls should be considered? Pawson and Pickett state that 'any management measures introduced for the purpose of controlling bass mortality must, therefore, not only take into account the wide variety of fishing methods used to catch them, but also the fragmented and opportunistic nature of a fishery which is also prosecuted by anglers. Such measures must be SIMPLE, ENFORCEABLE, and FAIR.' (my emphasis). But do those proposed bear up under close examination based on these criteria?

"Raising the MLS to 36 cm will only apply to UK fishermen; there is only a proposal to 'ask' the EC Commission to consider increasing the EC landing size from its present 32 cm to 36 cm. Particular significance must be given to this when one considers the extent of the French effort directed at the bass stock that winters in mid-Channel. Mesh size controls on nets will affect other, more important, fisheries. For example, the sole fishery along the Sussex coast is far more significant than that of the bass, and a large proportion of the trammels used for sole are under 100 mm. Just a small increase in mesh size will result in a large drop in the numbers of soles landed. The proposed exemption of drift nets for the purpose of catching mullet, is welcomed.

"In terms of enforcement, both larger trawlers and small inshore netting boats are subject to rigorous and efficient supervision by their local SFCs. Anglers, on the other hand, are both very numerous and widely dispersed, making effective enforcement almost impossible without a massive input of personnel and resources. European trawlers will be left untouched by the present proposals. Although subject to the increase in the MLS, our own larger trawlers' gear will be unaffected as there are no proposals to increase cod-end MMS to 100 mm. Measures like quota controls are also deemed to be inappropriate. Similarly, sport fishermen will not be excluded from nursery areas unless the local SFC brings in a local by-law, which in turn would give it a major enforcement headache.

"It would appear that the conservation measures, therefore, are almost exclusively aimed at the small inshore fisherman using trammels, gill nets, or drift nets. The retention curves for gill nets show how effective the proposed MMS of 100 mm will be against the targeted inshore fisherman. Is it not ironic that, by taking out the easily controlled inshore boats, exploitation of the bass stock will be left in the hands of larger (mainly European) trawlers and anglers – both groups beyond the effective control of Ministry powers and resources?

"The decision about these measures is clearly in the sphere of the politician when one considers the third element of the equation, FAIRNESS. As a part-time fisherman, I can only ask that our political representatives debate this policy in a fair and democratic fashion; and not legislate almost exclusively against the small full-time and part-time fishermen, who prove to be an easy target. We lack the capital investment, and consequent fishing flexibility, of the larger trawlers. We lack the organized political lobby of the angling fraternity, whose ultimate aim is to make the bass exclusively a game fish like the salmon and sea trout. The inshoreman would accept conservation if it were spread fairly across the board; but there is little mention of quota control, designation as a pressure stock (logical if the stock is really threatened), or specific restrictions on anglers.

"To conclude, these measures are based upon an unproven scientific need, and look to enforcement possibilities as the main criteria when lighting upon appropriate regulations. The result is that the burden falls disproportionately upon the inshore fisherman because he is most easily seen and regulated. If you want the bass to be exploited exclusively by larger mid-Channel trawlers or solely the preserve of the sports angler, let us be open about the decision and its consequences. They should not be allowed to 'masquerade under the cloak of conservation'."

It is difficult to reconcile the anglers' and commercial fishermen's opposing interpretation of the Government's proposals for managing the bass fishery, as illustrated above. Both points of view are entirely understandable, but they implicitly enshrine a policy of discrimination, recreation amenity against commercial use of the resource. Nevertheless, these and many other communications provided a wealth of information which was taken into consideration when redrafting the proposals to achieve a more acceptable, and effective, package.

Because, to be effective, any management control must be acceptable to and respected by those on whom it falls, whether implemented through local, national or international legislation, a more considered and objective-orientated rationale has to be formulated. Hence the role of fisheries science, and the importance of consultations, our experience of which carries a strong message for fisheries scientists: presentation is important. The most elegant research will not impress those to whom its conclusions are relevant unless it is put to them in an understandable and unambiguous form. Similarly, the exercise's objective must be made abundantly clear from the outset and, though it will not always be possible to enunciate the whole scope of management actions, the scientist should have these well in mind, both in giving advice to government officials and in consultations with the fisheries' participants. It is also important for scientific advice to be credible and for its direction of application to be consistent, though this should not be allowed to lead to accepted procedures and formats outstaying their usefulness when it is obvious that a new approach is required.

17.4 MANAGEMENT OPTIONS

Minimum landing size (MLS)

The most common tactic used to conserve fish stocks is to stipulate a size – usually a linear dimension – below which individual fish cannot be legally killed or landed. Such an MLS might be used to serve several purposes. It can be pitched at the size at which most female fish of a particular species first mature to spawn, thus helping to protect immature fish and preserve the stock's breeding potential. It can be aimed at preventing too many small fish being caught, so that yields to the fishery are not depressed by inadequate growth in the population as a whole. It can be fixed at the smallest marketable size for the species in question, thus avoiding excessive dumping and wastage by the fishery. In all cases, complementary restraints on fishing operations (e.g. mesh size controls, restricted access areas, catch-and-release arrangements), which help fishermen minimize the number of undersized fish they have to discard from their catch, are a necessary adjunct to an MLS regulation.

As far as bass are concerned, there is ample scope for a lively debate as to just how large (or small) the MLS should be. Because there is virtually no lower limit to the size at which bass have a market value (very small wild fish could be harvested for ongrowing), a commercial fisherman might be expected to resist strongly any attempt to increase the MLS above the size of fish most usually taken in his local fishery. This is a particular concern in those fisheries where juvenile bass predominate, and in view of the premium price often obtained for plate-sized bass. At the other extreme, the true sport angler is prepared to accept a much higher MLS; he is not too interested in small bass nor in their market value, and would welcome any measure which improved the chances of bass reaching a large size.

At a national level, the anglers' representative body (the now defunct National Anglers' Council) and BASS accepted the proposed MLS of 36 cm, but would have preferred 38 or 40 cm. In general, anglers had not complained about a shortage of large bass in the eastern English Channel or southern North Sea, but in the western English Channel and along the west coasts of England and Wales, there was genuine concern over the apparent deterioration of the stocks. Commercial fishermen along most of the south coast of England were opposed to any increase in MLS above 32 cm. Their main argument was that a higher MLS would cut catches (and earnings) drastically in these inshore fisheries, in which small fish predominated. They also claimed that the eventual benefits would accrue chiefly to the French fleet fishing for larger bass offshore, and which was constrained only by the EC MLS of 32 cm. They pointed out that there had been no decline locally in the number of small bass caught, but this observation is consistent with a fishing pattern targeted chiefly at juvenile fish and where there is as yet no recruitment failure owing to excessive depletion of the spawning stock. From Cornwall northwards, on the west coast, there was no resistance to an MLS of 36 cm.

It appeared, then, that a national MLS of 36 cm fell between the extremes of a call for 38 cm or higher by some of the more vociferous anglers' representatives, and a maximum of 32 cm demanded by many commercial fishermen along the English Channel coast. If the EC MLS for bass could also be raised to 36 cm, it seemed likely that there would be much less opposition from this latter quarter to the management package as a whole. It is worth noting that probings had suggested that those other countries having strong commercial bass fisheries interests within the EC would not easily countenance an MLS above 36 cm.

Mesh size controls

Having settled on a target MLS of 36 cm, it was necessary to determine complementary mesh size controls. From the beginning, it was clear that restrictions on mesh sizes in trawl nets were unlikely to be introduced specifically for bass fishing. First, trawl mesh sizes tend to be influenced by the need to comply with international rather than national requirements. In the English Channel – where nearly all bass trawling takes place – a general minimum mesh size of 80 mm is in force, and few boats are likely to use larger meshes in their cod ends. Second, inshore trawlers tend to tow their nets for short periods – usually less than 1 h – and the relatively high recapture rate of bass caught by trawl for tagging suggests that any undersized bass released alive have a good chance of survival. In addition, an increase in trawl mesh size might result in too many small bass becoming enmeshed in the netting, actually creating a problem that seldom occurs with the gear being used at present.

Catch data from gill nets used in the commercial fishery showed that, with a mesh size of 100 mm, very few bass under 36 cm would be caught (Fig. 17.1) (Reis and Pawson, 1992). Bearing in mind the primary objective, of raising the size of recruitment towards 38 cm, it seemed that this would be achieved by introducing a minimum mesh size (MMS) of 100 mm for all enmeshing gears.

It was necessary to restrict the use of small mesh sizes in all enmeshing nets because:

1. undersized fish caught in such gear have a low probability of survival

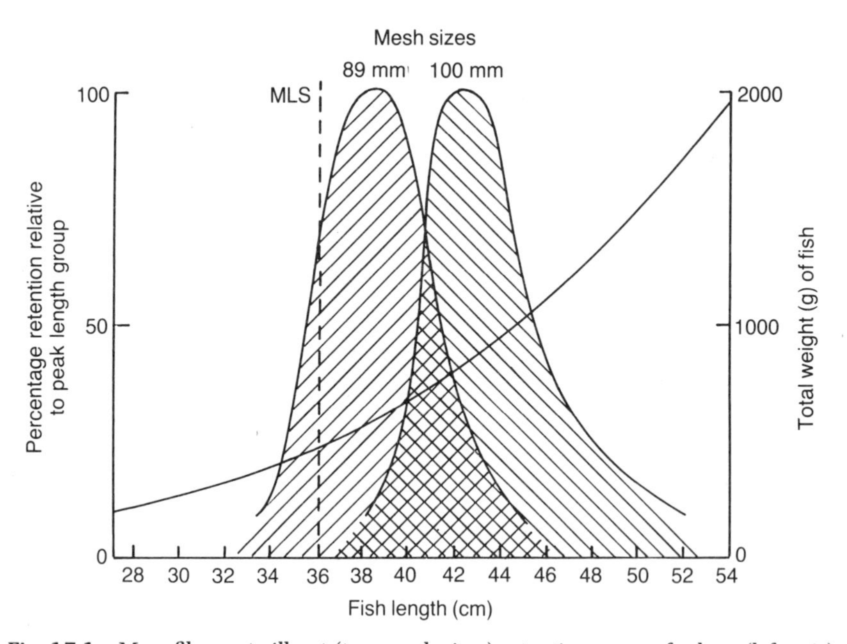

Fig. 17.1 Monofilament gill net (two mesh sizes) retention curves for bass (left axis), showing also the increase in bass weight with length (right axis). From Pawson and Pickett, 1992.

even if discarded, owing to the damage caused when extracting them from the meshes;

2. it is not possible to identify a 'bass gill net fishery', not least for the purposes of bringing prosecutions for infringements of regulations. The term 'gill net' is applied to a wide variety of enmeshing nets, which are often used to take a mixture of species, the relative proportions of which vary with the grounds being fished and the season;
3. about one-third of the commercial bass catch in the UK was taken by enmeshing nets; and
4. the method is commonly used by casual and part-time fishermen, with whom enforcement of any management measure (particularly specific ones such as an MLS) can be very difficult.

It was anticipated that introducing an MMS of 100 mm for enmeshing nets, to avoid the capture of immature bass, would also bring conservation benefits for sea trout, sole, grey mullet and several other species. From a management viewpoint, however, there are several fisheries around the English and Welsh coasts in which a 100 mm MMS could be considered to be a disadvantage, either because small but profitable fish would no longer be caught (e.g. sole, golden-grey mullet, red mullet (*Mullus surmuletus*) and the 'small pelagics' – sprat, herring, mackerel and pilchard), or through an unintentional diversion of effort towards fish which are more susceptible to meshes of over 100 mm (e.g. salmon). With respect to the latter, however, it should be borne in mind that there is no compulsion on fishermen to use smaller-meshed nets than those which most effectively catch the target species, i.e. those intent on actively pursuing salmon will already be using the most appropriate mesh size. Any minimum mesh size regulation is intended to reduce the chances of undersized fish being caught in nets and (probably) killed. To be effective, the controls must be enforceable, and therefore one comprehensive national regulation is to be preferred to local variations.

As expected, requests were made for exemptions from mesh size controls for gill net fisheries for other species in bass fishing areas. These fisheries could be assigned to one of two categories:

1. those using small-meshed gear which is unlikely to take undersized bass (i.e. drift nets for herring, sprat, mackerel and pilchard) and for which a *maximum* mesh size of 60 mm would be acceptable. A general, national derogation for such gear would probably be appropriate; or
2. those using enmeshing nets to catch sole, grey mullet, sea trout, etc., which employed mesh sizes below 100 mm, for which local derogations specific to each fishery might be permitted, provided:
   (a) that the imposition of an MMS of 100 mm would significantly reduce the economic viability of the fishery in question; and

(b) that the (discard) mortality of undersized bass (i.e. less than 36 cm), as a result of permitting the fishery to be pursued, would be insignificant compared with the value of the (non-bass) fishery.

Other criteria, based purely on fish stock conservation requirements (for example, the maintenance of a robust spawning stock or improvements in yield per recruit) are not really suitable for this purpose. Such information on the stocks in question tends to be patchy and, more importantly, it is not available to those seeking exemption from controls. Furthermore, a judgement on whether to grant an exemption would, in such a case, have to be made on biological grounds, whereas social and economic factors are probably more relevant.

This approach demonstrated the Government's recognition of the need to regulate gill net fisheries in general and to satisfy the requirements of the bass fishery management strategy, whilst allowing other fisheries to continue unhindered. In the event, the identification and designation of the fisheries which required exemption was quite straightforward, and it was not difficult to arrive at decisions as to which of these fisheries should be allowed to use mesh sizes under 100 mm. It was hoped that, once in place, these provisions would require only minor modification from time to time, in the course of monitoring the fisheries involved and/or in response to representation by fishermen.

The effectiveness of these measures in protecting juvenile bass relies upon certain assumptions about the behaviour of fishermen. With the increase of the MLS to 36 cm, it was expected that some bass netsmen would adopt larger mesh sizes or turn to other catching methods to maximize their catches of legal-sized bass. Gill nets are highly size selective, and the few undersized fish which these fishermen caught would probably be returned to the sea. Although netsmen fishing principally for species other than bass would probably continue to use their normal mesh size, by definition they do not catch many bass. Those undersized bass which are discarded and die will probably be a negligible proportion of the juvenile bass mortality that might ensue if gill net mesh sizes were not controlled.

The chief problem was that the less committed fishermen who, it was believed, accounted for a considerable mortality of bass even under an MLS of 32 cm, would continue to operate as before. The main concern was that gear which could be operated *in absentia* should not be permitted with mesh sizes less than 100 mm, especially where its use could result in large numbers of undersized bass being killed, even if they were discarded.

In part, this had already been achieved by provisions in the *Salmon Act 1986* for prohibiting fixed nets within the UK 6 mile zone (Great Britain – Parliament, 1986). There are, however, many areas in which such gear is authorized for catching sea fish by SFC enabling bye-laws. Similarly, unless SFCs were to introduce bye-laws controlling enmeshing net mesh sizes (these already existed throughout Wales and in the North-west of England), drift netting and other non-fixed styles of gill netting would not be subject to any controls, and may be used for any species, including bass.

The requirement, then, was for a general, national MMS of 100 mm for all types of enmeshing nets, with the possibility of specific derogations for smaller meshes in fisheries which would not be viable with 100 mm or larger mesh nets, but which do not operate to the detriment of MAFF's bass conservation strategy. This proposal was announced by the Minister on 24 July 1986, following which a series of consultation meetings took place around the coasts of England and Wales, where required by local interests.

The angling organizations supported the proposed introduction of an MMS of 100 mm for all enmeshing nets, though many anglers would rather have seen the use of the method banned or at least severely curtailed. In the early 1980s, sea anglers tended to regard gill netting as Public Enemy Number One. Commercial fishermen generally accepted that gill net mesh size controls should complement MLS regulations (not only for bass), though it was emphasized that a strict MMS of 100 mm on all enmeshing nets would have serious consequences for some local fisheries. In the main, however, these tended to be drift net fisheries for small, pelagic species (herring, sprat and mackerel), and grey mullet and small bass (in the Solent area only). These can be readily distinguished from fixed gill net fisheries because, in inshore waters, fishing boats have to attend drift nets (and ring nets and beach seines) at all times.

The main concern of the industry appeared to be the potential impact of an MMS of 100 mm on the south coast fisheries for grey mullet. Locally, and especially in Poole Harbour, this was regarded as being more important than any concomitant benefits to the bass fishery. The netsmen argued that mesh sizes as low as 75 mm were required for mullet, though this applied primarily to drift and ring nets. It was implied that a higher MMS may not be too unwelcome for fixed nets. We suspect, however, that the main (unspoken) requirement was to continue fishing for small bass in estuaries, though in most areas there were no claims that these were the only fish available to the fishery, such as were heard from the Solent fishery. It is worth noting that a 100 mm MMS for all enmeshing nets was already successfully enforced by the South Wales SFC, and that there was no opposition to the proposal from fishermen in Wales and East and North-west England, provided that suitable derogations were available for the mullet fishery.

It was concluded that there was little opposition and much support for the introduction of mesh size controls for enmeshing nets, though the exact mesh size level was in dispute. Whilst fishermen along the English Channel coast were not in favour of an MMS above 75–80 mm, they were chiefly concerned about fisheries other than those for bass, and especially ones using drift or ring nets rather than fixed gear. Elsewhere, a 100 mm MMS was acceptable, though a transition period (or interim MMS less than 100 mm) was requested to allow fishermen to adapt and use up existing gear which had smaller meshes.

Restrictions on fishing in nursery areas

It was easy to foresee that enforcement of an increased MLS for bass of 36 cm, and complementary controls for mesh sizes in gill nets, would be most important in areas where juvenile bass predominate in catches. These tend to be close inshore, where part-time fishermen and anglers account for most of the bass caught. Although it had proved difficult to monitor this part of the fishery, it was known from tagging exercises in the mid 1980s that exploitation rates in some estuaries can be as high as 50% each year and, from conversations with fishermen and merchants, that many bass under the MLS of 32 cm were being landed.

The key part of the package of management measures aimed at protecting juvenile bass from too high a level of exploitation was the proposal to restrict all fishing activities that are likely to take small bass, in areas in which they were particularly vulnerable to fishing. These areas were defined as those in which the majority of fish in the local bass population were below the legal MLS and vulnerable to exploitation, and they came to be known as the bass nursery areas. With an MLS of 36 cm, the bass in need of protection are predominantly fish up to 5 years old. This differs from the definition used by Kelley (1986) and others, who have described nursery areas as those containing only 0- and 1-group fish. The main task was to identify the most important bass nursery areas and to describe the fisheries there that might be affected if fishing was to be restricted.

The geographical range of adult bass extends as far north as Scotland on both east and west coasts of Britain, but, because of the susceptibility of 0-group bass to low temperatures, it is unlikely that potential nursery habitats north of the Dee Estuary on the west coast and the Thames Estuary in the North Sea, would contain sufficient small bass in most years to merit protective measures in addition to the MLS and MMS controls. The main exceptions are in those localities where warm water is discharged from coastal power stations, such as at Heysham, in Morecambe Bay, Lancashire. Sheltered or enclosed inshore areas (i.e. estuaries, inlets, rias and harbours) are the normal habitat of juvenile bass for most of the year.

At times – usually in winter – they may be found in the open sea, where they are then much less vulnerable to fishing and are mixed with larger bass; protection then has to rely on MLS and mesh size controls.

In the UK, the places normally frequented by juvenile bass can be grouped into five types, characterized by their general topographic and hydrographic features, as follows:

1. Lowland estuary (e.g. River Blackwater, Essex; River Severn). These usually have wide river mouths with sand banks, mudflats and saltings, and are often at the confluence of two or more rivers. They have a long inland penetration (16 km or more) of high-salinity water and, though generally less than 30 m deep, may retain 0–3-group bass throughout the winter. They often have adjoining tidal creeks through marshland, which tend to dry out at low tide and have very little freshwater input.
2. Downland estuary (e.g. River Arun, Sussex; River Itchen, Hampshire). These estuaries are narrow owing to geological constriction or man-made training banks, and there is usually a fast freshwater run off and relatively short inland penetration (less than 10 km) of salt water. They are strongly tidal and often turbid and do not provide such extensive feeding areas for bass as do some other types of estuary.
3. Sandy estuaries may vary from less than 1 km^2 (e.g. River Nevern, Pembrokeshire) up to 100 km^2 (e.g. Burry Inlet, Glamorgan), and are mainly produced by river deposits or long-shore drift creating a sand-bar across their mouth. They are associated either with areas having a mountain or moorland catchment (in Wales and South-west England) or gravel bank coasts (e.g. in Suffolk), and often have dunes or salt marshland along the extensive tidal reaches. Juvenile bass may predominate among the fish communities of such shallow (i.e. less than 10 m) habitats, and in the warmer months they can provide the main source of income for local fishermen.
4. Rias (e.g. Milford Haven; River Fal, Cornwall) are valleys which have been flooded by the sea, with little sedimentation, and are often associated with rocky areas. They are normally much deeper than true estuaries and some may not receive any significant freshwater input (e.g. Salcombe Harbour). They have large areas of deep water, even at low tide, and may contain both juvenile and adult bass throughout the year.
5. Natural harbours (e.g. Poole and Chichester Harbours) are tidal, non-estuarine arms of the sea bounded by bar-built islands or peninsulas. They are usually shallow with extensive intertidal zones, and almost completely drain at low tide, when young bass will be found in a deep-water channel near the entrance.

In addition, there are several artificial habitats where small bass congregate and which, therefore, attract considerable fishing effort, in particular around the warm-water discharges from coastal power stations.

The relative status of these nursery areas was assessed on the basis of (a) the probable proportional contribution of recruits to the adult bass stock as a whole, and (b) the significance of protecting juveniles there in terms of the benefits to local bass fisheries (Fig. 17.2). Nevertheless, the potential fishing mortality on undersized bass in unregulated nursery areas was the important consideration in the overall management package. In many cases, it was thought unnecessary to prohibit fishing in the whole nursery area and still be able to provide adequate protection to juvenile bass, and consideration was given to (a) the identification of specific locations in a nursery area and the time of year in which the bass there were particularly vulnerable to fishing, (b) the ease of delineating that area, and (c) the degree to which fishing restrictions could be enforced.

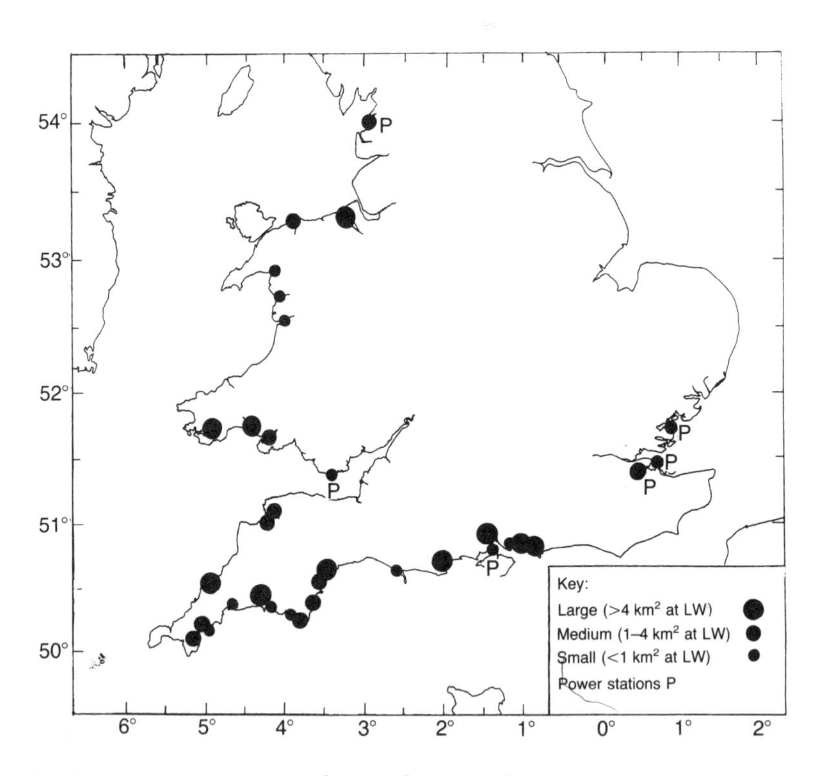

Fig. 17.2 Bass nursery areas in the UK, showing relative local benefits of closures to fishing, based on the size of the areas and estuaries closed to bass fishing. Nursery area sites at coastal power stations are indicated.

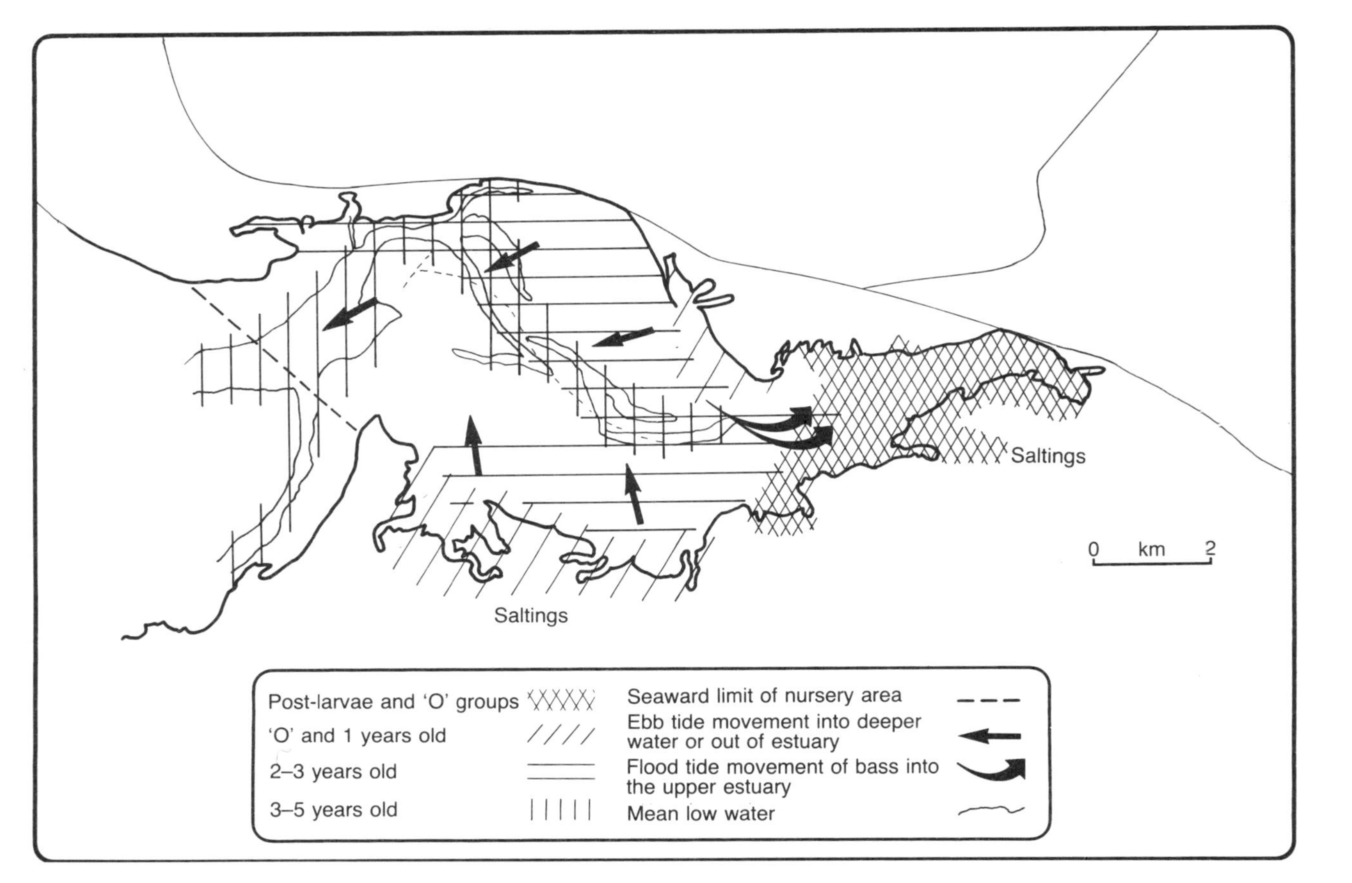

Fig. 17.3 Stylized description of the distribution and behaviour of bass in a typical estuarine nursery area (a medium-size sand-bar estuary).

Although the distribution of juvenile bass within each nursery area varies with tidal conditions, time of year and the relative abundance of the constituent year classes, the locality of the most vulnerable part of the population can usually be adequately delineated on the basis of the characteristic topography of each category of nursery area and information on local fishing areas. The distribution and movements of juvenile bass in a stylised estuarine nursery area are shown in Fig. 17.3.

Implications of nursery area restrictions for fisheries for species other than bass

It was the responsibility of DFR and administrators to develop guidelines with which decisions could be made as to which particular fisheries should be permitted to continue to operate in those nursery areas where restrictions were considered necessary to protect small bass. Ideally, these decisions should be based on an analytical assessment of the potential impact of (relatively) uncontrolled fishing on the local bass population and the benefits that regulation would have for yields in the fishery. But, in practice, this was not possible because of (a) the difficulty in determining the actual exploitation rates on juvenile bass – including fish under the MLS – and (b) the large, unpredictable variations in year class strength and the year-to-year changes in abundance and distribution of juvenile and adult bass in and around nursery areas.

A more pragmatic approach was therefore adopted, in which the vulnerability to capture of small bass in those areas known to harbour them was judged against the social and economic significance of local fisheries for bass and other species. Following the consultation exercise in 1988, the idea of prohibiting fishing for bass in nursery areas had largely been accepted by both sport and commercial interests. Considerable information on the various fisheries in 34 areas, identified as having a priority for control above and beyond the 36 cm MLS and enmeshing net mesh size controls, had been obtained during the consultation.

Angling is carried out to a varying extent throughout the year in all bass nursery areas. In some, e.g. the River Blackwater in Essex, several species, such as cod, whiting, flounders, rays and eels, are caught in addition to bass. Others are more noted for big fish of one or two species, e.g. the River Teign for flounders and Salcombe Harbour for plaice and gilthead bream. A small amount of sea trout angling takes place in tidal waters, though seldom from boats. It is not easy to assess the number of anglers who fish in these areas, but a great many would be severely affected should restrictions on fishing for bass include a complete prohibition of angling, as was initially proposed for some nursery areas.

Initially, the national angling organizations all welcomed the principle of closing nursery areas to fishing, particularly when it could be linked with protection for migratory salmonids. They were concerned, however, that agreement should be reached between all interested parties and that the possible high costs of enforcement were appreciated. It is not surprising that SFC and NRA responses on this issue were also cautious, and further consultation was requested over the extent of potential closed areas, the restrictions to operate within them, and their enforcement. There was no outright opposition from individual anglers (at whom the measure was aimed, in part) and sport angling organizations made it clear that they would be prepared to restrict their access to some areas in order to demonstrate their serious and responsible attitude regarding the called-for sport-fish status of bass. There was no doubt, however, that a closure of entire nursery areas to all angling would not be politically acceptable.

The main complaint on behalf of commercial fishermen was that closure of areas to protect juvenile bass – a principle which they also did not, generally, oppose – would deny to some people access to fisheries for other species which were important for their livelihood. It seemed, therefore, that the closure of designated bass nursery areas to all forms of exploitation was only acceptable if derogations were made for some important local fisheries for other species. It was intended that the details of such closures would be worked out with representatives of all interested parties, though some guidelines would have to be adopted around which local arrangements could be made.

It was possible, following consultation and by local arrangement, to identify some areas the closure of which would impinge little on existing fisheries, but still give sufficient protection to juvenile bass. In most cases, however, it was anticipated that exemptions for specified fisheries within the defined area would be necessary, provided that their continuance did not threaten the conservation requirements for juvenile bass. A knowledge of these fisheries provided valuable guidance as to their respective interactions with the bass fishery and the potential impact of management measures, but it was consultation at the local level that revealed the relative importance of each fishery and the participants' willingness to accept, or determination to oppose, the various aspects of the bass management proposals.

The degree to which fishing activity could be restricted in bass nursery areas depends largely on the local importance of fisheries for other species. There are commercial fisheries for grey mullet all round the coast within and adjacent to bass nursery areas. The catch of mullet taken in estuaries and harbours is not easy to estimate, because part-time fishermen predominate in these fisheries. For this reason, the fishing pressure on bass nursery areas is likely to be much greater in the vicinity of large conurbations (e.g. Portsmouth, Southampton, Poole, Plymouth) than in more remote areas. Mullet fisheries within harbours and estuaries employ similar methods in all regions, usually fixed and drifted gill nets and/or beach seines. It is likely that the proportion of the total UK mullet catch that is taken within harbours and estuaries is much larger than the corresponding proportion of the total (legal size) bass catch, and these areas are, therefore, relatively more important to mullet fishermen.

Requests were made on behalf of mullet fishermen in several regions, asking for them to be allowed to fish in potential bass nursery areas using gill nets with mesh sizes of around 80–85 mm. It is well known that such gear will catch bass of around 30–35 cm, but at Poole, for example, the fishermen claimed that despite this, few bass were caught by virtue of the fishing method employed – ring netting in this case. Decisions as to whether mullet fishing should be allowed in any particular bass nursery area, had to bear in mind the associated imposition of general controls on gill net mesh sizes.

With the exception of the English north-east coast (where few juvenile bass are caught), most licensed fishing for salmon around England and Wales occurs in estuaries in Wales and South-west England, and there is none in potential bass nursery areas to the east of Southampton Water. Sea trout are caught in most of the rivers that have salmon runs, and in some others which have few salmon. Occasional sea trout appear in estuaries in South-east England, but licensed fishing there tends to take place mainly on the open coast, using drift nets and beach seines to catch feeding fish well away from their natal rivers (Potter and Pawson, 1991). One particular problem was that some gill nets were licensed to fish for sea trout using mesh sizes below the national MMS (102 mm) for salmon and migratory trout. In some estuaries in South-west England and off the English south-east coast, derogations under the *Salmon and Freshwater Fisheries Act 1975* (Great Britain–Parliament, 1975) allowed mesh sizes down to 60 mm. It was agreed that there are benefits to both the sea trout and bass fisheries of increasing the MMS to at least 89 mm, and this has been implemented in the Anglian Region (ICES division IV c) through NRA bye-laws.

Other minor, though locally important, fisheries involve the use of small-meshed gill nets for sprat and herring in autumn and winter along the coasts of Suffolk, Essex and Kent, and in the Exe Estuary and Milford Haven for herring. There is also a whitebait (0-group herring and sprat) fishery in the Thames Estuary and sandeel seining along the English south coast for (bass) angling bait. Drift net fisheries using small meshes for pelagic species such as sprat, mackerel, pilchard and herring were easily identified, and it was considered that their continuance could be allowed, as they posed little threat to juvenile bass. Other inshore fisheries using small-meshed nets, such as fyke netting and pair trawling for eels, beach seining for sandeels, and beam trawls or push nets for shrimps, are known to catch and discard large numbers of undersized fish of many commercial species, including bass. As with the drift nets for small pelagic species, their continuing use in bass nursery areas had to be judged against the potential mortality of small bass, and from the viewpoint of the numbers of people whose livelihoods depend on their execution.

In most bass nursery habitats there is a small amount of part-time or casual, commercial fishing effort aimed at other marine species, e.g. cod, whiting, plaice and sole. Stake nets and lines are set from the shore at low tide and these areas may also be worked by small trawlers and fixed netters, especially when the weather is too rough for them to work off more exposed coastlines.

Fisheries that occur almost wholly in estuaries and harbours include those for eels (mainly fyke netting), oysters (*Ostrea edulis*) (dredging and cultivation), mussels (*Mytilus edulis*) and cockles (*Cardium edule*) (dredging and hand-raking) and shrimp (beam trawls and push nets). These fisheries are mainly pursued in a traditional, seasonal pattern with a fairly stable output and workforce, and in 1986 their total annual first sale value was estimated to be similar to that for bass alone: around £5 million. Their continuance is considered to be compatible with the requirement to close bass nursery areas to fishing methods likely to take undersize bass.

17.5 IMPLEMENTATION OF THE MANAGEMENT PACKAGE

As a regulatory measure, the concept of nursery areas is simple, but it is certainly the most radical part of the MAFF proposal for management of the bass fishery, and it may be helpful, therefore, to retrace its evolution. In the report to Ministers, recommending the package of measures for managing the bass fishery (Pawson and Pickett, 1987), it was pointed out that, to protect juvenile bass, restrictions on fishing in nursery areas were attractive both in biological conservation terms and in making management of the fishery more even-handed. MLS and mesh size restrictions have a much greater impact on commercial fishermen than they do on sport anglers. However, there were two major difficulties in attempting to introduce restrictions on fishing activity in nursery areas by Ministerial Order.

First, more work had to be done at a local level to identify more precisely the areas in which fishing would need to be controlled. In 1986, it was not clear whether closures should be on a permanent or seasonal basis, though this would obviously vary from area to area. Nor had any particular locations been identified, though it was envisaged that the size of such areas could vary from a hectare to 20 km^2, depending on the local distribution of juvenile bass and their vulnerability to the fishery. Within these areas there was a choice between closing the entire area while providing exemptions for other fisheries to continue, or closing only those parts of the area where small bass were thought to be particularly vulnerable, and banning all fishing there; the intention being to minimize opportunities to catch small bass.

The proposed nursery areas were specified and described in a consultative paper issued in March 1989. When this was published, it was emphasized that the extent of the areas, the length of closed seasons and the restrictions on fishing activities for species other than bass were not sacrosanct, and that the Fisheries Department was prepared to consider variations and modifications on the basis of evidence and informed comment from fishermen, anglers and other interests.

There was also another reason for these further detailed consultations at local level. To be successful, the restrictions applying in nursery areas would need the cooperation of fishing interests and local regulatory bodies such as SFCs and the NRA. All those concerned were, therefore, invited to study the scientific report, and to consider whether it would be appropriate to propose local arrangements which could be effectively implemented through local bye-laws.

A further consideration was the legislative framework within which bass nursery areas were to be implemented. Whilst section 13(5) of the *Sea Fisheries Regulation Act 1966* (Great Britain–Parliament, 1966) provides that 'any local fisheries committee may, within their district, enforce any Act relating to sea fisheries', it is an obvious prerequisite that the enforcing body must look for breaches of the legislation to bring prosecutions. Unfortunately, the 1966 Act contains none of the ancillary powers of entry, inspection, search and seizure needed for that purpose, and the power conferred by section 13(5) can only be used to take proceedings. Similarly, the powers conferred by section 10(2) of the 1966 Act are given only for the purpose of enforcing bye-laws made by a local sea fisheries committee, and cannot be used for the purpose of enforcing a Parliamentary Act relating to sea fisheries.

It must be borne in mind that the UK's strategy for bass conservation encapsulated management plans for the international fishery for adult fish outside territorial waters, in addition to the protection of juvenile bass in extremely parochial nursery areas. Consequently, it was apparent that, despite the attendent limitations, nursery areas and the other measures would best be designated by a national order, thus conferring upon them a status which would assist with future international negotiations. To this end, section 13(1) of the *Sea Fish (Conservation) Act 1967* (Great Britain – Parliament, 1967) provides that 'a local fisheries committee may take proceedings in respect of any contravention of section 1, 2 or 3 of this Act occurring within the district of the committee'. Sections 1 and 2 contain prohibitions on landing, selling and carrying undersized fish and on their possession for business purposes; section 3 creates offences relating to nets and other fishing gear. The supporting powers for the power conferred by section 13(1) (that is, the powers lacking in the 1966 Act) are to be found in sections 16(1) and 17 of the 1967 Act. The 36 cm minimum landing size regulation, and mesh size controls on enmeshing nets, were made under sections 1 and 3, respectively (Great Britain–Parliament, 1989a and 1989b), and 34 nursery areas (Anon., 1990) were established under section 5 of the 1967 Act (Great Britain–Parliament, 1990). For the purposes of enforcement, sea fisheries committee personnel can be appointed as British Sea Fishery Officers under section 7 of the *Fisheries Act 1968* (Great Britain–Parliament, 1968).

Unlike almost any previous sea fisheries legislation in the UK, the bass regulations were expected to affect both commercial fishermen and anglers to a similar degree. The impact would vary, depending on, for example, the extent to which fishermen had previously focused their effort on small bass and whether they habitually worked in those nursery areas for which restrictions on fishing were being proposed. Ironically, perhaps, it turned out to be the anglers who would face the greatest restrictions if nursery areas were closed to all fishing activity.

Under the provisions of the *Sea Fisheries (Conservation) Act 1967* (Great Britain–Parliament, 1967), enmeshing nets with a mesh size between 65 and 89 mm were prohibited within British fishery limits lying south of lines of latitude drawn through Haverigg Point in Cumbria on the west coast and Donna Nook in Lincolnshire on the east coast. These restrictions were confined to nets carried by registered fishing boats, but restrictions on fishing for bass in nursery areas were to apply only to any fisherman or angler working from a boat, whether registered or not. However, it was hoped that local SFCs, using their powers under the *Sea Fisheries Regulation Act 1966* (Great Britain–Parliament, 1966) would, if necessary, introduce bye-laws applying restrictions on mesh sizes of nets worked from the shore and, similarly, restrictions on fishing and angling from the shore in nursery areas. Although the mesh size restrictions would also affect net manufacturers and suppliers, it was intended that there would be at least 2 years' warning of these requirements. Whilst the proposed restrictions could not, under European Community law, be applied to fishermen from other member states, such fishermen are not permitted to fish within UK coastal waters out to 6 miles. Although the boats of some European countries have limited access to waters between 6 and 12 miles offshore, bass under 40 cm are seldom encountered there. Nevertheless, the Government pressed for an early increase in the EC minimum landing size for bass from 32 to 36 cm, in order to strengthen the regulation by applying it across North-west Europe and increasing the legislation's powers. This was accomplished in December 1989.

For fishermen to comply with the regulations, the main change in fishing practices would be an adjustment in gear and fishing grounds to avoid taking juvenile bass. It was not anticipated that any of these changes would be widely welcomed, but inshore fishermen are adaptable, and they often change their methods and choice of fishing grounds to counter fluctuations in the availability of fish resources. By a fate of nature, three successive year classes (1984–86) of bass were relatively weak, and this had resulted in a dearth of fish between 32 and 36 cm in the late 1980s. By the time that the regulations were introduced early in 1990, many bass fishermen had already become used to directing their effort at older, larger bass of the 1982 and 1983 year classes (Fig. 15.1).

It was also recognized that the regulations would affect market and consumer choice, particularly in view of recent demand for plate-sized bass in response to increased supplies in the mid 1980s and improved marketing facilities throughout Europe. Despite this trend, the unit price for bass above and below 36 cm was higher than that of most other marine fish species. Nevertheless, the measures were not expected to have a significant impact on the market for small bass, which was increasingly receiving farmed fish from Mediterranean countries.

It was anticipated that some fishermen operating in nursery areas might elect to forgo fishing altogether rather than change their ways. There would, however, be no disruption of activities that did not interfere with the bass fishery, such as potting for shellfish, drift and ring netting for grey mullet (for which mesh size controls were exempted within the 3 mile limit between Beachy Head in Sussex and Rame Head in Cornwall), trawling and licensed salmon netting. Although it was inevitable that in the short term, there would be a loss of earnings for some people, this would probably be reversed within 1 to 2 years, as those bass benefiting from protection grew and recruited to the fishery.

In addition to providing greater bass fishing opportunities for anglers than the previous free-for-all, when unrestricted gill netting in some nursery areas had greatly lessened their attraction to anglers, it was estimated that within five years of the regulations coming into full effect, the yield to the fishery should increase by up to 25%. Because the measures were aimed at the long-term management of the bass fishery, it would be necessary to review the effects of the restrictions on mesh sizes and on fishing in nursery areas during DFR's monitoring of the fishery.

All this might seem complicated to the layperson, but the lesson is that the best resource management plans are made alongside appropriate assessments of their impact and legislation that enables them to be enforced effectively.

17.6 INTERNATIONAL REGULATION OF BASS FISHERIES

Whilst the bass fishery around England and Wales is prosecuted close inshore by small boats, which target both adult and small bass, there are also many offshore pair trawlers which have been fishing the previously underexploited adult stock(s) during winter and early spring. Although France, England and Wales contribute the majority of bass landings from Biscay northwards, the Danes, Dutch, Scots and Iberian fishermen have been looking for a foothold in this fishery.

Implementation of the EC MLS of 36 cm, coupled with the national UK measures to increase protection of juvenile bass, was expected to help safeguard the stock. But these measures were associated with effort and exploitation patterns in the late 1980s and may not in future be adequate to protect either the stock, or the economic viability of the UK fishery, in the event of a continued expansion in fishing effort on bass.

There has been no international assessment of the stocks involved, and though, by 1992, the UK had much improved landings statistics, annual catch-at-age data by ICES Division, and estimates of recruitment and mortality levels in the inshore fishery, it will not be possible to determine exploitation levels in the population as a whole until the offshore international fishery is well sampled. Nevertheless, some limitation on the international offshore fishery might eventually be desirable to maintain an adequate spawning stock and increase the numbers of adult bass available inshore in summer, so that the inshore fisheries of the UK and other countries which implement conservation measures to protect small bass can benefit from their own restraint.

Management units

Whilst the fisheries for most international marine fish resources are centred offshore, with the exploitation in the coastal margins being a relatively small part of the total, the reverse has been true for bass in the UK and Ireland, with the coastal fishery predominating over the offshore activity. This presents a considerable difficulty in arriving at satisfactory analytical assessments and catch forecasts for conventional TAC regulation.

Inshore fishery

The UK inshore fishery operates mainly from April to October with small boats which often land directly to retailers. The majority take bass in a seasonal fishery, and though bass is a valuable species, bass landings generally do not by themselves constitute the mainstay of fishermen's earnings. Assessment and monitoring of the catches of the inshore fishery is based on a voluntary sampling approach (the fishermen's logbook scheme, Chapter 12) and not a complete census of all major landings of the species, as in the major marine fisheries.

Offshore fishery

The offshore fishery is conducted between November and May by vessels of 10 m or more in length, from which landings could be monitored through EC logbooks or at markets, though they tend not to be reported internationally because there is no international agreement on assessing or managing the bass stock and its fishery. The directed fishery (as it is conducted at present) is easily identified.

Management strategy

No matter how the stocks or fisheries of a species are defined, resource management has two aspects: the need to regulate the size and age of the fish caught, and control of the proportion of the stock caught each year. The first is achieved by technical measures (e.g. mesh size controls, MLS, closed areas), the second by catch or effort limitation. For bass, the gill net mesh size and nursery area measures introduced at the UK national level could also be applied through the EC to augment the existing MLS control, or they could be applied nationally on the UK model to any of the effectively independent north European coastal fisheries which predominantly exploit juvenile rather than adult bass. As long as these measures were effectively enforced, the inshore fisheries by vessels under 10 m may then be said to be soundly managed, as far as the exploitation pattern (age-related mortality) is concerned. It might also be argued that the exploitation pattern in the offshore fishery is already satisfactory, by virtue of the natural distribution of the fish which are principally adults in this fishery. There would, therefore, be no need for technical measures in addition to those applied in the coastal zone.

Management that relies on technical measures alone is only adequate so long as fishing remains stable at an appropriately modest level. But bass is now a high-value international commodity and remains outside the EC's quota management system, so there is justifiable concern that exploitation could escalate to undesirable levels within the time it will take to establish the scientific basis necessary for an analytical assessment and TAC or effort controls to be implemented. Moreover, the unusual mixture of coastal and offshore interests, and an absence of internationally maintained catch records, may present a considerable difficulty in achieving national and international agreement on allocations and quotas.

Because data are not available with which to assess the quantitative relationships between the various national coastal and offshore stocks of bass and the yields to their respective fisheries, it is not possible to determine how an analytical TAC could be varied and allocated annually on the basis of geographical components of the stock. Provided that the exploitation pattern is directed towards adult fish, however, it might be feasible to use a more general TAC, covering the North Sea, Irish Sea, Celtic Sea and, possibly, Biscay, based on steady-state calculations that also take into account recruitment, because this can be forecast with confidence. Whichever method is used, and although the catches of the offshore fishery could easily be monitored, estimating or monitoring the catches of the coastal fishery, in order to manage quotas throughout the year, would be very difficult indeed for reasons given earlier.

The regulation of fishing effort has somewhat analogous difficulties. The offshore fishery is conducted by relatively few, easily identified vessels, but the coastal fishery is conducted by a host of small commercial vessels and anglers, whose level of bass fishing activity simply could not be controlled other than by opening and closing the fishery nationally. Given the seasonal variations in the availability of bass around the coasts owing to migrations, this approach would be extremely disruptive.

17.7 THE WAY FORWARD

The overall position, then, is that the current package of UK technical measures should provide adequate control of exploitation patterns on bass, and could be extended to the coastal fisheries of other countries, but there is a foreseeable requirement for international control of the level of exploitation. The assessments required for analytical forecasts of catch options, and their implications for bass stocks, are not yet available. Even a precautionary TAC, which would allow for variation in recruits to the stock and take account of catches in both offshore and inshore fisheries, would probably have to be set so high that it could not provide an effective control of the level of exploitation in most years.

On the other hand, there is as yet no international acceptance of the way in which direct regulation of fishing can be achieved, i.e. through days at sea controls. Whilst this might be appropriate to the offshore fishery for bass, however, it would pose enforcement difficulties inshore. It is evident that the inshore and offshore sectors could be considered as separate management units, subject to regulatory regimes that would be closely related but would differ in detail, especially with respect to annual adjustments.

The offshore fishery could be controlled by regulation of fishing effort, through, for example, a limited number of boats being licensed to exploit the stock during a restricted season. Alternatively, a TAC could be applied to this sector of the fishery, adjusted annually in response to variations in indices of recruitment (to the offshore fishery), obtained from analysis of the coastal fisheries catch-at-age data. Quantitative catch or effort regulations are unlikely to be effective in the coastal fisheries because month-by-month catch monitoring is difficult, and effort monitoring, especially of the angling component, would be impossible. Because of this enforcement difficulty, any quantitative limit would only be regarded as a broad target to keep the managers informed as to whether the level of inshore fishing was being held within reasonable bounds, rather than the enactment of a close control on the fishery backed by legal sanctions.

It could be argued, instead, that the technical measures introduced in 1990 will suffice to stabilize the UK coastal fishery within reasonably liberal catch limits. The fishery in its present form is not expected to be subject to a rapid increase in directed fishing for bass, though pair trawling teams have, since 1991, become established in the English inshore zone of the English Channel. They fish for bass in summer as well as winter, and local line fishermen have claimed concomitant declines in their catches in some areas. However, such specialized bass fishing vessels are readily identifiable and could be included in the 'offshore' controls.

These various characteristics of the bass fishery put severe limitations on the management options for regulating the international level of fishing when this becomes necessary, as it will, sooner or later. There must be a fairly liberal limit on the catch of the coastal fisheries, within which fishing effort might be stablized by technical measures, but which would be capable of responding to variation in stock size so that actual catches provide guidance for adjusting the precautionary-cum-analytical TAC for the offshore fishery. As with the coastal fishery, the offshore sector TAC could only provide an imprecise guideline for monitoring the performance of, in this case, a fleet of licensed vessels, the licensing being the effective control of fishing capacity intended to exclude the intrusion of mobile fleets and a consequent foreshortening of the fishing season. It might be argued that a catch allocation which does not restrict the coastal fishery discriminates in its favour, but this can be refuted on the grounds that this fishery's effort is subject to very restrictive technical measures that protect the resource basis of the offshore fishery. The inshore fishery also employs many more units than the offshore fishery, but these are small operators and do not have the scope and geographical flexibility to exploit alternative resources should the seasonal bass catch fail to materialize.

Chapter eighteen

Progress and prognosis

It will be evident that we have not attempted to give an entirely dispassionate academic account of what is known about the biology of the European sea bass, nor have we merely provided a description of its fishery and a discussion of the problems it faces and the theoretical remedies for them. In fact, we have not been bystanders in this matter, but a small part of a movement which is intensely interested in bass – not as an angler's quarry nor a commercial resource, but as a scientific subject. We hope that evidence of this enthusiasm for the species is not missing from these pages.

The opportunity to work on a fish which had been so little studied prior to the 1970s arose because many people were concerned for the future of the bass stocks around the coasts of England, Wales and Ireland. What little was known of its biology in the late 1970s was, nevertheless, sufficient to indicate that protection of local juvenile populations was bound to assist conservation of bass. The same, of course, can be said for any fish stock. Quantitative assessments of the size of the bass population – even in relative terms – and of its productivity and mortality rates were entirely lacking, and so any restraint on exploitation (if this really were to be necessary) could not be justified on the basis of an analytical assessment, nor could it take the form of direct catch controls. Management had to take the form of technical measures, aimed at directing exploitation away from juvenile bass. Unfortunately, at that time, so little was known about the fishery in which bass were caught, how, when and where it operated and the nature of its interactions with fisheries for other species, that the only management measure that could be considered was a minimum landing size regulation, specific to bass. For the purposes of protecting stocks it was easy to see that an MLS could be usefully set at around the size of first maturity in female fish. This was already well known to be around 40 cm for bass around British coasts. But what effect would such a measure have on the commercial fishery, on yields, and on the livelihoods of all the people who caught, sold or bought bass, and would a sufficiently high MLS (for effective conservation of the stocks) prove so unacceptable that it would be ignored? The government has to be mindful of a fisheries policy which includes considerations of non-discrimination and enforceability.

The lack of knowledge about the bass fishery and its participants, and the decision by the Minster of State for Fisheries that it would be premature to increase the bass MLS beyond 32 cm before at least a preliminary report had been prepared on the outcome of relevant investigations, prompted DFR to set up the bass fishery research programme in 1981.

The first task was to find out just who was catching bass, when and where and by what methods, and also to examine the distribution of stocks and their movements so that an understanding of the relationship between exploiters and resource could be gained. Despite considerable advances in the technology of remote sensing and telemetry, there is only one way to do this, go out and work in the fishery itself. Consequently, 1982 and 1983 were devoted to tagging studies in which commercial fishermen were chartered, hired or otherwise persuaded into letting members of a small intrepid team occupy their boats as they fished. There were four important achievements during this period. Information was gathered about the fishery, bass were tagged and released into populations all around the UK coasts (complementing the work of Don Kelley, with whom we had soon made mutually beneficial contact), scale samples and other biological data (including gill net retention information) were obtained and, probably most important to us, we met the interested parties, shared their pots of tea, and generally established a working relationship.

We were not long in discovering – as we were told in no uncertain terms more than once – that official Ministry figures for the quantities of bass landed in England and Wales were grossly under-estimated. In the early 1980s, all important commercial marine fisheries were monitored using data on catches landed at 60–70 of the most important ports in England and Wales. At some of these, the landings were also sampled by species for length and age analysis. Bass, however, was not an important species at that time, and very few catches actually went to the ports routinely sampled by MAFF's SFI. An alternative sampling strategy had to be devised and implemented.

The logbook scheme (Chapter 12) has produced a better time series of catch and effort data for the bass fishery than is available for almost any other UK inshore fishery. Quota management under the European Commission's CFP management regime has disrupted traditional fishery patterns and led to misreporting in many catch returns, and assessment data for many stocks are becoming less reliable. In the bass fishery, by contrast, we know which part of the fleet is being sampled, how its activities relate to unsampled vessels, and there is no tangible gain to the fishermen in misreporting; the scheme is voluntary, data are confidential and there are no quotas to be observed or track records of catches to be established.

Adequate biological samples have been less easy to obtain. Other marine species, such as cod, plaice or herring, have been regularly supplied to particular ports, in season, where MAFF regional staff can easily intercept and measure fish and collect material for ageing (usually otoliths). Additional measurements, otoliths and biological material such as stomach contents and gonad tissues are routinely collected on research vessel cruises using trawl gear. With bass, however, not only are landings often remote from regional staff, and relatively unpredictable, but because the fish are sold ungutted and whole, sampling beyond weight, length and a few unnoticed scales is very expensive. Adult bass are seldom caught on routine charter or research vessel surveys. Considerable time, energy and ingenuity have therefore been spent in amassing sufficient biological data on bass to enable catch estimates to be expressed as length or age distributions, and for basic biological studies (e.g. feeding patterns, fat, maturity and condition cycles and genetic identity) to be completed. On the other hand, juvenile bass are relatively easy to find and catch, once you know where and how. Routine sampling of several coastal power stations' cooling-water intake screens, and charter vessel surveys of nursery areas, have provided reliable data on the relative abundance of bass year classes before they recruit to the commercial fishery.

Much of the work described in this book has been conducted and supervised from the Fisheries Laboratory at Lowestoft, though many people outside DFR have been involved in data collection. Limits on resources, and the rather urgent need to establish a sound scientific basis for management advice, meant that DFR's work programme had to concentrate on those aspects of the biology of bass and its fishery that were most relevant to the provision of that advice. Extramurally, however, every encouragement was given to those who were interested in investigating other matters, in particular the species' early life history. In addition to his tagging work, Don Kelley's nursery studies on the west and south-west coasts were a most valuable contribution, pre-empting much of the environmentalists' present concerns for the threatened status of estuaries. A series of postgraduate research studentships at Swansea and Plymouth Universities focused on the reproductive biology and the ecology and population dynamics of bass during their first year of life. It is also possible that DFR's promotion of bass research elicited publications by other fish biologists who had collected appropriate data on bass incidental to their main research subjects, but who otherwise had little incentive to analyse or publish the results.

In appraising the progress that has been made in increasing our knowledge of the biology of bass and assessment of its stocks and fishery, it is difficult to decide whether a comparison is best made with that other contentious resource and recreation species, the Atlantic salmon, or with some commercial marine species.

The salmon has been studied intensively and extensively for well over 100 years, and information on its life history, with the exception of its salt-water sojourn, is abundant and comprehensive. There is a very special curiosity about a fish that is born and lives most of its life in clear freshwater streams, only to disappear from view for one or more years before reappearing in them as a large, conspicuous, fighter of waterfalls and anglers. This freshwater phase is not only relatively easy to observe, but its continued success is critical to the salmon's survival. And *that* is economically and politically important. Our knowledge of the reproductive and juvenile biology of the bass has improved considerably with developments in its aquaculture, but we have only been able to sketch the outline of the behaviour of wild spawners, the mechanism of recruitment of larvae to nursery areas and the general ecology of the species' early life history. Nevertheless, the sea phase of salmon is probably no better understood than that of the bass.

Many commercial marine fish have also been extensively studied, in a difficult environment, chiefly with a view to identifying and assessing the impact of exploitation on stocks rather than elucidating their basic biology. It is likely, therefore, that in ten years the bass has been brought well up the league of marine fish knowledge.

It is useful to make a distinction between knowledge that is linked to a time and a place, and soon becomes dated, and that which will stand the test of time. The first category includes descriptions of fisheries, the state of stocks and, so it seems, reasons for managing them. Basic biological knowledge is more enduring. This might explain why data on commercial fish species are less widely disseminated or used in scientific papers than those collected to answer biological questions, and why there are more formal publications on gobies and sticklebacks than on haddock or plaice. Our own publishing record on bass during the 1980s is slight, preoccupied as we were with assessment, provision of management advice and the consultation exercise leading to implementation of regulations for the bass fishery in the UK. In fact, most of the papers dealing with bass around England and Wales have been written by researchers in whose investigations the species figured only briefly. Nevertheless, it is to their credit that information published on the biology of bass has improved considerably during the last 6 or 7 years. Prior to 1983, only three published papers dealt mainly with the biology of *D. labrax* around England and Wales; now there are more than 30.

For as long as DFR continues with its bass assessment programme, information on the fishery around England and Wales is likely to be maintained at its present detailed level. Should the requirement for catch and effort data in the inshore sector lapse, however, as might occur if management of the fishery solely concerned effort or catch quota controls on boats over 10 m in length, our current knowledge of the whole fishery for bass (and several other 'minor' species) will deteriorate. For that reason alone, the information given in Chapters 9 and 12 might never be improved upon. Similarly, the bass pre-recruit surveys in the Solent have provided a robust time series of data on year class strength, and on the distribution and growth of juvenile bass. This has enabled us to examine the influences of climate on reproductive success and juvenile survival, and gives a reliable advanced warning of good or poor recruitment to the bass spawning stock and fishery in the English Channel.

Looking further into the future, it is highly unlikely that there will be no attempts to increase the level of exploitation of bass stocks, nor is it likely that coastal development can be held in check in bass nursery areas. It is important, therefore, that research continues on the ecological impact of tidal barrages, marinas, aggregate extraction and any other perturbation of the inshore environment upon which the bass appears to be so utterly dependent. For once, in dealing with a marine fish species, we are in a strong position to assess the impact on stocks of degradation of nursery grounds, and it is readily apparent when recruitment to a bass fishery in any particular area falls below the regional norm.

As to fisheries management, with the bass we have seen clear evidence that priorities and objectives are changing. The introduction of TACs and quota management regimes for most of the more important fisheries exploiting shared stocks in North-west Europe in 1984 confirmed and emphasized the need to regulate catching power in the various national fleets. The EC's Common Fisheries Policy was aimed at an acceptable share-out of the available yields, sustained employment in the industry and providing legitimate access for national fishing fleets across median lines between countries, but it was also aimed at conserving fish stocks and the fisheries themselves (Wise, 1985). As never before, it became startlingly obvious just what quantities of the various assessed species' stocks were available to be caught, focusing fishermen's minds on how best to guarantee their share and maximize its value. Such calculating thinking in a monetarist society must have led many people to consider how good a living could be made from a high-value and relatively unregulated species like the bass. As a consequence, it has become necessary to introduce management measures which go as far as the limits of legislation and enforcement allow, to safeguard spawning stocks and yields to the fishery, without recourse to direct controls on catch or effort. That this has come about signals, to us, the recognition that small-boat fisheries –

and recreational angling – are important in England and Wales (as they are in much of coastal Europe), socially, economically and, therefore, politically.

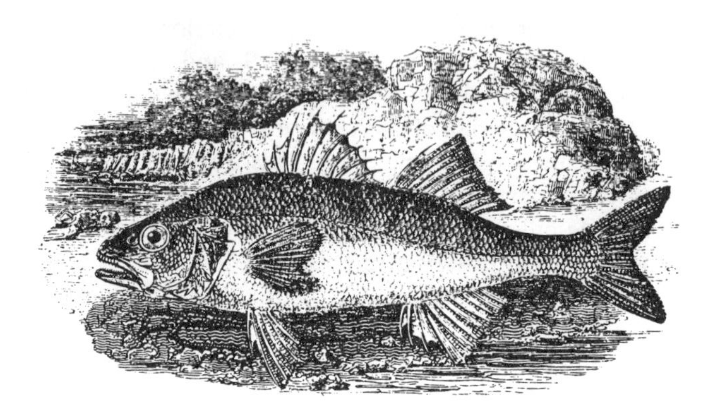

Basse. From Yarrell, 1836.

Bibliography

Aflalo, F.G. (1891) *Sea-Fishing on the English Coast*, L. Upcott Gill, London.

Aflalo, F.G. (1898) *Sea Fish* (The Anglers' Library, II) (ed. The Right Hon. Sir Herbert Maxwell), Bart. Lawrence and Bullen Ltd, London.

Aflalo, F.G. (1904) *British Salt-water Fish*. Woburn Library of Natural History, London, 328 pp.

Aflalo, F.G. (1910) *Sea Fishing in Ottoman Waters*. (British Sea Anglers Society Qtrly, Vols II & III), G. Barber, London.

Albuquerque, R.M. (1956) *Poissons du Portugal. Clefs de Leur Détermination*. 1164 pp. + 274 figs.

Alderson, R. and Howell, B.R. (1973) The effect of algae on the water conditions in fish rearing tanks in relation to the growth of juvenile sole *Solea solea* (L.). *Aquaculture*, **2**, 281–8.

Alliot, E., Pastoureaud, A. and Thebault, H. (1983) Influence de la température et de la salinité sur la croissance et la composition corporelle d'alevins de *Dicentrarchus labrax*. *Aquaculture*, **31**, 181–94.

Alliot, E., Pastoureaud, A., Palaez-Hudlet, J. and Metailler, R. (1979) Utilisation des farines végétales et des levures cultivées sur alcanes pour l'alimentation du bar (*Dicentrarchus labrax*), in *Finfish Nutrition and Fishfeed Technology*, Vol. 2 (eds J.E. Halver and K.Tiews), pp. 229–35.

Andrade, J.P. (1983) Contribution to the biological knowledge (age determination and growth study) of the bass *Dicentrarchus labrax* (L.) (Pisces, Moronidae) from Ria de Aveiro, in *Proceedings of two meetings on Iberian Ichthyology, Barcelona*, 1983, 21 pp.

Anon. (1990) Bass nursery areas and other conservation measures. MAFF Pubs, London, 15 pp.

Aprahamian, M.W. and Barr, C.D. (1985) The growth, abundance and diet of 0-group sea bass, *Dicentrarchus labrax*, from the Severn Estuary. *J. Mar. Biol. Ass. U.K.*, **65**, 169–80.

Arias, A. (1980) Crecimiento, regimen alimento y reproduction de la dorada *Sparus aurata* y del robalo *Dicentrarchus labrax* en los esteros de Cadiz, *Invest. Pesq.*, **44**(1), 59–83.

Arnold, G.P. (1969) A flume for behaviour studies of marine fish. *J. exp. Biol.*, **51**, 671–80.

Arnold, G.P., Greer Walker, M. and Holford, B.H. (1990) Fish behaviour: achievements and potential of high-resolution sector-scanning sonar. *Rapp. P.-v. Réun Cons. Int. Explor. Mer*, **189**, 112–20.

Audousset, D. (1978) Le bar, *Dicentrarchus labrax*: zoologie et élevage, thése pour le Doctorat Veterinaire, Ecole Nationale Veterinaire d'Alfort.

Bagenal, T.B. and Tesch, F.W. (1978) Age and growth, in *Methods for the Assessment of Fish Production in Fresh Waters* (ed. T. Bagenal), Blackwell Scientific, Oxford, pp. 101–36.

Barahona-Fernandes, M.H. (1978) Effect of aeration on the survival and growth of sea bass (*Dicentrarchus labrax*) larvae: a preliminary study. *Aquaculture*, **14**, 67–74.

Barahona-Fernandes, M.H. (1979) Some effects of light intensity and photoperiod on the sea bass larvae (*Dicentrarchus labrax* (L.)) reared at the Centre Oceanologique de Bretagne. *Aquaculture*, **17**, 311–21.

Barahona-Fernandes, M.H., Girin, M. and Metailler, R. (1977) Expériences de conditionnement d'alevins du bar (Pisces, *Dicentrarchus labrax*) a differents aliments composés. *Aquaculture*, **10**, 53–63.

Barnabé, G. (1972) Contribution à l'étude de la biologie du loup (*Dicentrarchus labrax* L.) de la région de Sète, thèse 3ème Cycle, Univ. Sc. Techn. Languedoc, Montepellier, 160 pp.

Barnabé, G. (1973). Contribution à la connaissance de la croissance et de la sexualité du loup (*Dicentrarchus labrax*) de la région de Sète. *Annls. Inst. Océanogr.* Paris, **49**(1), 49–75.

Barnabé, G. (1974) Mass rearing of the bass *Dicentrarchus labrax* L., in *The Early Life History of Fish* (ed. J.H.S. Blaxter), Springer-Verlag, Berlin, pp 749–53.

Barnabé, G. (1976a) Contribution à la connaissance de la biologie du loup *Dicentrarchus labrax* (L.) de la région Sète. *Thèse Univ. Sc. Tech. Languedoc Montpellier*, 426 pp. multicop.

Barnabé, G. (1976b) Elevage larvaire du loup (*Dicentrarchus labrax* (L.); Pisces, Serranidae) à l'aide d'aliment sec composé. *Aquaculture*, **9**, 237–52.

Barnabé, G. (1978) Etude dans le milieu naturel et au captivité de l'écoéthologie du loup *Dicentrarchus labrax* (L.) (Poisson, Serranidae) à l'aide de nouvelles techniques. *Ann. Sci. Nat. (Zool.)*, **20**, 423–502.

Barnabé, G. (1980) Exposé synoptique des données biologiques sur le loup ou bar, *Dicentrarchus labrax*. *Synop. F.A.O. Pêches*, no. 126 (70 pp).

Barnabé G. (1990) Rearing bass and gilthead bream, in *Aquaculture*, Vol. 2. (ed. G Barnabé), Ellis Horwood, N. Y., pp. 647–86.

Barnabé, G. and Barnabé-Quet, R. (1985) Avancement et amélioration de la pointe induite chez le loup *Dicentrarchus labrax* (L.) à l'aide d'un analogue de LHRH injecté. *Aquaculture*, **49**, 125–32.

Barnabé, G. and Billard, R. (1984) Edition des Actes du Colloque 'L'Aquaculture du bar (loup) et des Sparides'. Sète, 15–17 Mars, 1983.

Barnabé, G. and Rene, F. (1972) Reproduction contrôlée du loup *Dicentrarchus labrax* (Linné) et production en masse d'alevins. *C. r. hebd. Seanc. Acad. Sci. Paris*, **275**D, 2741–4.

Barnabé, G. and Tournamille, J. (1972) Expériences de reproduction artificiele du loup, *Dicentrarchus labrax* (L). *Rev. Trav. Inst. Pêches Marit., Nantes* **36**(2), 185–9.

Bedier, E. (1981) Pilot scale production of sea bass (*Dicentrarchus labrax*) fry. *Rapp. P.-v. Réun. Cons. Int. Explor Mer*, **178**, 530–2.

Bell, M.V., Henderson, R.J. and Sargeant, J.R. (1986) The role of polyunsaturated fatty acids in fish (mini review). *Comparative Biochemistry and Physiology*, **83B**, 711–9.

Ben Tuvia, A. (1974) On the occurrence of the mediterranean serranid fish *Dicentrarchus punctatus (Bloch)* in the Gulf of Suez. *Copeia*, 1974, 741–3.

Bertignac, M. (1986) Etude de la pêcherie du Bar dans le Morbras – etat d'avancement des travaux. *Cah. Morbras*, no. 18, pp. 1–14.

Bertalanffy, L. von (1938) A quantitative theory of organic growth (Inquiries on growth laws, II). *Human Biol.*, **10**(2), 181–213.

Bertignac, M. (1987) L'exploitation du bar (*Dicentrachus labrax*) dans le Monbras (Bretagne Sud). Thèse de Doctorat en sciences, Ecole National Supcricure Agronomique de Rennes, 236 pp.

Bertolini, F. (1933) Serranidae. In uova, larve e stadi giovonili di teleostei. Fauna u Flora Golf. *Neapel Monogr.*, **38**, 310–3.

Beverton, R.J.H. and Holt, S.J. (1957) *On the dynamics of exploited fish populations.* Fishery Invest., Lond., (Ser. 2), **19**, 533 pp.

Bickerdyke, J. (1887) *Angling in Salt Water.* L. Upcott Gill, London.

Bigelow, H.B. and Schroeder, W. C. (1953). Striped bass *Roccus saxatilis* (Walbaum) 1792, in *Fishes of the Gulf of Maine, U.S. Fish Wildl. Serv., Fish Bull.*, **53**, 389–404.

Bini, G. (1968) *Atlante dei Pesci della Coste Italiane*, Vol. IV, Edit. Mondo Sommerso, 161 pp.

Bloch, M.E. (1792) *Naturgeschichte der auslandischen Fische*, Berlin 6: 53, pl 305.

Bolla, S. (1989) Fatty acid composition of Atlantic halibut larvae fed on enriched *Brachionus, Artemia* and collected plankton. *Rapp. P.-v.s Réun., Cons. Int. Explor. Mer*, **191**, 475.

Bonn, E.W., Bailey, W.M., Bayless, J.D., Erickson, K.E. and Stevens, R.E. (1976) *Guidelines for Striped Bass culture.* Am. Fish. Soc. Striped Bass Committee of the Southern Division, Amer. Fish. Soc., Bethesda, MD, 103 pp.

Bou Ain, A. (1977) Contribution à l'étude morphologique, anatomique et biologique de *Dicentrarchus labrax* et *Dicentrarchus punctatus* des côtes tunisiennes, Thèse de Doctorat de spécialité, Faculté des Sciences, Tunis, 115 pp.

Boulenger, G.A. (1895) *Catalogue of the Perciform fishes in the British Museum.* 2nd ed. Trustees Br. Mus. (Nat. Hist.), London, 391 pp.

Boulineau-Coatanea, F. (1968) Etude anatomique et descriptive des otolithes (*sagitta*) du bar (*Morone labrax*). *Bull. Mus. nat. Hist. Nat., Paris*, 2ème Série, **40**(3), 474–84.

Boulineau-Coatanea, F. (1969) Contribution à l'étude biologique du bar *Dicentrarchus labrax* (Linné). Thèse 3ème cycle, Fac. Sci., Univ. Paris., Océanogr. Biol, 121 pp.

Boulineau-Coatanea, F. (1970) Régime alimentaire du bar *Dicentrarchus labrax (Serranidae)* sur la côte atlantique bretonne. *Bull. Mus. Nat. Hist. Nat., Paris*, 2ème Série, **41**(5), 1106–22.

Brander, K.M. and Bennett, D.B. (1986) Interactions between Norway lobster (*Nephrops norvegicus*) and cod (*Gadus morhua*) and their fisheries in the Irish Sea, in (eds G.S. Jamieson and N. Bourne), *North Pacific workshop in stock assessment and management of invertebrates, Can. Spec. Publ. Fish. Aquat. Sci.*, **92**, pp. 269–81.

Bromage, N., Carrillo, M. and Zanuy, S. (1988) Light controls spawning in sea bass. *Fish Farming Int.*, **15**, 22.

Brunel, G. and Fuzeau, P. (1990) Aquaculture in heated water, in *Aquaculture*, Vol. 2. (ed. G. Barnabé), Ellis Horwood, Chichester, pp 801–19.

Truslé, J. and Roblin, C. (1984) Sexualité du loup *Dicentrarchus labrax* en condition d'élevage contrôlé, in *L'Aquaculture du Bar et des Sparidés* (eds G. Barnabé and R. Billard), INRA, Paris, pp. 33–43.

Buller, F. (1981) *Pike and the Pike Angler*, Stanley Paul & Co. London, 288 pp.

Bye, V.J. (1984) The role of environmental factors in the timing of reproductive cycles, in *Fish Reproduction: Strategies and Tactics* (eds G.W. Potts and R.J. Wootton), Academic Press, London, pp. 187–205.

Cadenat, J. (1935) Les Serranidés de la côte Occidentale d'Afrique du Cap Spartel au Cap vert. *Revue Trav. Inst. Pêches Marit.*, **8**(4), 377–422.

Carlander, K.D. (1982) Standard intercepts for calculating lengths from scale measurements for some centrarchid and percid fishes. *Trans. American Fish. Soc.*, **111**, 332–36.

Carlisle, D.B. (1961) Inter-tidal territory in fish. *Journal of Animal Behaviour*, **9**, 106–7.

Carrillo, M., Bromage, N., Zanuy, S., Serrano, R. and Prat, F. (1989) The effect of modifications in photoperiod on spawning time, ovarian development and egg quality in sea bass (*Dicentrarchus labrax* L.). *Aquaculture*, **81**, 351–65.

Carter Platts, W. (1940) *The Young Angler*. Adam and Charles Black, London, 227 pp.

Chervinski, J. (1974) Sea bass, *Dicentrarchus labrax* Linne (Pisces, Serranidae) a 'police-fish' in freshwater ponds and its adaptability to various saline conditions. *Bamidgeh*, **26**, 110–13.

Chervinski, J. (1975) Sea basses (*Dicentrarchus labrax* (Linne) and *D. punctatus* (Bloch)) (Pisces, Serranidae) a control fish in freshwater. *Aquaculture*, **6**, 249–66.

Chervinski, J. and Lahav, M. (1979) Fresh water feed of young European sea bass (*Dicentrarchus labrax* L.). *Bamidgeh*, **31**, 44–8.

Chevalier, C. (1980) Contribution a l'étude de la croissance des juveniles de *Dicentrarchus labrax* L. en Bretagne Nord. *Int. Coun. Explor. of the Sea*, CM 1980/L:35, 9 pp.

Chevey, P. (1925) Recherches sur la perche et le bar, *Thèse Fac. Sc. Paris*, Série A, no. 1011, no. d'ordre 1847, 226–92.

Child, A.R. (1992) Biochemical polymorphisms in bass, *Dicentrarchus labrax*, in the waters around the British Isles. *J. Mar. Biol. Ass. U.K.*, **72**, 357–64.

Claridge, P.N. and Potter, I.C. (1983) Movements, abundance, age composition and growth of bass, *Dicentrarchus labrax*, in the Severn Estuary and inner Bristol Channel. *J. Mar. Biol. Ass. U.K.*, **63**, 871–9.

Clark, J.R. (1968) Seasonal movements of striped bass contingents of Long Island Sound and the New York Bight. *Trans. Am. Fish. Soc.*, **97**, 320–43.

Clement, O. (1990) Aquaculture in marshes: the salt marshes of the French Atlantic coast, in, Aquaculture, Vol. 2 (ed. G. Barnabé), Ellis Horwood, Chichester, pp. 787–800.

Coombs, S.H., Nichols, J.H., Conway, D.V.P., Milligan, S. and Halliday, N.C. (1992) Food availability for sprat larvae in the Irish Sea. *J. Mar. Biol. Ass., U.K.*, **72**, 821–34.

Cooper, E. (1950) *Modern Sea Fishing – from bass to tunny*. 2nd ed. The Sportsmans' Library, A. & C. Black, London, 186 pp.

Corps, M.H.V. (1992) Cannibalism in juvenile bass, *Dicentrarchus labrax*. Porqupine Newsletter, **5**(5).

Couch, J. (1862) *A history of the fishes of the British Islands*, Vol. 1, Groombridge & Sons, London.

Cox, B. (1985). *Uptide and Boatcasting*, A. & C. Black Ltd, London, 104 pp.

Craig, J.F. (1987) *The Biology of Perch and Related Fish*, Croom Helm, London, 333 pp.

Cuvier, G., and Valenciennes, A. (1828). *Histoire Naturelle des Poissons*, II, Paris, 490 pp.

Dando, P.R. and Demir, N. (1985) On the spawning and nursery grounds of bass, *Dicentrarchus labrax*, in the Plymouth area. *J. Mar. Biol. Ass. U.K.*, **65**, 159–68.

Daubert, J.T. and Young, R.A. (1981) Recreational demands for maintaining instream flows: a contingent valuation approach. *Amer. J. Agric. Econ.*, **63**(4), 665–76.

Davedjank, K. (1926) *Pêche et Pêcheries de Turquie Adm. Dette*, Publ. Ottomane, Constantinople, 486 + 169 pp.

Davey, J.T. (1980) Spatial distribution of the copepod parasite *Lemanthropus kroyeri* on the gills of bass, *Dicentrarchus labrax* (L.). *J. Mar. Biol. Ass. U.K.*, **60**, 1061–7.

Davidson, N.C., Laffoley, D.d'A., Doody, J.P., Way, L.S., Gordon, J., Key, R., Pienkowski, M.W., Mitchell, R. and Duff, K.L. (1991) *Nature conservation and estuaries in Great Britain*. Nat. Cons. Coun., Peterborough, 422 pp.

Davies, J.K. (1988) A review of information relating to fish passage through turbines: implication to tidal power schemes. *J. Fish Biol.*, **33**(A), 111–26.

Day, F. (1884) *The Fishes of Great Britain and Ireland*, Vol. 1 (1880–1884), Williams and Moorgate, London, 336 pp.

Dendrinos, P. and Thorpe, J.P. (1985) Effects of reduced salinity on growth and body composition of the European bass *Dicentrarchus labrax* (L.). *Aquaculture*, **49**, 333–58.

Desauney, Y., Perodou, J.B. and Beillois, P. (1981) Etude des nurseries de poissons du littoral de la Loire Atlantique. *Sci. pêche*, 319, 26 pp.

Devauchelle, N. (1984) Reproduction decalée du bar (*Dicentrarchus labrax*) et de la daurade (*Sparus aurata*), in *L'Aquaculture du Bar et des Sapridés* (eds G. Barnabé and R. Billard), Paris, INRA, pp.53–61.

Dieuzeide, R., Novella, M. and Roland, J. (1954) Catalogue des poissons des côtes algériennes. II Ostéoptértygiens. *Bull. Stn. Agric. Pêche Castiglione, N.S., Alger*, 258 pp.

Do Chi, T. and Lam Hoai Thong. (1971) Croissance différentielle de *Dicentrarchus labrax* (L.). Etude préliminaire du phénomène de la région des Sables d'Olonne (Vendée). *Trav. Lab. Biol. Halieutique, Univ. Rennes*, **5**, 29–43.

Dunn, M., Potten, S., Radford, A. and Whitmarsh, D. (1989) *An economic Appraisal of the Fishery for Bass (Dicentrarchus labrax L.) in England and Wales*, a report to the Ministry of Agriculture, Fisheries and Food, Vol. 1, Marine Resources Research Unit, Portsmouth Polytechnic, 217 pp.

Establier, R. and Gutierrez, M. (1980) Acumulación de cadmio a partir del agua de mar por el rôbalo, *Dicentrarchus labrax*, y la dorada, *Sparus aurata* y sus efectos histopatológicos. *Invest. Pesq.*, **44**(1), 43–54.

Establier, R., Gutierrez, M. and Arias, A. (1978) Acumulación de mercurio inorgánico a partir del agua de mar por el rôbalo, *Dicentrarchus labrax* L., y sus efectos histopatológicos. *Invest. Pesq.*, **42**(2), 471–83.

Fahy, E. (1981) The Wexford commercial sea bass, *Dicentrarchus labrax* (L.), fishery. *Fish. Bull.* (Dublin), 3, 10 pp.

FAO (1992) *FAO Year Book. Fisheries. Fisheries Statistics*, Vol. 70, 1990. Food and Agriculture Organisation of the United Nations. Rome 1992, 396 pp.

Farrugio, H. and Le Corre, G. (1985a) Les pêcheries de lagune en Méditerranée, Rapport final. IFREMER, DRV-85-1/PE/SETE: 250 pp.

Farrugio, H. and Le Corre, G. (1985b) Stratégie d'échantillonnage des pêches aux petits métuers en méditerranée. Rapport Coinvention CEE. XIV-B.1 83/2/MO9P1: 120 pp.

Farrugio, H. and Le Corre, G. (1986) Interactions entre pêcheries de lagunes, pêche au chalut dans le Golfe du Lyon. Rapport IFREMER, DRV-86.003/RM/SETE: 208 pp.

Ferrari, I. and Chieregato, A.R. (1981) Feeding habits of juvenile stages of *Sparus auratus* L., *Dicentrarchus labrax* L. and mugilidae in a brackish embayment of the Po River Delta. *Aquaculture*, **25**, 243–57.

Fisher, W. (1973) *Fiches FAO d'Identification des Espèces pour les Besoins de la Pêche Méditerranée et Mer Noire (Zone de pêche 37)*, I et II, FAO, Rome, pag. var.
Fitzmaurice, P. (1978) Some observations on the life history of the bass, *Dicentrarchus labrax* (L). Irish Specimen Fish Committee. Report for 1978, pp. 36–52.
Floyd, K. (1985) *Floyd on Fish*, BBC, London, 112 pp.
Fouché, M. (1986) Impact du marquage sur la biologie et le comportement de bars (*Dicentrarchus labrax*) immatures, *Thèse de Doctorat en sciences agranomiques. Ecole national supérieure agronomique de Rennes*. 131 pp.
Gammon, C. (1967) *Fishing: A Pictorial Guide*, Fredrick Muller, London, 128 pp.
Gatesoupe, F.J. and Robin, J.H. (1982) The dietary value for sea-bass larvae (*Dicentrarchus labrax*) of the rotifer *Brachionus plicatilis* fed with or without a laboratory cultured alga. *Aquaculture*, **27**, 121–7.
Geoffroy St Hilaire, L. (1809) Histoire naturelle des poissons de la Mer Rouge et de la Méditerranée, in *Description d l'Egypt*, Vol. 1, Paris, pp. 311–40.
Gill, T.N. (1861) Monograph on the genus *Labrax* of Cuvier. *Proc. Acad. Nat. Sci. Phil.*, 108–19.
Gmelin, J.F. (1788) *Linnaei Systema Naturae* 1 (3)–Pisces, 1126–515.
Goode, G.B. (1884) The food fishes of the United States, in *The Fisheries and Fishery Industries of the United States*, Section 1, *Natural History of Useful Aquatic Animals*, (eds G.B. Goode and a Staff of Associates), U.S. Govt Print. Off., Washington, DC, pp. 163–682.
Gravier, R. (1961) Les bars (loups) du Maroc atlantique, *Morone labrax* (L.) et *Morone punctatus* (Bloch). *Revue Trav. Inst. Pêches Marit.*, **25**(3), 281–92.
Great Britain–Parliament (1966) *Sea Fisheries Regulation Act, 1966*. Chapter 38, Her Majesty's Stationery Office, London, 12pp.
Great Britain–Parliament (1967) *Sea Fish (Conservation) Act 1967*. Chapter 84, Her Majesty's Stationery Office, London, 23 pp.
Great Britain–Parliament (1968) *Sea Fisheries Act 1968*. Chapter 77, Her Majesty's Stationery Office, London, 26 pp.
Great Britain–Parliament (1975) *Salmon and Freshwater Fisheries Act 1975*. Chapter 51, Her Majesty's Stationery Office, London, 43 pp.
Great Britain–Parliament (1981) *The Immature Bass Order 1981*. Her Majesty's Stationery Office, London, 2 pp., (Statutory Instrument 1981/535).
Great Britain–Parliament (1986) *Salmon Act 1986*. Chapter 62, Her Majesty's Stationery Office, London, 46 pp.
Great Britain–Parliament (1989a) *The Sea Fish (Specified Sea Area) (Regulation of Nets and Prohibition of Fishing Methods) Order 1989*. Her Majesty's Stationery Office, London, 4 pp., (Statutory Instrument 1989/1284).
Great Britain–Parliament (1989b) *The Undersized Bass Order*. Her Majesty's Stationery Office, London, 3 pp., (Statutory Instrument 1989/1285).
Great Britain–Parliament (1990) *The Bass (Specified Areas) (Prohibitions of Fishing) Order*. Her Majesty's Stationery Office, London, 6 pp., (Statutory Instrument 1990/1156).
Guérin-Ancey, O. (1973) Contribution à l'étude de la croissance des jeunes de *Dicentrarchus labrax* L. du Golfe de Marseille. *Cah. Biol. Mar.* **14**, 65–77.
Gulland, J. (1956) On the fishing effort in English demersal fisheries. *Fish. Invest. London*, Ser. 2, 20(5), 41 pp.
Gulland, J.A. (1965) Estimation of mortality rates. Annexe to the Report of the Arctic Fisheries Working Group, *Int. Coun. Explor. of the Sea*, C.M. 1965/3: 9 pp.
Gulland, J.A. (1966) Manual of sampling and statistical methods for fisheries biology, part 1, sampling methods, FAO Manuals in Fisheries Science, Rome.

Gulland, J.A. (1969) Manual of methods for fish stock assessment. Part 1. Fish population analysis, FAO Manuals in Fisheries Science, No. 4, Rome, 154 pp.

Gunther, A. (1859) *Catalogue of the Acanthopterygian Fishes in the Collection of the British Museum*, Vol. 1, Trustees Br. Mus. (Nat. Hist.), London, 524 pp.

Gunther, A. (1863) On the European species of the genus *Labrax*. *Ann. Mag. Nat. Hist.*, **12**(3), 174–5.

Harden Jones, F.R. (1963) The reaction of fish to moving backgrounds. *J. Exp. Biol.*, **40**(3), 437–46.

Hartley, P.M.Y. (1940) The saltash tuck-net fishery and the ecology of some estuarine fishes. *J. Mar. Biol. Ass. U.K.*, **24**, 1–68.

Hickey, C.R., Jun., Young, B.H. and Bishop. R.D. (1977) Skeletal abnormalities in striped bass. *N.Y. Fish Game J.*, **24**, 69–85.

Holbrow, E.J. (1909) *The Bass*. (British Sea Anglers Society's Quarterly, Vol. 2), G. Barber, London 123–6.

Holcombe, F.D. (1921) *Modern Sea Angling*, Frederick Warne & Co. Ltd, London.

Holden, M.J. and Williams, T. (1974) The biology, movements and population dynamics of bass, *Dicentrarchus labrax*, in English waters. *J. Mar. Biol. Ass. U.K.*, **53**, 91–107.

Holland, B.F., Jun. and Yelverton, G.F. (1973) Distribution and biological studies of anadromous fishes off-shore North Carolina. Div. Commer. Sport Fish., NC Dept Nat. Econ. Resour. Spec. Rep. no. 24, 132 pp.

Howell, B.R. (1973) Marine fish culture in Britain VIII. A marine rotifer, *Brachionus plicatilis* Muller, and the larvae of the mussel *Mytilus edulis* L. as foods for larval flatfish. *J. Cons. Int. Explor. Mer.*, **35**, 1–6.

Howell, B.R. (1979) Experiments on the rearing of larval turbot, *Scophthalmus maximus*, L. *Aquaculture*, **18**, 215–25.

Hudson, J. (1963) The splendour of the bass, *Creel*, **1**(2), 24–5.

Jackman, A. (1954) The early development stages of the bass *Morone labrax* (L.). *Proc. Zool. Soc. Lond.*, **124**(3), 531–4.

Jennings, S. (1990) Population dynamics of larval and juvenile bass *Dicentrarchus labrax* (L.), PhD thesis, University of Wales, 266 pp.

Jennings, S. (1992) Potential effects of estuarine development on the success of management strategies for the British bass fishery. *Ambio*, **21**, 468–70.

Jennings, S. and Pawson, M.G. (1991). The development of bass, *Dicentrarchus labrax*, eggs in relation to temperature. *J. Mar. Biol. Ass. U.K.*, **71**, 107–16.

Jennings, S. and Pawson, M.G. (1992) The origin and recruitment of bass, *Dicentrarchus labrax*, larvae to nursery areas. *J. Mar. Biol. Ass. U.K.*, **72**, 199–212.

Jennings, S., Lancaster, J.E., Ryland, J.S. and Shackley, S.E. (1991) The age structure and growth dynamics of young-of-the-year bass, *Dicentrarchus labrax*, populations. *J. Mar. Biol. Ass. U.K.*, **71**, 799–810.

Jordan, D.S., and Eigenmann, C.H. (1887) Notes on the specific names of certain North American fishes. *Proc. Acad. Nat. Sci. Phil.*, **295**.

Jordan, D.S., and Eigenmann, C.H. (1890) A review of the genera and species of Serranidae found in the waters of America and Europe. *Bull. U.S. Natl. Comm.*, **8**, 329–441.

Katavic, I. (1986) Diet involvement in mass mortality of sea bass (*Dicentrarchus labrax*) larvae. *Aquaculture*, **58**, 45–54.

Katavic, I., Jug-Dujakovic, J. and Glamuzina, B. (1989) Cannibalism as a factor affecting the survival of intensively cultured sea bass. *Aquaculture*, **77**, 135–43.

Katavic, I., Tudor, M., Komljenovic, J. and Kuzic, N. (1985) Changes in the biochemical composition of *Artemia salina* (L.) in relation to different feeding conditions. *Acta Adriatica*, **26**, 123–34.

Kelley, D.F. (1979) Bass populations and movements on the west coast of U.K. *J. Mar. Biol. Ass. U.K.*, **59**, 896–936.

Kelley, D.F. (1986) Bass nurseries on the west coast of the U.K. *J. Mar. Biol. Ass. U.K.*, **66**, 439–64.

Kelley, D.F. (1987) Food of bass in U.K. waters. *J. Mar. Biol. Ass. U.K.*, **67**, 275–86.

Kelley, D.F. (1988a) The importance of estuaries for sea-bass *Dicentrarchus labrax* (L.). *J. Fish Biol.*, **33** (Suppl. A), 25–33.

Kelley, D.F. (1988b) Age determination in bass and assessment of growth and year-class strength. *J. Mar. Biol. Ass. U.K.*, **68**, 179–214.

Kelley, D.F. and Reay, P.J. (1988) The shallow creek fish communities of south-west England and West Wales estuaries. *J. Fish Biol.*, **33** (Suppl. A), 221–2.

Kennedy, M. and Fitzmaurice, P. (1968) Occurrence of eggs of bass (*Dicentrarchus labrax*) on the Southern coast of Ireland. *J. Mar. Biol. Ass. U.K.*, **48**, 585–92.

Kennedy, M. and Fitzmaurice, P. (1972) The biology of the bass *Dicentrarchus labrax* in Irish Waters. *J. Mar. Biol. Ass. U.K.*, **52**, 557–97.

Kentouri, M. (1980) Elevage des larves de loup (*Dicentrarchus labrax* L) a l'aide d'organismes du zooplankton congelé: resultats préliminaires. *Aquaculture*, **21**, 171–80.

Kerr, J.E. (1953) Studies on fish preservation at the Contra Costa Steam Plant of the Pacific Gas and Electric Company. *Calif. Dept Fish Game, Fish. Bull.*, **92**, 66 pp.

Klein, J.T. (1749) *Historiae Piscium Naturalis: de piscibus per branchias apertas spirantibus* (2nd ser.), Gedani, 102 pp.

Labourg, P. and Stequert, B. (1973) Régime alimentaire du bar, *Dicentrarchus labrax* des réservoirs à poissons de la région d'Arcachon. *Bull. Ecol.*, **4**(3), 187–94.

Ladle, M. and Vaughan, A. (1988) *Hooked on Bass*. The Crowood Press, Ramsbury (Wilts.).

Ladle, M., Casey, H. and Gledhill, T. (1983) *Operation Sea Angler*. A. & C. Black, London.

Lam Hoai Thong. (1970) Contribution à l'étude des bars de la région des Sables d'Olonne. *Trav. Fac. Sci. Rennes, Ser. Océanogr. Biol.*, **3**, 39–68.

Lancaster, J. (1991) The feeding ecology of juvenile bass, *Dicentrarchus Labrax*, PhD thesis, University of Wales, 281 pp.

Langford, T.E.L. (1987) The effects of a thermal discharge on the growth and feeding of bass, *Dicentrarchus labrax* (Linnaeus, 1758) in the Medway Estuary, England. C.E.G.B. C.E.R.L. TPRD/L/3126/R.87. 21 pp. + figs.

Le Cren, E.D. (1951) The length–weight relationship and seasonal cycle in gonad weight and condition in the perch (*Perca fluviatilis*). *J. Anim. Ecol.*, **20**, 201–19.

Le Danois, E. (1913) Contribution à l'étude systématique et biologique des poissons de la Manche orientale. *Annls Inst. Océanogr.*, Paris, **5**, 58–9.

Le Mao, P. (1985) Peuplements piscicoles et teuthologique du basin maritime de la Rance: l'impact de l'amenagement marimateur, these de Doctorat Ingenieur en Sciences agronomiques, Option Halieutique, Rennes, 125 pp.

Le Masson, V. (1981) La pêche du bar à Etel: Introduction à une gestion du stock. Association pour le developpment de l'aquaculture en Moubihan, France (thesis, Ecole National Superieure Agronomique de Rennes), 66 pp.

Lea, E. (1910) On the methods used in the herrings investigations. *Publ. Circ. Cons. Int. Explor. Mer.*, **53**, 7–175.

Lee, R.M. (1920) A review of the methods of age and growth determination in fishes by mean of scales. *Fishery Invest., Lond.*, **2**(4), 32 pp.

Linneaus, C. (1758) *Systema naturae*, Editio decima Holmiae, 482 pp.

Lockley, P. (1991) 10 tonne bass haul nets record £66,000. *Fishing News*, Emap, London, (4 Jan. 1991).

Loomis, J., Sorg, C. and Donnelly, D. (1986) Economic losses to recreational fisheries due to small-head hydro-power development: a case study of the Henry's Fork in Idaho. *J. Env. Mgmt.*, **7**, 1–19.

Lumare, F. and Villani, P. (1973) Ricerche sulla riproduzione artificiale ed allevamento delle larve in *Dicentrarchus labrax* (L.). *Aquaculture*, **28**, 71–5.

Mansueti, R.J. (1958) Eggs, larvae and young of stiped bass, *Roccus saxatilis*. *Chesapeake Biol. Lab.*, Contrib. 113, 12 pp.

Mansueti, R.J. (1960) An unusually large pugheaded striped bass, *Roccus saxatilis* from Chesapeake Bay, Maryland. *Chesapeake Sci.*, **1**, 111–13.

Mansueti, R.J. (1961) Age, growth and movements of striped bass, *Roccus saxatilis*, taken in size-selective fishing gear in Maryland. *Chesapeake Sci.*, **2**, 9–36.

Maranuacht (1990) Newsletter of the Department of the Marine, Issue No. 5 (Sept. 1990), Eire.

Mayer, I. (1987) Reproductive biology of the bass *Dicentrarchus labrax* L., PhD thesis, University of Wales, 144 pp.

Methot, R.D. (1981) Spatial covariation of daily growth rates of larval northern anchovy, *Engraulis mordax* and northern lampfish, *Stenbranchius leucopsarus*. *Rapp. P.v. Reun. Cons. Int. Explor. Mer.*, **178**, 424–31.

Mills, D.H. (1989) *Ecology and Management of Atlantic Salmon*. Chapman & Hall, London, 351 pp.

Mitchill, S.L. (1815) The fishes of New York, described and arranged. *Trans. Lit. Philos. Soc. N.Y.*, **1**, 355–492.

Moore, A., Pickett, G.D. and Eaton, D.R. (1994) A preliminary study on the use of acoustic transmitters for tracking juvenile bass (*Dicentrachus labrax*) in an estuary. *J. Mar. Biol. Ass. UK*, **74**(2).

Morales-Nin, B. (1985) Daily growth increments in the otoliths of *Dicentrarchus labrax*. *Rapp. P.-v. Réun. Comm. Int. Explor. Scient. Mer Medit.*, **29**, 95–7.

Morgan, R.P. II, Koo, S.Y. and Krantz, G.E. (1973) Electrophoretic determination of populations of the striped bass, *Morone saxatilis*, in the Upper Chesapeake Bay. *Trans. Am. Fish. Soc.*, **102**, 21–32.

Mosneron Dupin, J. and Lagardere, J.-P. (1990) Behavioural responses of sea bass *Dicentrarchus labrax* (Linne, 1758) to low temperatures. Preliminary results recorded in saltmarshes by means of acoustic telemetry. *C. r. Hebd. Séanc. Acad. Sci.*, (111), Paris, **310**, 279–84.

Muyard, J. (1978) Le bar et sa pêche dans les pertuis charentais. Rapport de D.A.A. en Halieutique, E.N.S.A. de Rennes/I.S.T.P.M. de la Rochelle, 18 pp.

National Opinion Poll Market Research Ltd (1970) National Angling Survey, 1969–70. A report prepared for the Natural Environment Research Council Steering Committee, 4 (Sea Angling).

National Opinon Poll Market Research Ltd (1980) National Angling Survey, 1980 (short report). Water Research Centre, Marlow, Bucks.

New, M., Insull, D., Ruckes, E. and Spagnolo, M. (1987) The markets for the prime Mediterranean species – sea bass, sea bream, mullets and eels – and their links with investment. U.N. Dev. Prog, FAO, U.N. ADCP/REP/87/29. Rome. 46 pp.

Niall, I. (1964) Bass at point north in, (ed. B. Venables), *Creel*, **1**(7), 18–19.

Ohno, A. and Okamura, F. (1988) Propagation of calanoid copepod, *Acartia tsuensis*, in outdoor tanks. *Aquaculture*, **70**, 39–51.

Ottaway, E.M. and Simkiss, K. (1979) A comparison of traditional and novel ways of estimating growth rates from scales of natural populations of young bass (*Dicentrarchus labrax*). *J. Mar. Biol. Ass. U.K.*, **59**(1), 49–59.

Paling, J.E. (1966) The attachment of the monogenean *Diplextanum aequans* (Wagener) Diesing to the gills of *Morone labrax* L. *Parasitology*, **56**, 493–503.

Parsons, C. (1982) Biology of bass (*Dicentrarchus labrax*) from South-west Wales, unpub. part II honours project, University of Wales, 66 pp.

Pawson, M.G. (1990) Using otolith weight to age fish. *J. Fish. Biol.*, **36**, 521–31.

Pawson, M.G. (1992) Climatic influences on the spawning success, growth and recruitment of bass (*Dicentrarchus labrax* L.) in British waters. *ICES Mar. Sci. Symp.*, **195**, 388–92.

Pawson, M.G. and Benford, Teresa, E. (1983) Coastal fisheries of England and Wales, Part 1: A review of their status in 1981, MAFF Direct. Fish Res., Lowestoft, Int. Rep., no. 9, 54 pp.

Pawson, M.G. and Pickett, G.D. (1987) The bass and management of its fishery in England and Wales. MAFF Lowestoft Laboratory Leaflet no. 59, 37 pp.

Pawson, M.G. and Pickett, G.D. (1988) Assessment and management of the U.K. bass (*Dicentrarchus labrax* L.) fishery. ICES CM/H:71, 18 pp.

Pawson, M.G. and Pickett, G.D. (1992) Management of bass (*Dicentrarchus labrax* L.) fisheries in the UK, in *Fisheries in the Year 2000* (ed. K. O'Grady). (Inst. Fish. Manage. 21st Anniv. Conf. Proc., pp. 189–98.

Pawson, M.G. and Rogers, S. (1989) Coastal Fisheries of England and Wales. Part II: A review of their status in 1988, MAFF Direct. Fish. Res. Lowestoft, Int. Rep., no. 19, 76 pp.

Pawson, M.G., Kelley, D.F. and Pickett, G.D. (1987) The distribution and migrations of bass, *Dicentrarchus labrax* L., in waters around England and Wales as shown by tagging. *J. Mar. Biol. Ass. U.K.*, **67**, 183–217.

Persoone, G., Soregeloos, P., Roels, O. and Jaspers, E. (1980) *The Brine Shrimp: Artemia*, Universal Press, Wetteren, 3 vols, 318 pp., 636 pp. and 428 pp.

Petersen, C.G.J. (1891) Eine methods zur bestimmung des alters und des wuches der fische. *Mitt. Dtsch. Seefisherei Ver.*, **11**, 226–35.

Pickett, G.D. (1989) The sea-bass. *Biologist*, **36**, 89–95.

Pickett, G.D. (1990) Assessment of the UK bass fishery using a log-book based catch record system. MAFF Direct. Fish. Res. Lowestoft Tech. Rep. no. 90, 33 pp.

Pickett, G.D. and Pawson, M.G. (1992) A log-book scheme for catch and effort assessment data; an example from the UK bass fisher, in *Catch Effort Sampling Strategies; their application in freshwater fisheries management* (ed. I. Cowx), pp 262–74.

Pickett, G.D., Pawson, M.G. and Eaton, D.R. (1988) Colostomy in a bass (*Dicentrarchus labrax* L.)? *J. Cons. Int. Explor. Mer.*, **45**, 105–6.

Pickett, J.F. (1979) Pugheadedness in a largemouth bass. *N.Y. Fish Game J.*, **26**(1), 98–9.

Pitcher, A.J. and Hart, P.J.B. (1982) *Fisheries Ecology*. Chapman & Hall, London, 414 pp.

Poll, M. (1947) *Faune de Belgique. Poissons Marins*, Mus. Roy. Hist. Nat. Belg., 452 pp.

Pope, J.G. (1972) An investigation of the accuracy of virtual population analysis using cohort analysis. *ICNAF, Res. Bull.*, No. 9, pp. 65–74.

Potter, E.C.E., and Pawson, M.G. (1991) *Gill Netting*. MAFF Direct. Fish. Res., Lowestoft, Lab. Leafl. no. 69, 34 pp.

Pullen, G. (1990) *Go Fishing for Bass*. Oxford Ill. Press, Yeovil, 102 pp.

Radford, A. (1985) The economics and value of recreational salmon fisheries in England and Wales: an analysis of the rivers Wye, Mawddach, Tamar and Lune. CEMARE, 102 pp.

Rae, B.B. and Shearer, W.M. (1965) Seal damage to salmon fisheries. *Mar. Res.*, **2**, 33 pp.

Rafail, S.Z. (1971) Investigation on Sciaenidae and Moronidae catches and on the total catch by beach Seine on the U.A.R. Mediterranean coast. *Stud. Rev. Gen. Fish. Coun. Mediterr.*, **48**, 1–26.

Rafinesque, C.S. (1820) Ichthyologia Oklensis, a Natural History of the Fishes Inhabiting the River Ohio and its Tributary Streams, Preceded by a Physical Description of the Ohio and its Branches. Lexington, KY, 23 pp.

Ré, P., Rosa, H.C. and Dinas, M.T. (1986) Daily microgrowth increments in the sagittae of *Dicentrarchus labrax* larvae under controlled conditions. *Invest. Pesq.*, **50**, 397–402.

Reis, E.G and Pawson, M.G. (1992) Determination of gill-net selectivity for bass (*Dicentrarchus labrax* L.) using commercial catch data. *Fish. Res.*, **13**, 173–87.

Ricker, W.E. (1958) Handbook of computations for biological statistics of fish population. *Bull. Fish. Res. Bd Can.*, **119**, 119–300.

Riley, J.D., Andrews, M.J., Aprahamian, M.W. and Claridge, P.N. (1986) Bass (*Dicentrarchus labrax*) year-class strength size variation as shown by sampling the 0-group on power station cooling water intake screens, English coast 1972–1983. *Annls Biol.*, **40**, 181.

Roblin, C. (1980) Etude comparée de la biologie du développement (gonadogénèse, croissance, nutrition) du loup (*Dicentrarchus labrax*) en milieu naturel et en élevage contrôlé, thèse 3ème cycle, Université de Perpignan, 150 pp.

Roblin, C. and Bruslé, J. (1984) Le régime alimentaire des alevins et juvéniles du loup (*Dicentrarchus labrax* L.) des lagunes littorales du Golfe du Lion (étangs Roussillonnais, France). *Vie et Milieu*, **34**, 195–207.

Rosenberg, A.A. and Beddington, J.R. (1988) Length-based methods of fish stock assessment, in *Fish Population Dynamics* (2nd edn), (ed. J. Gulland), John Wiley & Sons Ltd, pp. 83–103

Rounsefell, G.A., and Everhart, W.H. (1953) *Fishery Science. Its methods and applications*. John Wiley & Sons Inc. New York; Chapman & Hall Ltd., London, 444 pp.

Russell, F. S. (1935) On the occurrence of post larval stages of the bass, *Morone labrax* (L.) in the Plymouth area. *J. Mar. Biol. Ass. U.K.*, **20**, 71–2.

Sabriye, A.S. (1983) Age and growth of Bass (*Dicentrachus labrax*) from Plymouth waters, Devon, UK, 1982. Unpublished. Part III honours project, Plymouth Polytechnic.

Sabriye, A.S. (1986) Reproduction and early life-history of bass *Dicentrarchus labrax*, in Plymouth Waters, MSc. thesis, Plymouth Polytechnic, 102 pp.

Santulli, A. (1985) La distribuzione dei livelli di L-carnitina nei tessuti dei pesci ed il suo effetto sulla crescita degli stadi giovanili di *Dicentrarchus labrax*. *Oebalia*, **11**, 69–71.

Santulli, A., Modica, A., Cusenza, L., Curatolo, A. and D'Amelio, V. (1993) Effects of temperature on gastric evacuation and absorbtion and transport of dietary lipids in sea bass (*Dicentrarchus labrax*, L.). *Comp. Biochem. Physiol*, **105A**, 363–7.

Smitt, F.A. (1892) *Scandinavian Fishes*, Part 1, pp. 45–46.

Soriano, M., Moreau, J., Hoenig, J.M. and Pauly, D. (1990) New functions for the analysis of two-phase growth of juvenile and adult fishes, with application to Nile Perch. ICES CM 1990/D:16. Statistics Cttee. 13 pp.

Stequert, B. (1972) Contribution à l'étude du bar *Dicentrarchus labrax* (L.) des réserégion d'Arcachon, Thèse de 3ème cycle Univ. Bordeaux I. no. d'ordre 1009, 149 pp. multicop.

Stirling, H.P. (1972) Feeding, growth and proximate composition of the European bass, *Dicentrarchus labrax*, PhD thesis, University of Southampton, 151 pp.
Stirling, H.P. (1977) Growth, food utilization and effect of social interaction in european bass *Dicentrarchus labrax*. *Marine Biology*, **40**, 173–84.
Svetovidov, A.N. (1964) *Les poissons de la Russie*. 560 pp.
Tesch, F.W. (1968) Age and growth. International Biological Programme, *IBP Handbook*, no. 3, Oxford, pp. 93–123.
Tesseyre, C. (1979) Obtention de loups (*Dicentrarchus labrax*) portions en 20 mois d'élevage intensif avec recyclage de l'eau, in *Finfish Nutrition and Fishfeed Technology* (eds J.H. Halver and K. Tiews), Heenemann, Berlin.
Thompson, B.M. and Harrop, R.T. (1987) The distribution and abundance of bass (*Dicentrarchus labrax*) eggs and larvae in the English Channel and southern North Sea. *J. Mar. Biol. Assoc. U.K.*, **67**, 263–74.
Tortonese, E. (1973) Serranidae, in *Check-list of the Fishes of the Northeastern Atlantic and of the Mediterranean*, Vol. 1, UNESCO, Paris (eds C. Hureau and T. Monod), pp. 355–62.
Tournamille, J. (1975) Contribution à l'étude électrophorétique des protéines seriques et cristalliniennes chez le loup (*Dicentrarchus labrax* L. et *D. punctatus* B.), thèse 3ème cycle, Univ. Sc. Tech. Languedoc, Montpellier, 129 pp. multicop.
Turnpenny, A.W.H. (1988) Fish impingement at estuarine power stations and its significance to commercial fishing. *J. Fish Biol.*, **33**(Suppl.A), 103–10.
Utrecht, W.L. Van. (1979) Remarks on the anatomy and ontogeny of scales of teleosts. *Aquaculture*, **17**(2), 159–74, 20.
Van den Broek, W.L.F. (1979) A seasonal survey of fish populations in the lower Medway estuary, Kent, based on power station screen samples. *Est. Coastal Mar. Sci.*, **9**, 1–15.
Venables, B. (1964) Bernard Venables goes fishing – Portugal. *Creel*, **1**(7), 8–10.
Walbaum, J.J. (1792) *Petri Artedi Suici Genera Piscium in Quinbus Systema Totum Ichthyologiae Proponitur cum Classibus Ordinibus Genereum Characteribus, Specierum Differentiis, Observationaibus Plurimis. Ichthyologiae*, Pars III, 2nd edn, Grypeswaldiae, 723 pp.
Waldman, J.R. (1986) Diagnostic value of Morone dentition. *Trans. Am. Fish Soc.*, **115**, 900–7.
Waldman, J.R., Dunning, D.J. and Mattson, M.T. (1990) A morphological explanation for size-dependent anchor tag loss from stiped bass. *Trans. Am. Fish. Soc.*, **119**, 920–3.
Waldock, M.J., Thain, J.E., and Waite, M.E. (1987) The distribution and potential toxic effects of TBT in UK estuaries during 1986. *Appl. Organometallic Chem.*, **1**, 287–301.
Wallace, P.D. and Hulme, T.J. (1977) The fat/water relationship in the mackerel, *Scomber scombrus* (L.), pilchard, *Sardina pilchardus* (Walbaum) and sprat, *Sprattus sprattus* (L.), and the seasonal variations in fat content by size and maturity. Fish. Res. Tech. Rep., MAFF Direct. Fish. Res., Lowestoft, no. 35, 10 pp.
Watanabe, T. (1982) Lipid nutrition in fish. *Comp. Biochem. Physiol.*, **73B**, 3–15.
Weatherley, A.H. (1972) *Growth and Ecology of Fish Populations*, Academic Press, London, 293 pp.
Weatherley, A.H. and Gill, H.S. (1987) *The Biology of Fish Growth*, Academic Press, London, 443 pp.
Wheeler, A. (1969) *The Fishes of the British Isles and North-west Europe*, Macmillan, London, XVII 613 pp.
Whitehead, P.J. and Wheeler, A.C. (1966) The generic names used for the sea bass of

Europe and N. America (Pisces Serranidae). *Annali Mus. Civ. Stor. Nat. Genova.*, **76**, 23–40.

Wilcocks, J.C. (1868) *The Sea Fisherman*, 2nd edn, Longmans, Green & Co., London.

Winch, J. (1983) The biology of *Atrispinum labracis* n. comb. (Monogenes) on the gills of bass, *Dicentrarchus labrax*. *J. Mar. Biol. Ass. U.K.*, **63**, 915–27.

Wise, M. (1984) *The Common Fisheries Policy of the European Community*, Methuen & Co., New York, 316 pp.

Woolcott, W.S. (1957) Comparative osteology of serranid fishes of the genus *Roccus* (Mitchill). *Copeia*, **1957**, 1–10.

Yarrell, W. (1836) *A History of British Fishes*. Vol. 1. John vanVoorst, London, 408 pp.

Young, A. (1955) *Bass. How to Catch Them*. Herbert Jenkins Ltd, London, 96 pp.

Zanuy, S. and Carrillo, M. (1985) Annual cycles of growth, feeding rate, gross conversion efficiency and haematrocrit levels of sea bass (*Dicentrarchus labrax* L.) adapted to different osmotic media. *Aquaculture*, **44**, 11–12.

Author index

Species index

Subject index

Fish Ecophysiology

J C Rankin, and **F B Jensen**, both at Institute of Biology,
Odense University, Odense, Denmark

The range of environments inhabited by fishes is vast, and the successful maintenance of fish populations in challenging environments requires responsive adjustments in physiology. Ecophysiology forms the interface between ecology and physiology, overlapping with other disciplines such as behaviour and morphology. *Fish Ecophysiology* describes how the physiology of fish is affected by and regulated in response to environmental changes.

There is great concern, internationally, about the impact of human activities on the environment , as well as the potential environmental changes which may be brought about by predicted global climate changes. A full understanding of the effects of environmental and potential habitat changes on fish physiology is vital to all those working in this area.

Fish Ecophysiology contains 15 chapters, written by internationally-acknowledged experts, each chapter giving a vital insight into a particularly important aspect of the subject area. As such, the book will be of great value to upper level students, researchers and professionals working in fish biology, fisheries, aquaculture, environmental sciences, ecotoxicology, physiology and ecophysiology.

Contents: Preface - fish ecophysiology: The comparative physiologists's viewpoint - *J C Rankin and F B Jensen;* Bioenergetics: Feed intake and energy partitioning - *M Jobling;* Biochemical correlates of growth rate in fish - *D F Houlihan, E Mathers and A Foster;* Growth, reproduction, and death in lampreys and eels - *L Olesen Larsen and S Dufour;* Salmonid smolting: A pre-adaptation to the oceanic environment - *G Boeuf;*. Role of peptide hormones in fish ormoregulation - *Y Takei;* Environmental perturbations of oxygen transport in teleost fishes; Causes, consequences and compensations - *F B Jensen, M Nikinmaa and R E Weber;* Cardiovascular and ventilatory control during hypoxia - *R Fritsche and S Nilsson* Acid-base regulation in response to changes of the environment: Characteristics and capacity - *N Heisler;* Environmental effects on fish gill structure and functions - *S F Perry and P Laurent;* Effects of water pH on gas and ion transfer across fish gills - *D J Randall and H Lin;* Endocrine responses to environmental pollutants - *J A Brown;* Branchial mechanisms of acclimation to metals in freshwater fish - *D G McDonald and C M Wood;* Phenotypic plasticity of fish muscle to temperature change - *I A Johnston;* Recent advances in the ecophysiology of Antarctic notothenioid fishes: Metabolic capacity and sensory performance - *J C Montgomery and R G M Wells;* The ecophysiology of intertidal fish - *C R Bridges.*

Fish and Fisheries Series 9

(Series Editor: T J Pitcher College, UK)

December 1992: 234x156: c.408pp, c.108 illus

Hardback: 0-412-45920-5: £55.00

On the Sex of Fish and the Gender of Scientists

A collection of essays in fisheries science

D Pauly, International Center for Living Aquatic Resources Management (ICLARM), Philippines

Daniel Pauly is the most widely cited fisheries scientist of his generation. *On the Sex of Fish and the Gender of Scientists* comprises an edited and updated collection of 27 of Daniel Pauly's essays, spanning a great range of exciting and sometimes contriversial topics, many of them breaking new scientific ground.

The book is divided into four sections covering tropical and other fisheries, patterns in fish biology, overfishing and how fisheries science is implemented. There is a Foreword to the book written by Ray Hilborn and a Series Foreword by Series Editor Tony Pitcher.

There is something of interest in this book for everyone involved in fisheries science and fish biology. The book should find a place on the shelves of all fisheries workers, professionals and students alike.

Key Benefits

* numerous illustrations aid in reader understanding and enjoyment
* author is the most widely cited fisheries worker of his generation - provides you with information from a respected source

Contents: Foreward. Preface and acknowledgements. On tropical and other fisheries and on learning from each other. Patterns in fish biology: beyond complexity. Overfishing, or why more is less. Fisheries science: how people make it happen. Getting closer to books. Back matters.

Fish and Fisheries Series 14
(Series Editor: T Pitcher)

June 1994: 234x156: c.264pp, 56 line illus
Paperback: 0-412-59540-0: c. £29.95